Hydrogen Energy System

NATO ASI Series

Advanced Science Institutes Series

A Series presenting the results of activities sponsored by the NATO Science Committee, which aims at the dissemination of advanced scientific and technological knowledge, with a view to strengthening links between scientific communities.

The Series is published by an international board of publishers in conjunction with the NATO Scientific Affairs Division

A **Life Sciences** B **Physics**	Plenum Publishing Corporation London and New York
C **Mathematical and Physical Sciences** D **Behavioural and Social Sciences** E **Applied Sciences**	Kluwer Academic Publishers Dordrecht, Boston and London
F **Computer and Systems Sciences** G **Ecological Sciences** H **Cell Biology** I **Global Environmental Change**	Springer-Verlag Berlin, Heidelberg, New York, London, Paris and Tokyo

PARTNERSHIP SUB-SERIES

1. **Disarmament Technologies**	Kluwer Academic Publishers
2. **Environment**	Springer-Verlag / Kluwer Academic Publishers
3. **High Technology**	Kluwer Academic Publishers
4. **Science and Technology Policy**	Kluwer Academic Publishers
5. **Computer Networking**	Kluwer Academic Publishers

The Partnership Sub-Series incorporates activities undertaken in collaboration with NATO's Cooperation Partners, the countries of the CIS and Central and Eastern Europe, in Priority Areas of concern to those countries.

NATO-PCO-DATA BASE

The electronic index to the NATO ASI Series provides full bibliographical references (with keywords and/or abstracts) to more than 50000 contributions from international scientists published in all sections of the NATO ASI Series.
Access to the NATO-PCO-DATA BASE is possible in two ways:

– via online FILE 128 (NATO-PCO-DATA BASE) hosted by ESRIN, Via Galileo Galilei, I-00044 Frascati, Italy.

– via CD-ROM "NATO-PCO-DATA BASE" with user-friendly retrieval software in English, French and German (© WTV GmbH and DATAWARE Technologies Inc. 1989).

The CD-ROM can be ordered through any member of the Board of Publishers or through NATO-PCO, Overijse, Belgium.

Series E: Applied Sciences - Vol. 295

Hydrogen Energy System

Production and Utilization of Hydrogen and Future Aspects

edited by

Yuda Yürüm
Department of Chemistry,
Hacettepe University,
Ankara, Turkey

Springer Science+Business Media, B.V.

Proceedings of the NATO Advanced Study Institute on
Hydrogen Energy System, Utilization of Hydrogen and Future Aspects
Akçay, Turkey
August 21–September 3, 1994

A C.I.P. Catalogue record for this book is available from the Library of Congress.

ISBN 978-94-010-4053-2 ISBN 978-94-011-0111-0 (eBook)
DOI 10.1007/978-94-011-0111-0

Printed on acid-free paper

Originally published by Kluwer Academic Publishers in 1995
Softcover reprint of the hardcover 1st edition 1995

TABLE OF CONTENTS

Storage and Transmission

Applications

Case Studies

FOREWORD

This volume contains the lectures of the Advanced Study Institute on "Hydrogen Energy System, Utilization of Hydrogen and Future Aspects" which was held at Akçay, Edremit, Turkey, between August 21 and September 3, 1994. The book includes 21 chapters originally prepared for the meeting by distinguished scientists of the field. Company of these colleagues contributed to the success of the Institute and our understanding of the Hydrogen Energy System increased immensely by their efforts.

The course consisted of four parts: 1) Production of Hydrogen; 2) Storage and Transmission of Hydrogen; 3) Applications; and 4) Case Studies on Hydrogen Utilization and Hydrogen Energy System Related Subjects. Courses in relation with these subjects were presented throughout two weeks period of the Institute. I expect that the written material presented in the book will be of great benefit both for the beginners of the subject and experienced people working on the hydrogen energy system.

Many people contributed to the success of the Institute on which this volume was based. Professor Nejat Veziroğlu is deeply appreciated for introducing us to the subject of hydrogen energy system. I take this occasion to convey my gratitude to my colleagues who lectured in the Institute both for their efforts during the two weeks and their expertly prepared lecture notes that reached to me in time.

I thank to some special people who contributed to the organizational aspects of the Institute. I include in this list Dr. Emin Arca, Dr. Murat Azık and Dr. A. Kerim Karabakan and my graduate students Nurşen Öztaş, Silmer Özvatan, Dilara Bozkurt, Sara Uzun and Mehtap Gündoğan for their efforts during the Institute and hosting all of the participants. Last but not least my thanks and appreciation go to my wife Perla Yürüm for her endless support and encouragement through the realization of the Institute.

The Institute was generously sponsored by the Scientific and Environmental Affairs Division of the NATO and their contribution is deeply acknowledged.

I should also extend my appreciation to the publishers themselves for their customary efficiency in bringing out this volume.

Prof. Yuda Yürüm
Director
NATO ASI on
Hydrogen Energy System, Utilization of Hydrogen and Future Aspects

Ankara, March 1995

TWENTY YEARS OF THE HYDROGEN MOVEMENT: 1974-1994

T.N.VEZİROĞLU
Clean Energy Research Institute
University of Miami
FL 33124
U.S.A.

ABSTRACT. It is now the twentieth anniversary of the beginning of the Hydrogen Energy Movement. Over the twenty years, there have been accomplishments in every front-from the acceptance of the concept as an answer to energy and environment related global problems, to research, development and commercialization. The Hydrogen Energy System has now taken firm roots. Activities towards the implementation are growing.

1. Introduction

It is the 20th Anniversary of the beginning of the Hydrogen Movement. It was just a little over 20 years ago, during the first international conference on Hydrogen Energy, the Hydrogen Economy Miami Energy Conference or the THEME Conference for short, which was held in March of 1974 in Miami Beach, Florida, U.S.A. In the afternoon of the second day, a small group, later to be named "Hydrogen Romantics", got together: Cesare Marchetti, John Bockris, Tokio Ohta, Bill Van Vorst, Anibal Martinez, Walter Seifritz, Hussein Abdel-Aal, Bill Escher, Bob Zweig, late Kurt Weil, myself and a few other enthusiasts, whose names escape me. There was a passionate, yet deliberate, debate. It was agreed that the Hydrogen Energy System, Figure 1, was an idea, whose time had arrived. It was the permanent solution to the depletion of conventional fuels, it was the permanent solution to the global environmental problems.

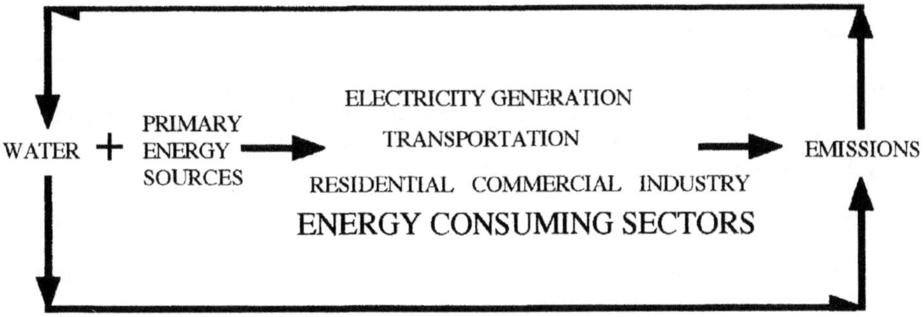

Figure 1. A Schematic diagram of the hydrogen energy system.

Y. Yürüm (ed.), Hydrogen Energy System, 1–14.

Then the discussion turned to whether there was a need for a formal organization. It was Anibal Martinez of Venezuela, incidentally one who took part in setting up the petroleum cartel OPEC, urged the founding of a society dedicated to crusade for the establishment of (what it seemed to the gathering and later proven to be so) the inevitable and the universal energy system. It was ironic that he was proposing the establishment of an organisation, which would make OPEC obsolete. The rest is history. IAHE was established by the end of that year and started working in earnest.

2. Accomplishements

In spite of the fact that, the competition, the conventional fuels are subsidized, since the damage they cause to the health of the people, to the well-being of this planet and to the integrity of everything living and/or standing on it, are paid for not by the conventional fuels but by the society; in spite of the fact that equivalent subsidies are not yet extended to hydrogen, for example, though a carbon tax and/or a CO_2 tax; in spite of these handicaps; in the twenty years since the Hydrogen Movement began, hydrogen has made significant inroads in several directions in the energy field-because of its unmatched superior properties and characteristics as an energy carrier.

In the twenty years, a lot of ground has been covered. There has been progress in every front. All those who took part in the Movement, all those who are taking part in this Movement, can be very proud. There is every reason why we should be satisfied, why we should celebrate this milestone, the 20th Anniversary of the Hydrogen Movement. Let us survey the happenings and achievements of the last twenty years.

2.1. CONFERENCES

One of the first activities of the International Association for Hydrogen Energy was to organize the biennial World Hydrogen Energy Conferences (WHECs) to provide a platform for the Hydrogen Energy community, for the scientists, energy engineer, environmentalists, decision makers and the thinkers of the future of the planet Earth.

WHEC Conferences have been to most of the major countries around the world. The first WHEC Conference was held in Miami in 1976, and the others followed in two year intervals in Zurich, Tokyo, Pasadena, Toronto, Vienna, Moscow, Honolulu, Paris, and in 1994 in Cocoa Beach, Figure 2.

I would like to add that in parallel with the WHEC Conferences, there have been organized several other conferences dedicated to specific applications of hydrogen, such as on transportation, on fuel cells, on hydrogen-metal interactions, etc. Also the number of national and international conferences devoted to hydrogen energy or having sessions on hydrogen energy, are increasing in number.

2.2. CONCEPT

Before the THEME Conference, little attention was paid to hydrogen as an energy carrier. The words "hydrogen energy", "hydrogen economy", "hydrogen energy system" were unknown even to most of those well-versed in energy. Today, these words are well-known and accepted, Figure 3. Not only the scientists and engineers, but also the public at large is becoming exposed to the concept. We see more and more articles and news items in popular press about the environmental benefits of hydrogen, about hydrogen as the fuel of the future and about happenings in hydrogen energy technologies. The expressions hydrogen energy, hydrogen economy, hydrogen energy system have entered the scientific literature, newspapers and into everyday vocabulary.

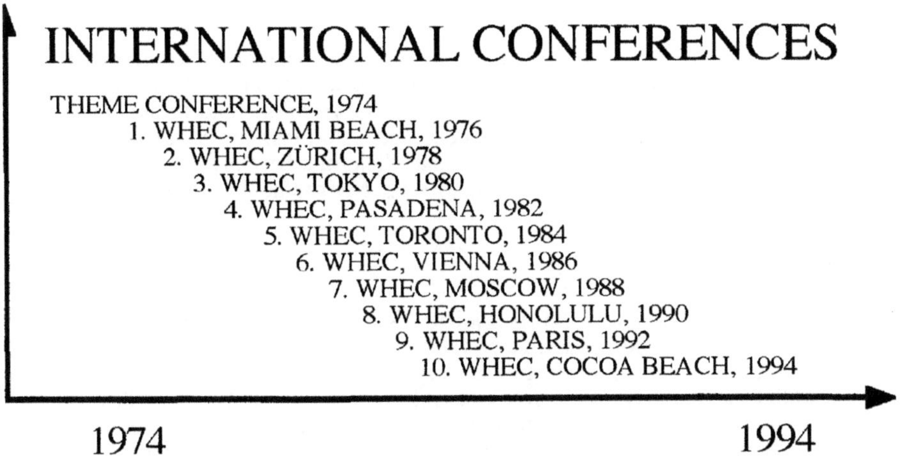

Figure 2. International conferences on hydrogen energy.

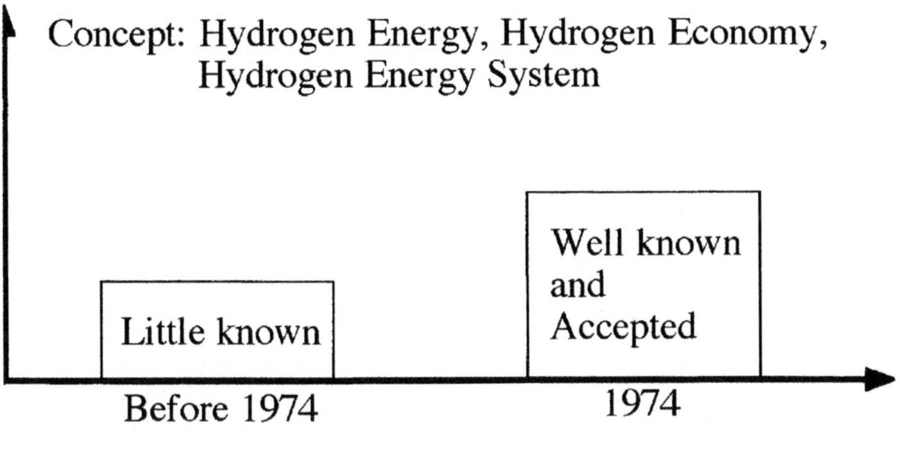

Figure 3. Acceptance of "hydrogen energy", "hydrogen economy", "hydrogen energy system" concepts.

2.3. ORGANIZATIONS

Twenty years ago, there was no organization dedicated to hydrogen energy. Today, national and international organizations devoted to hydrogen energy or to the applications of its unique and particular properties, cover the globe from one end to the other, from the United States to Japan, from Korea to Sweden. As it can be seen from Figure 4, there are at least thirteen such organizations and their numbers are growing. Many of these organizations are also forming alliances with environmental groups, establishing chapters and educating the uninitiated.

2.4. PERIODICALS

Twenty years ago, there was no periodical dedicated to hydrogen energy. The International Journal of Hydrogen Energy (IJHE), which is the official journal of the International Association for Hydrogen Energy (IAHE), is in it's nineteenth year. In 1975, it started as a quarterly. Three years later, it became bimonthly; in 1982, it became monthly. We just had to increase the page budget because of the growing number of the papers being received. In addition to IJHE, there are now several other periodicals-not only in English, but also in many other languages, Figure 5.

2.5. BOOKS

The number of hydrogen energy related books has shown an exponential growth over the last twenty years. The THEME and the WHEC Conference proceedings now stand at 40 volumes. In addition, dozens of volumes have been published in hydrogen energy, hydrogen fuelled transportation, hydrogen fuel cells, hydrogen-metal interactions in all the major languages of the world, Figure 6.

2.6. VISUAL PROGRAMS

Twenty years ago, we heard little or nothing in the visual media about hydrogen, with the exception that it was the fuel for the rockets in space programs. Since then, many imaginative and popular programs and documentaries have been produced by major television organizations and companies working on hydrogen energy technologies, Figure 7.

2.7. ELECTRIC POWER GENERATION

Hydrogen is a unique fuel with unmatched properties. One of its unique properties is that it can be converted to electricity electro-chemically in fuel cells with high efficiencies. It is not subject to the limitations of the Carnot Cycle, which is the case with the present day thermal power plants-whether they burn fossil fuels or nuclear fuels. Because of this high utilization efficiency advantage of hydrogen, electric utilities, electric power equipment manufacturers, power industry research organizations have taken a particular interest in electric power generation through hydrogen fuel cells. Tokyo Electric Utility started experimenting with a 4.5 MW Pratt and Whitney fuel cell years ago. Now, they have a second, an 11 MW fuel cell, in line. Major power generating equipment manufacturers have become involved in research, development and marketing of hydrogen fuel cell power plants. Several new companies have been formed specifically to work on fuel cells, Figure 8.

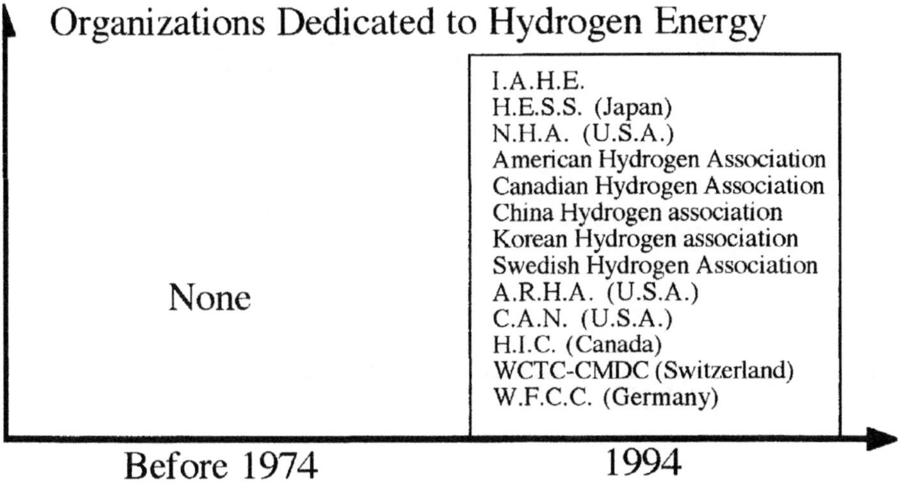

Figure 4. Organizations dedicated to hydrogen energy.

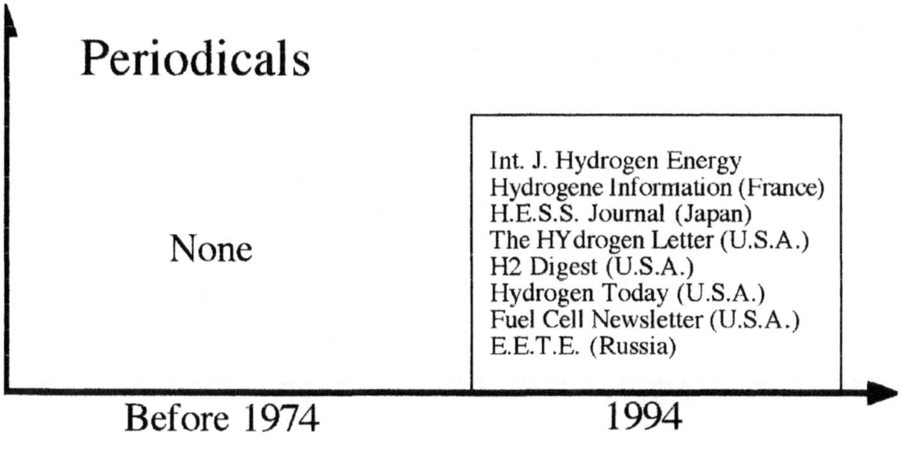

Figure 5. Periodicals on hydrogen energy.

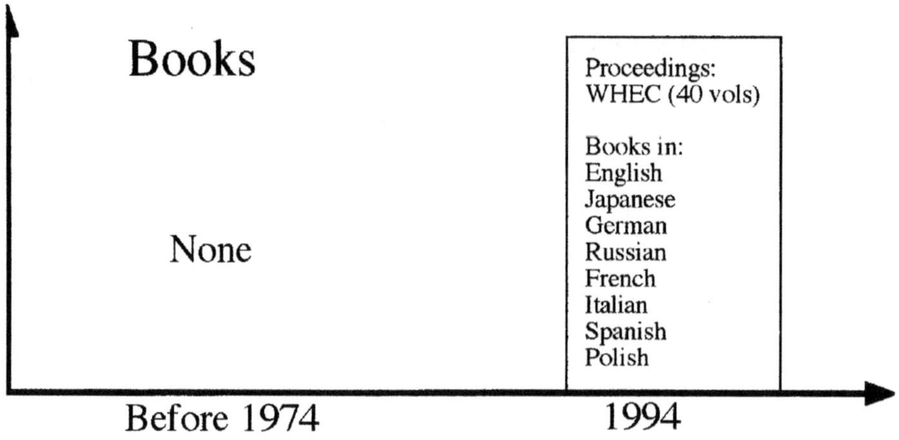

Figure 6. Books on hydrogen energy.

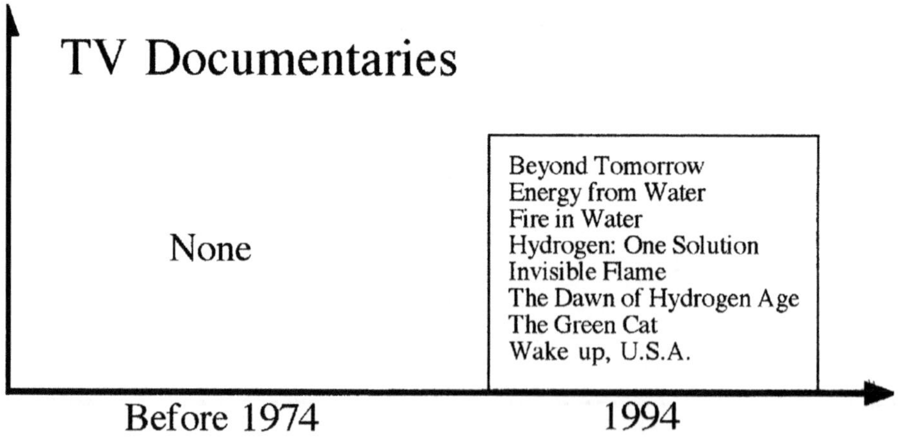

Figure 7. Visual programs on hydrogen energy.

There are many types of fuel cells. They have different and unique properties making each type suitable for a particular application: phosphoric acid fuel cells, alkaline fuel cells, proton exchange membrane fuel cells, molten carbonate fuel cells and solid oxide fuel cells. Some of these have already been commercialized, demonstration projects have been started for some and the others are in research and developmental stage.

In addition to having high utilization efficiencies, hydrogen fuel cells are clean, only by-product being water and quiet, i.e., no moving parts. They are versatile; they can be used for large scale power generation in central power plants, as well as for small scale electricity production in distributed mode. Because of all these unique characteristics, there is no doubt, that hydrogen fuel cell generating capacity will grow fast.

2.8. SURFACE VEHICLES

Hydrogen's unique properties make it suitable as a fuel for motive power, both for I.C. engine powered vehicles and electric powered vehicles. In addition to its unsurpassed environmental characteristics, lean burning property of hydrogen make it a suitable and efficient fuel for stop-and-go type city driving. Mazda Corporation of Japan has reported that hydrogen is the best fuel for Wankel Engines.

The efficiency advantage of hydrogen fuel cells is being put into use in electric cars, in which hydrogen fuel cells provide the motive power rather than electric batteries. Hydrogen fuel cells can and do overcome the short-driving range, small acceleration, low velocity handicaps of battery powered electric cars. As it can be seen from Figure 9, all the major car companies of the world are now involved in research and development of hydrogen fuelled motor vehicles. Some of these companies are preparing to offer hydrogen fuelled cars in 1998 in California and in other states and countries, which may pass zero emission vehicle laws.

2.9. NAVAL APPLICATIONS

Because it does not produce any harmful chemicals, because of its stealth characteristics, because its higher efficiency will provide longer under water cruising range, the German Navy has decided to have its next generation of submarines to have hydrogen fuel cell power plants. Also, the Australian Navy is experimenting with hydrogen fuel cells in their submarines. The Canadian Navy will do likewise, Figure 10. Of course, because of the aforementioned characteristics, hydrogen is the right fuel for sea surface transportation, as well.

2.10. SPACE PROGRAMS

Before 1974, hydrogen was used in rockets by the Soviet and the United States space programs. Now, the other countries, which have space programs, have joined them and are using hydrogen as the staple fuel of their space programs, because of another unsurpassed, unmatched property of hydrogen-that of being the lightest fuel, Figure 11.

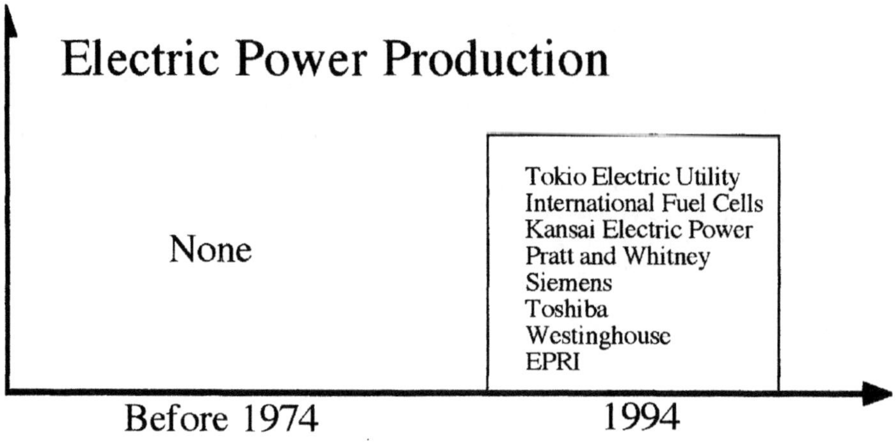

Figure 8. Companies and organizations involved in hydrogen fuel cell electric power generation.

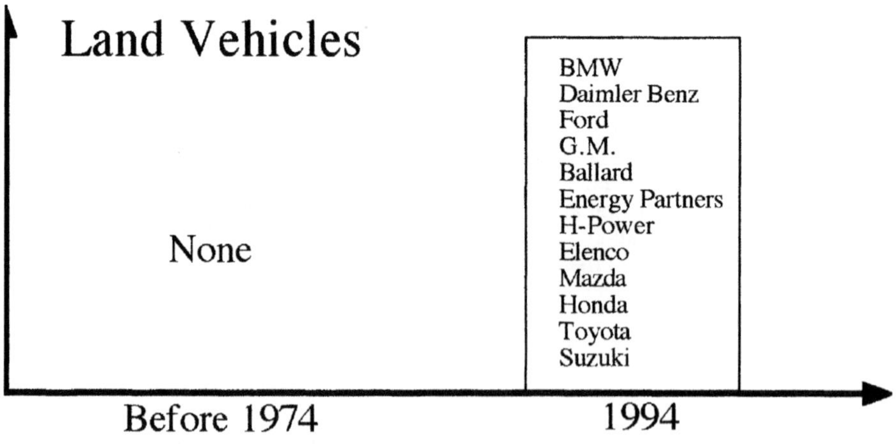

Figure 9. Companies involved in hydrogen fuelled vehicles.

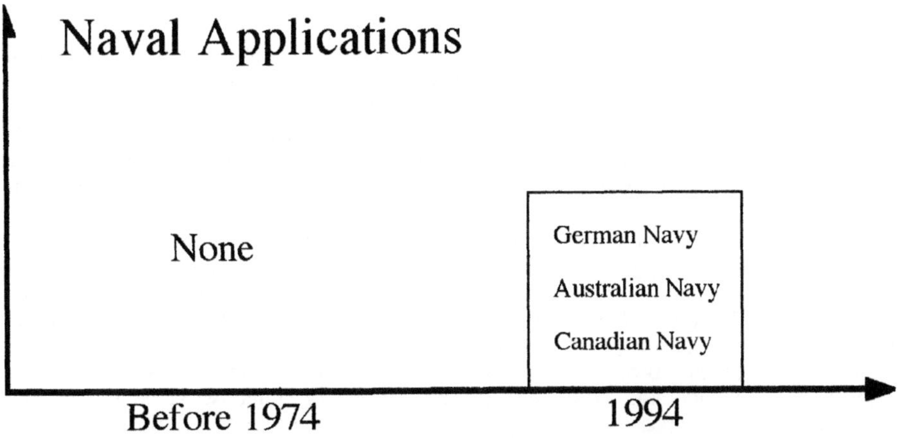

Figure 10. Naval applications of hydrogen.

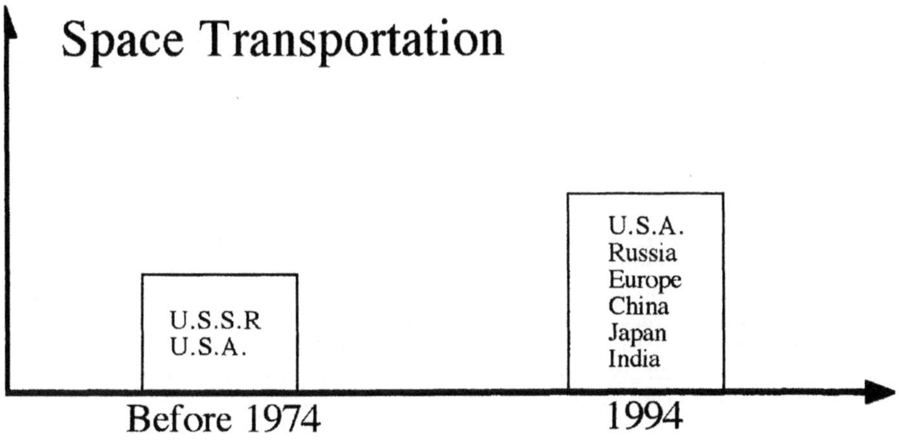

Figure 11. Hydrogen in space programs.

2.11. AEROSPACE PLANES

Before 1974, there were no aerospace planes, which of course would have used hydrogen as fuel, because it is the fuel of choice for space programs. Today, we have the American shuttle visiting space, putting communication satellites, observation satellites in orbit and conducting various scientific experiments. Russia has built a shuttle which can land automatically, without a pilot being in charge. The European two plane Sanger System is on the drawing board and studies are being conducted as to its practicality.

Extensive research and developmental work has been done on the engines, materials and structure of the U.S. National Aerospace plane with the involvement of the major aerospace companies, Figure 12. The fuel selected for this plane is the so-called "slush hydrogen", a mixture of liquid and solid hydrogen, which makes use of another unique property of hydrogen, resulting in reduction of storage size. The latest decision is that not to build the prototype, but continue with research and development work. There is no doubt that the experience gained will be of immense value in building tomorrow's hypersonic passenger transport-of course, to be fuelled by hydrogen.

2.12. AIRPLANES

In 1956, a Pratt and Whithey developed hydrogen fuelled turbo-jet engine was mounted on one side of a B-57 bomber and some in-flight data were collected. After 1974, hydrogen fuelled airplane activities have increased, Figure 13. In 1988, retired Pan American pilot Bill Conrad converted a four seater Grumman Cheetah to hydrogen. The power plant was a liquid hydrogen fuelled I.C. engine. It became the first airplane to take-off, cruise and land on hydrogen power alone, at the Fort Lauderdale International Airport, Florida, U.S.A.

This was followed by the testing of a hydrogen jet engine mounted on a Tupolev 154 Transport, which was renumbered 155, in Russia. Germany and Russia has signed an agreement of cooperation for the development of hydrogen fuelled air transportation.

European Airbus Company has initiated a program of research and development work for a hydrogen fuelled air transport. Their studies indicate that, although hydrogen costs more than jet fuel, the airfares for hydrogen fuelled air transportation would be competitive with today's airfares, because of the great weight and energy savings with the much lighter hydrogen fuel. Japan has initiated research and development work on a hypersonic transport, for which hydrogen is expected to be the fuel of choice, because of its excellent combustion properties, light weight and environmental compatibility.

2.13. HYDRIDE APPLICATIONS

One of the unique properties of hydrogen is that it will combine with certain metals and alloys easily, in large amounts, forming hydrides in exothermic chemical reactions. When hydrides are supplied with heat, hydrogen is released. The temperature and pressure characteristics vary for different metals and alloys. The advantage is being taken of these properties for many electrochemical and thermochemical applications, Figure 14. Smaller size hydrogen hydride batteries are already commercial. Bigger batteries for electric cars are being developed.

There are demonstration project for hydrogen-hydride air conditioning, refrigeration and heat pumps. They do not need chloroflorocarbons and as such they will not damage the ozone layer. Conversion to hydrogen-hydride air conditioning and refrigeration systems will put a definite stop to ozone layer depletion.

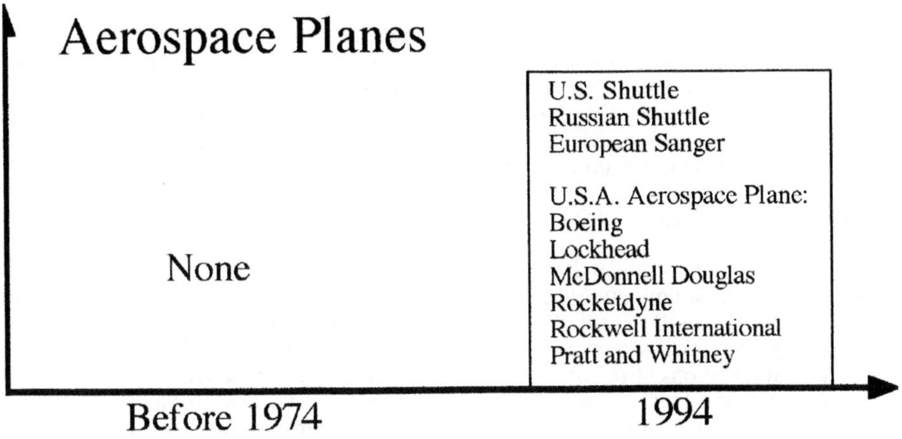

Figure 12. Hydrogen in aerospace planes.

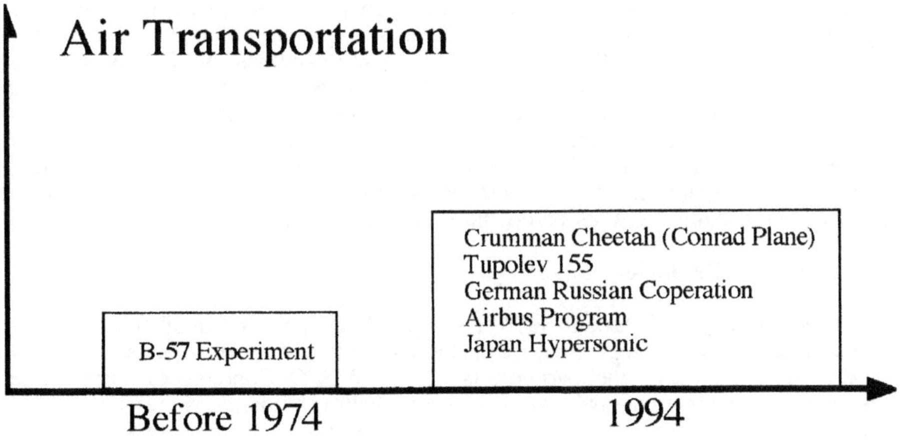

Figure 13. Hydrogen in air transportation.

2.14. CATALYTIC COMBUSTION

Another unique property of hydrogen is the flameless combustion or the catalytic combustion in the presence of small amounts of catalysts, such as platinum or palladium. Catalytic combustion applications have many advantages over those of flame combustion applications: They are safer, they have higher second law efficiencies, as well as being environmentally compatible. Many residential and commercial appliances have been developed using this unique property of hydrogen, Figure 15.

2.15. INTERNATIONAL PROGRAMS

Today German/Saudi Arabian Hy-Solar Project is producing solar hydrogen in the world's biggest petroleum country. Saudi Arabians expect eventually to be the permanent exporters of energy, in the from of solar hydrogen and they are preparing for it.

Euro-Quebec Project is in its fifth year and moving ahead in all fronts. They are looking into applications of relatively inexpensive hydro power produced liquid hydrogen imported to Europe from Canada-applications such as city bus transportation and smelting of iron, as well as the development of infrastructure for overseas transportation and storage of liquid hydrogen. Norway and Germany are planning to initiate a similar program.

Japanese WE-NET Program is the most ambitious and most comprehensive hydrogen program in the world. Japan expects to spend about $3 billion over the next 28 years to achieve -what amounts to- in a deliberate and planned way conversion to hydrogen. The Space station, which will use hydrogen for transportation, hydrogen fuel cells to provide electricity and potable water, is another important international program based on hydrogen energy. Planning for the United Nations Industrial Development Organization's International Centre for Hydrogen Energy Technologies (to be established in Istanbul, Turkey) is moving ahead, Figure 16.

2.16. INTERNATIONAL STANDARDS

Of course, no technology can take roots without standards and no universal technology can be established without international standards. In 1990, the International Standards Organization, based in Geneva, Switzerland (which is now an affiliated organization of the United Nations), with the initiative and representations of Gustov Grob of Switzerland, had decided that the time had arrived for preparing international standards for hydrogen energy technologies. They established, a committee, ISO/TC-197 Committee, to prepare such standards. In its first meeting, the Committee formed 10 sub-committees, listed in Figure 17, to work on the standards of hydrogen energy technologies. It is gratifying to know that the Committee's work is moving ahead and it had its fourth meeting the week before the 10th WHEC Conference.

3. Concluding Remarks

In the twenty years, 1974 through 1994, Hydrogen Energy moved in all fronts, making in-roads in all areas of energy. Because of the unrelenting work of scientists, engineers and dreamers, such as the participants of the WHEC Conferences, such as the membership of the International Association for Hydrogen Energy, we can all rejoice in the progress made in the last twenty years and this progress has been substantial.

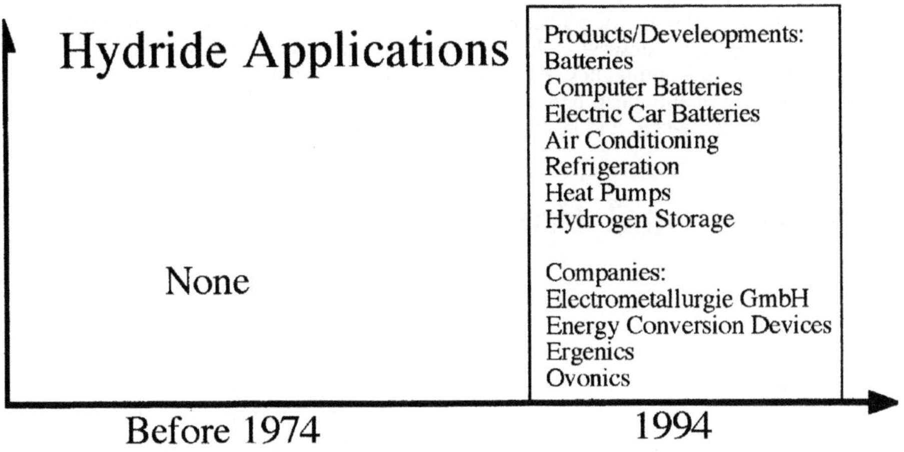

Figure 14. Hydrogen hydride applications.

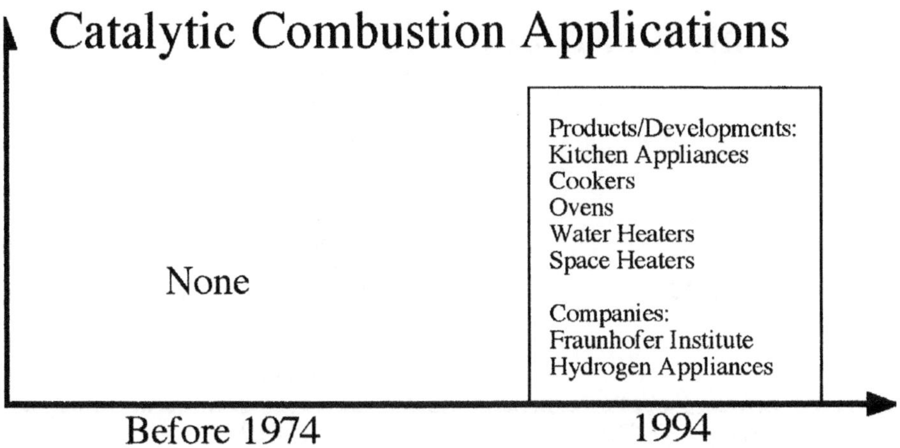

Figure 15. Hydrogen catalytic combustion applications.

In the next twenty years, the progress will be many folds greater and Hydrogen Energy System will provide the planet Earth, the only one known to be hospitable to life, with the energy system it deserves: clean, efficient and abundant energy.

My friends, my colleagues, pioneers of the hydrogen age, your efforts will expedite the spread of the clean and abundant energy system, enhance the quality of life for the peoples of the world and help preserve our biosphere. I wish you Godspeed.

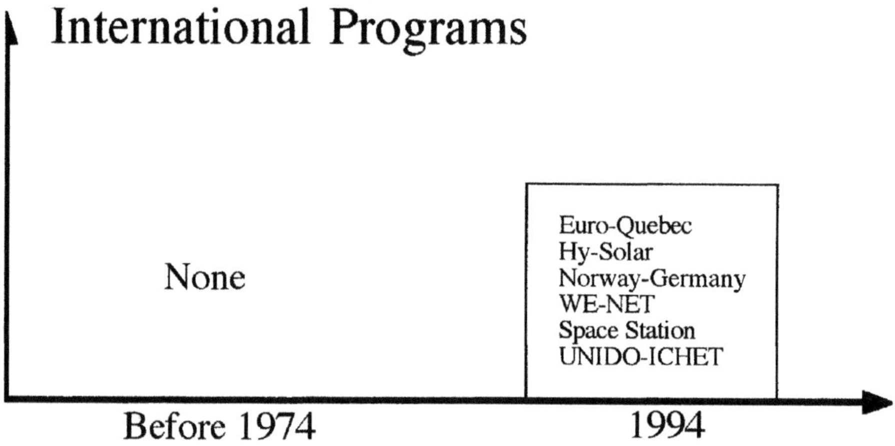

Figure 16. Hydrogen energy international programs.

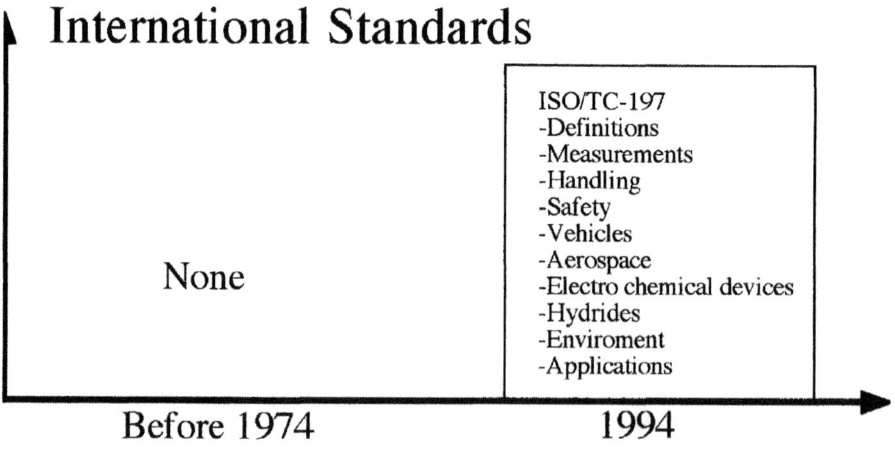

Figure 17. International standards for hydrogen energy technologies.

HYDROGEN PRODUCTION METHODS

Y. YÜRÜM
Department of Chemistry
Hacettepe University
Beytepe
Ankara 06532
Türkiye

ABSTRACT. Commercially available hydrogen production methods such as steam reforming of natural gas, partial oxidation of heavy oil, coal gasification, coal carbonization, steam iron process that are based on fossil hydrocarbons and methods in the stage of development, like thermolysis, radiolysis and photolysis of water, thermochemical cycles, hydrogen from hydrogen disulfide and biomass are discussed.

1. Introduction

At the present time, about 80 percent of the world energy demand is met by fossil hydrocarbons-coal, petroleum and natural gas (Davis, 1990). Hydrocarbons are the main source also for the production of hydrogen industrially. Another source for hydrogen production is water. Water is the most abundant available source of hydrogen and can be regarded as an unlimited source of hydrogen. With the utilization of an external energy source, hydrogen can be produced from either fossil hydrocarbons or water alone or from a mixture of both. Since the energy demand for hydrogen production using hydrocarbons is much lower than that of electrolysis of water, methods based on fossil hydrocarbons are preferred. Therefore the major portion of the hydrogen for industrial uses is produced from natural gas, oil and coal and a smaller percentage is manufactured by electrolysis of water. Hydrogen can also be produced using different sources and methods, however most of these are not economical and have not been proven to be used industrially. A variety of hydrogen production methods using solar energy (and its derivatives) have been or are currently being developed. They are direct thermal decomposition (thermolysis), radiolysis, thermochemical cycles, photolysis which includes photoelectrolysis, catalytic photolysis, biophotolysis.

The yields actually achieved in industrial hydrogen production from fossil hydrocarbons are quite different from the maximum possible theoretical yields. This is due to the fact, that the energy needed to produce hydrogen is usually supplied by the raw material itself. For the economic aspects of production not only the yield but also the prices of the feedstock and the investment costs for the plant are important, because there are only small differences in the cost of hydrocarbons based on the heating value, the investment costs for the production of hydrogen from light hydrocarbons are lower, and yields from light hydrocarbons are higher (Stull et al., 1969; Seider et al. 1980).

In the present chapter methods mentioned above will be discussed in general. Hydrogen production from electrolysis will be presented in a different chapter.

Y. Yürüm (ed.), Hydrogen Energy System, 15–30.

2. Hydrogen Production from Fossil Hydrocarbons

2.1. HYDROGEN FROM STEAM REFORMING OF NATURAL GAS

Steam reforming of hydrocarbons (mainly natural gas) has been the most efficient, economical and widely used process for hydrogen and hydrogen/carbon monoxide production. Hydrocarbons and steam catalytically are converted to hydrogen and carbon oxides. The process is applicable only to hydrocarbons ranging from natural gas to naphtha and light fuel oil (Minet and Desai, 1983). A simplified basic flow diagram of the conventional steam reforming process is shown in Figure 1.

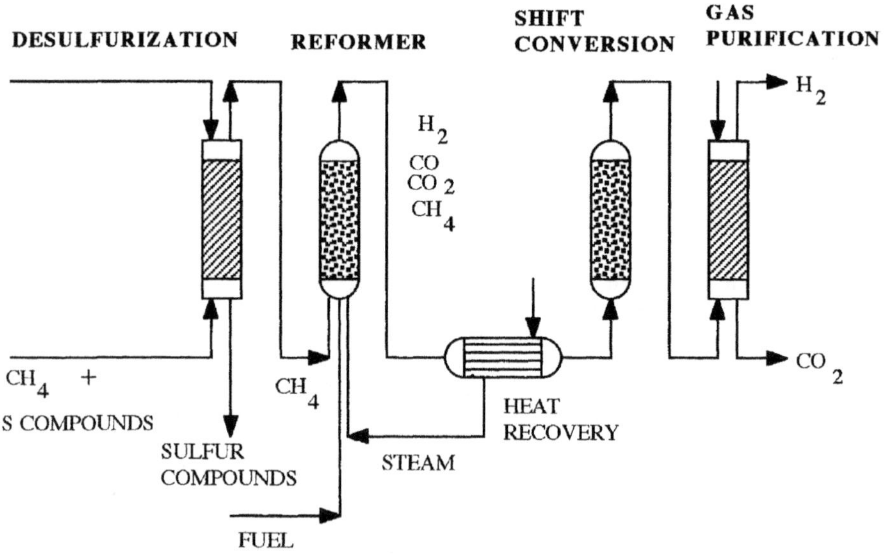

Figure 1. Process flow sheet for hydrogen production from steam reforming of natural gas.

Hydrocarbons react with steam in an endothermic reaction to form carbon monoxide and hydrogen. The most important raw material for the catalytic steam reforming process is natural gas. Other raw materials that can be used for the same purpose are propane, butane, liquefied petroleum gas and some naphtha fractions. The choice is usually made on the readiness and the price of the raw material. Disposition is given to lighter feedstock, if the price on a heating value basis is equivalent. The intention for this is the better operability of plants using lighter raw materials as well as the easier pretreatment operations and hence the lower investment costs for comparable plants.

The process essentially consists of three main steps: 1. synthesis gas generation, 2. water-gas shift, and 3. gas purification (Steinberg and Cheng, 1988). To preserve the catalysts, natural gas has to be desulphurized before being fed to the reformer. The desulphurized raw material is then combined with process steam and reacted over a nickel based catalyst present inside of a system of alloyed steel tubes. The following reactions occur in the reformer (Stull et al., 1969; Seider et al. 1980):

$$C_n H_m + n H_2O \rightleftharpoons n CO + (n+m/2) H_2$$
$$CO + H_2O \rightleftharpoons CO_2 + H_2$$
$$CO + 3 H_2 \rightleftharpoons CH_4 + H_2O$$

When n = 1 and m = 4 the hydrocarbon is methane and for n = 1 and m = 2.2 the hydrocarbon used is naphtha.

The reforming reaction is strongly endothermic, energy for this reaction is supplied by combustion of natural gas or fuel oil. The metallurgy of the tubes usually confines the reaction temperature to 700-925°C. A synthesis gas exit the catalytic reformer is typically a blend of H_2 (74%, by vol.), CO (18%, by vol.), CO_2 (6%, by vol.) and CH_4 (2%, by vol.) (Moore, 1983). After the reformer, the gas mixture passes through a heat recovery step cooling down to about 350°C and is transferred into the water-gas shift reactor to produce superfluous H_2.

The shift reaction, where carbon monoxide formed reacts with excess water, occurs concurrently to this reaction. Extra hydrogen is produced by this exothermic reaction:

$$CO + H_2O \rightleftharpoons CO_2 + H_2$$

The heat of reaction of the overall endothermic process is provided by indirect heating. The equilibrium of the hydrocarbon-steam reaction and the water gas-shift reaction states that high temperature at the end of reactor, high excess of steam and low pressure must be encountered to accomplish maximum hydrogen yields. The process conditions for the endothermic reforming reaction occurring in a fired tubular reactor are temperatures of 800-900°C and pressures varying between 1.5 and 3.0 MPa. The accurate choice of the process conditions depends on reaction kinetics, mechanical and economical settlement.

The amount of steam added to the hydrocarbon feedstock is dictated by the molar steam-to-carbon ratio. The steam-to-carbon ratio varies from 2.5 to 6 depending on the raw material and changes according to process optimization conditions. The following step is the high-temperature shift reaction occurring at 340-450°C which in certain cases is followed by a low- temperature shift reaction at 200-250°C.

Steam reforming and the coexisting shift reaction are catalyzed by supported nickel catalysts. The catalysts contain 15-25 wt % nickel oxide on mineral carriers of α-alumina, aluminosilicates, cement and magnesia. Nickel oxide must be reduced to metallic nickel before startup. Reduction is done generally with hydrogen but also with natural gas or even with the raw material gas itself. In some types of catalysts uranium oxide and chromium oxide are used as promoters. The use of promoters is claimed to give a higher resistance to catalyst poisoning by sulphur compounds and a lower inclination to make carbon deposits. For certain industrial operations such as low-temperature reforming some noble metal catalysts which have not been used commercially so far have been raised.

Nickel catalysts with alkaline carriers or alkaline-free catalysts with magnesium oxide as additive can be used for the steam reforming of liquefied petroleum gases and naphtha though both types of catalyst are less active than the traditional nickel catalyst. Thus, a slower decomposition of the hydrocarbons is attained. The reaction of water with any carbon formed is catalyzed at the same time (Bridger, 1972; Catalyst Handbook, 1972). The main catalyst poison in steam reforming plants is sulphur, which is present in most of the raw materials. Concentrations as low as 0.1 ppm produce a deactivating layer on the catalyst surface (Rostrup-Nielsen, 1977). Activity loss of the poisoned catalyst can be offset by raising the reaction temperature. This helps to reconvert the inert nickel sulphide formed by the reaction of metallic nickel with sulphur compounds to active nickel sites. If sulphur concentrations in the feedstock is in the forms of pulses of high concentrations permanent deactivation of the catalyst may occur together with mechanical problems caused by carbon deposits. Desulfurizing units therefore are used to clean the feedstock from sulphur components prior the raw material is fed to the catalytic converters. Chlorine and other halogen compounds as well as lead, arsenic and vanadium are the other catalyst poisons.

Nickel-free catalysts have been suggested for raw materials heavier than naphtha. These catalysts contain mostly strontium, aluminum and calcium oxides and are satisfactorily resistant to coking and operate with high-sulphur feedstocks. Moreover, the methane concentration in the exit gas is high with these catalysts even at reaction temperatures as high as 1000°C. Thus a secondary reformer has to be installed and higher hydrocarbons than methane are not usually found in the exit products of the steam reformer. The equilibrium composition can be calculated from equilibrium data as a function of the temperature, pressure and steam to carbon ratio.

The cold gas next passes through gas purification units in which CO_2, CO and other impurities present in the gas are removed to produce purified hydrogen. Several industrial operations such as wet scrubbing and pressure swing adsorption can be used to remove CO_2.

Efficiency of the steam reforming process, defined as the ratio of the heating value of produced hydrogen to the energy input as raw material, fuel and electricity is about 65-75%. The price of produced hydrogen is about $6/GJ, but this value depends on the cost and availability of the natural gas or other hydrocarbon feedstock (Steinberg and Cheng, 1988).

2.2. HYDROGEN FROM PARTIAL OXIDATION OF HEAVY OIL

Steam reforming is not applicable to convert hydrocarbons heavier than naphtha to hydrogen, thus partial oxidation of heavier feedstocks is used to produce hydrogen. Another objective to use partial oxidation for the conversion of gaseous fuel is the likelihood to produce synthesis gas with a high carbon monoxide content. Heavy residues from petrochemical processes are the preferred feedstocks for the production of hydrogen and carbon monoxide-rich gases. In principle, hydrogen can be produced by partial oxidation processes from all hydrocarbons including oil fractions with high sulphur and/or high heavy metal contents which are difficult to handle in upgrading processes and not considered as fuels for environmental reasons. The production, however, is only economically viable if heavy residues are used. A simplified flow diagram of the process is shown in Figure 2.

The process basically proceeds at moderately high pressures with or without catalyst depending on the raw material and process selected. The catalytic partial oxidation which occurs at about 600°C, uses feedstock ranging from methane to naphtha. The noncatalytic partial oxidation operates at 1150-1315°C and can use hydrocarbons ranging from methane, heavy oil to coal.

There are three main steps: 1. synthesis gas generation, 2. water-gas shift reaction, and 3. gas purification. The shift reaction and gas purification either by the conventional wet scrubbing or by the pressure swing adsorption operation are similar to those used in steam reforming. In the synthesis gas generation step, the hydrocarbon feedstock is partially oxidized with oxygen and the carbon monoxide produced is shifted with steam to CO_2 and H_2. Pure oxygen is used in the process because of the difficulties of separating nitrogen to produce pure hydrogen. The partial oxidation reactions are as follows :

$$C_n H_m + n/2\ O_2 \rightleftharpoons n\,CO + m/2\,H_2 + heat$$
$$C_n H_m + n\,H_2O + heat \rightleftharpoons n\,CO + (n + m/2)\ H_2$$
$$CO + H_2O \rightleftharpoons CO_2 + H_2 + heat$$

where n = 1 and m = 1.3 for residual fuel oils.

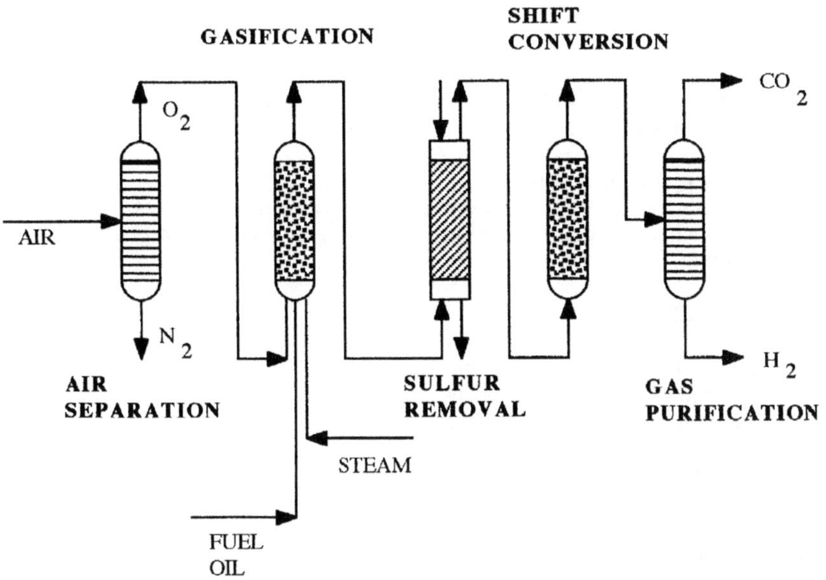

Figure 2. Process flow sheet for hydrogen production from partial oxidation of heavy oil.

The hydrocarbon-oxygen reaction is exothermic, the additional energy required for the endothermic hydrocarbon-steam reaction, similar to steam reforming, is supplied by burning additional fuel. The reactions occur mainly in the flame and in the empty brick-lined reactor. The residence time is usually < 5 s at temperatures 1250-1500°C. Pressure is often adjusted to 3-12 MPa to accommodate succeeding processes. The shift reaction occurs along with the oxidation-gasification reactions at temperatures greater than 1200°C without the presence of catalysts. Higher hydrogen yields, nevertheless, can be accomplished by attaining equilibrium at about 600°C in the presence of steam and suitable catalysts. The product is a very hot mixture of H_2 (46%, by vol.), CO (46%, by vol.), CO_2 (6%, by vol.), CH_4 (1%, by vol.), N_2 (1%, by vol.), (Moore, 1983). Molten ash or slag is cooled with water as it exits the reactor to solidify the slag. The sulphur contained in the feedstock is converted mainly into H_2S and only a small portion to COS. After sulphur removal, the gas is treated in the same way as the product gas of the steam reforming process.

The efficiency of partial oxidation is generally lower than for steam reforming, usually about 50%. Hydrogen can be produced for approximately \$10/GJ if cheap oil cheap oil with a cost of less than \$4/GJ is available, (Steinberg and Cheng, 1988).

2.3. HYDROGEN FROM COAL GASIFICATION

Coal gasification is defined as a process, in which coal is converted to gaseous products. When coal is fed into a gasification reactor and heated up, it mainly undergoes pyrolysis reactions and gases, volatile tar components and char are formed. It is known that (Hüttinger, 1988), not coal itself but the volatile pyrolysis products and char are gasified. Coal gasification generally done with oxygen of >95 % purity, at high temperatures and elevated pressures is used for the production of hydrogen or hydrogen-rich gas products. Gasification reactions can be presented as

$$2\,C + O_2 \rightleftarrows 2\,CO + \text{heat}$$
$$C + H_2O + \text{heat} \rightleftarrows CO + H_2$$

where C stands for pyrolysis products.

Coal gasification industrially is used to produce (van Krevelen, 1993),

a. Synthesis gas, rich in carbon monoxide and hydrogen suitable for Fischer-Tropsch synthesis of gasoline,

$$CO + 2\,H_2 \rightleftarrows (\text{-}CH_2\text{-}) + H_2O$$

b. Synthesis gas for the production of acetic anhydride,
c. Synthesis gas used in ammonia synthesis,
d. Synthesis gas used in methanol synthesis,

$$CO + 2\,H_2 \rightleftarrows CH_3OH$$

e. Synthesis gas convenient for the production of substitute natural gas with CH_4 concentration greater than 90 % CH_4,

$$CO + 3\,H_2 \rightleftarrows CH_4 + H_2O$$

f. Medium-and low-calory fuel gas for combined cycle processes which contain hydrogen, carbon monoxide, methane, carbon dioxide,
g. Hydrogen gas with concentrations greater than 95 %.

All of the coal gasification processes can be used to produce hydrogen. If pulverized coal is gasified with oxygen at high temperatures higher yields are achieved as it is done in the entrained-bed gasification reactor, commercialized by Koppers-Totzek and the Texaco processes (Clauson and Marion, 1980). The pulverized coal is quickly partially oxidized with oxygen and steam at essentially atmospheric pressure under slagging conditions. The raw gas is then cooled to recover waste heat, followed by quenching with water to remove entrained ash particles before going through the steps of compression, shift and gas purification by conventional wet scrubbing or pressure swing adsorption operations. Since hydrogen compression is highly energy consuming and most of the hydrogen users prefer hydrogen at higher pressures, it is significant to gasify coal at elevated pressures. The Texaco coal gasification process which operates at high pressures, around 5.5 MPa, utilizes a high steam content together with the synthesis gas in the shift reaction for hydrogen production. The raw gas is then desulphurized, shifted, and purified. The product hydrogen is at about 4 MPa with a purity higher than 97%. The simplified flow sheet of the process is given in Figure 3.

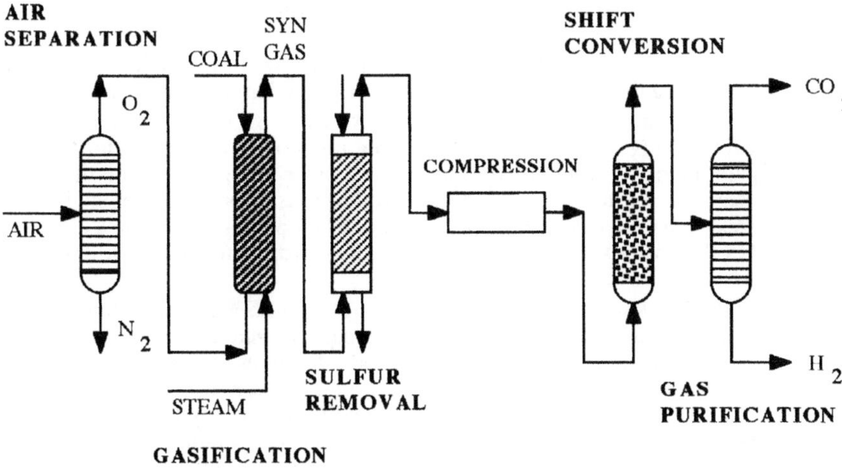

Figure 3. Process flow sheet for hydrogen production from coal gasification.

If coal is gasified at temperatures greater than 1300°C the chemical equilibrium at these temperatures produces a gas containing mainly hydrogen and carbon monoxide and concentrations of methane and other hydrocarbons in the synthesis gas are negligible. Carbon monoxide can be shift converted to hydrogen and thus the yield of hydrogen is increased.

Basically, all types of coal can be gasified to produce hydrogen, although certain limitations associated with the gasification process or the characteristics of the coal have to be thought. The gasification of petroleum coke in a fluidized bed gasifier has been proposed and the gasification of coke from tar sands has been successfully tested in an entrained bed gasifier. Gasification of other solid fuels such as peat, wood, refuse or sludge from waste water treatment plants leads to a low calory gas, which is less convenient for the production of hydrogen. Because of the high moisture and oxygen content of the raw material these gases are generally used on the plant site for heat or electricity production.

The coal gasification process is complicated because of the need of solids handling including transport, size reduction and removing large amounts of ash. The solids handling operations significantly affect cost of hydrogen production. Usually, the cost of coal does not counterbalance the higher capital cost of coal gasification systems, compared to steam reforming. The cost of hydrogen produced in coal gasification changes from \$12 to \$14/GJ (Steinberg and Cheng, 1988).

2.4. HYDROGEN FROM COKE OVEN GAS

Coke oven gas was the major source of hydrogen in the first half of this century. The hydrogen for ammonia synthesis and for coal liquefaction was obtained from coke oven gas and water gas, until the other methods were developed. New iron production techniques in which iron oxide is reduced with synthesis gas needs no metallurgical coke at all. Such new developments decreased the demand for metallurgical coke and therefore the production of coke oven gas also decreased significantly. In modern technology production of hydrogen from coke oven gas has now been superseded by more effective petroleum based processes and by coal gasification. However coke oven gas is still an important and convenient source for hydrogen, in locations where available. Similar aspects apply when gas from caking operations in refineries or from modern pyrolysis processes of hard coal, lignite, organic materials are considered as a source of hydrogen.

In modern coke oven batteries from one ton of coal about 350 m^3 of coke oven gas is produced (Bertling and Nashan, 1981), 50 % of the raw coke oven gas manufactured is used to energize the coke oven batteries and supply other energy needs of the coke oven plant and the rest of the coke oven gas produced is processed for hydrogen separation. Provided that better heat management and developments are achieved in the plant the amount of the coke oven gas that can be used as hydrogen source will increase accordingly.

Coke oven gas leaving the coke oven batteries at 750-850°C and atmospheric pressure contains impurities which have to be separated. The main components of the impurities are several aromatic heavy to medium oil fractions of the tar produced, hydrogen cyanide, ammonia, sulfur compounds and water. The amount of the impurities depends on the rank of coal utilized and process conditions employed in the coking process. The coke oven gas is first cooled to ambient temperatures and oil and naphthalene fractions and hydrogen cyanide, ammonia and sulfur compounds present are separated. Hydrogen content of the pretreated coke oven gas changes from 50 to 60 %. The separation of hydrogen from the pretreated coke oven gas can be done by two methods (Dalluge and Polster, 1981; Bernard and Vidal, 1984):

a. Low temperature condensation method is used generally to separate hydrogen from coke oven gas. Condensation of methane, ethylene and other compounds with a higher boiling point is carried out at pressures of 0.5-1 MPa and low temperatures of 60-70 K. The hydrogen obtained by this method is about of 95% purity. Hydrogen can be further be purified by utilizing a liquid nitrogen scrubber where the residual methane, carbon monoxide and oxygen content can be decreased to smaller levels. The gas obtained from the liquid nitrogen scrubber can be used in ammonia synthesis.

b. Pressure swing adsorption method is used to purify hydrogen at much lower investment costs. Higher hydrocarbons and sulfur compounds are separated after compression of the raw coke oven gas by a preadsorption stage. Hydrogen is separated from the other gas components in an adsoption train with four or more beds. Oxygen remaining in the product gas is converted catalytically to water in a reactor. Water produced is separated in an adsorption tower to yield hydrogen of a purity greater than 99.9%.

If the hydrocarbons in the raw or pretreated coke oven gas are converted to hydrogen by catalytical processes higher hydrogen yields can be obtained. Catalytic processes used for this purpose are steam-reforming and autothermic processes. Performance of both of these processes is increased when the sulfur content of the coke oven gas is low and unsaturated and cyclic hydrocarbons are absent in the raw or pretreated coke oven gas. To separate these, a multi stage precleaning operation is used (Meckel and Flockenhaus, 1981).

2.5. HYDROGEN FROM STEAM IRON PROCESS

The steam-iron process is one of the oldest ways of producing hydrogen. Steam iron process is a coal-based process, the synthesis gas generated from coal is mainly used for the reduction of iron oxide to iron in the iron generator. Hydrogen is obtained from the interaction of steam with iron oxide. It consists of four main steps: 1. coal gasification, 2. iron regeneration, 3. hydrogen generation and 4. purification. A basic block flow diagram for the process is presented in Figure 4.

In principle, it is a cyclic process for water cleavage, whereby coal is consumed. In the first step, coal is gasified with steam and air to produce a reducing gas containing carbon monoxide and hydrogen.

$$2\,C + O_2 \rightleftharpoons 2\,CO$$
$$C + H_2O \rightleftharpoons CO + H_2$$

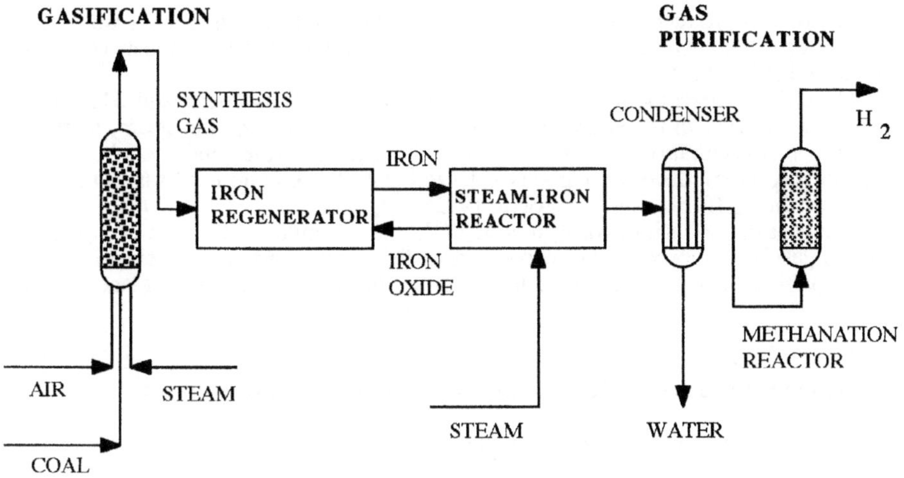

Figure 4. Process flow sheet for hydrogen production from steam iron process.

This gas is reacted with iron oxides in the second stage, to produce reduced forms of iron oxides.

$$Fe_3O_4 + H_2 \rightarrow 3\,FeO + H_2O$$
$$Fe_3O_4 + CO \rightarrow 3\,FeO + CO_2$$
$$FeO + H_2 \rightarrow Fe + H_2O$$
$$FeO + CO \rightarrow Fe + CO_2$$

The reducing synthesis gas is produced externally in a coal steam-air gasification stage. The main disadvantage of the steam-iron process is that only 60% of the reduction potential of the gas is utilized. The synthesis gas, actually CO and H_2, are not completely converted in reducing iron oxide. Most of the remaining heating value plus sensible heat at 825°C in the gas exiting from the iron regenerator can be used to generate power with a gas-steam combined cycle turbine generator. The synthesis gas, actually CO and H_2, are not completely converted in reducing iron oxide. Most of the remaining heating value plus sensible heat at 825°C in the gas exiting from the iron regenerator can be used to generate power with a gas-steam combined cycle turbine generator. In a sense, this can be considered as a cogeneration plant producing hydrogen and electrical power.

The reduced species are then reoxidized in the third step of the process with water to form the original oxides at the same time producing hydrogen. The regenerated iron then enters the steam-iron reactor where it is oxidized by steam to produce Fe_3O_4 and a hydrogen-rich gas.

$$Fe + H_2O \rightarrow FeO + H_2$$
$$3\,FeO + H_2O \rightarrow Fe_3O_4 + H_2$$

Continuous production of hydrogen can be achieved if reduction and oxidation of the solids occur in two separate reactors and the solid is transported between them (Tarman and Biljetina, 1978).

The effluent from the steam-iron reactor contains 37% hydrogen and 61% steam and small amounts of nitrogen and carbon oxides. Condensation of the steam, increases the hydrogen concentration to about 96% with 2% carbon oxides and 2% nitrogen. A methanation reactor reduces the carbon oxides to 0.2%. There is no need for CO shift or acid gas scrubbing (Barbir and Veziroğlu, 1992).

The cost of produced hydrogen is $12/GJ, this is similar to the cost of hydrogen from coal gasification process. The economics of this process depends on electric power produced which amounts to 25 % of the gross production costs (Steinberg and Cheng, 1988).

3. Hydrogen Production from Non-fossil Hydrocarbon Sources

3.1. HYDROGEN FROM WATER THERMOLYSIS

Theoretically the simplest method available is the direct thermal cleavage of water. Water can be divided thermally at temperatures greater than 2000 K. The components present after thermolysis reactions are: H_2O, H_2, O_2, H, O and OH. Appreciable quantities of hydrogen can only be obtained at temperatures significantly above 2000 K. The overall thermal dissociation reaction of water can be shown as (Ohta, 1978; Baykara and Bilgen, 1989):

$$H_2O \rightarrow a\,H_2O + b\,OH + c\,H + d\,O + e\,H_2 + g\,O_2$$

The thermolysis product of water is a mixture of different gases. The equilibrium composition depends on pressure and temperature of the reaction. The degree of water dissociation as a function of temperature is only 1.0 % at 2000 K, 2.7 % at 2300 K, 8.5 % at 2500 K, 10.3 % at 2700 K, 34.3 % at 3000 K and 57.4 % at 3500 K (Lede et al. 1983; Baykara and Bilgen, 1988). The main problems with the direct thermal decomposition of water are related to materials needed at extremely high temperatures. Other problems are associated with the recovery or separation of hydrogen from the hot mixture, conservation and recovery of the high-temperature energy. An economic utilization of this process is difficult, because of the recombination of the reaction products at the high temperatures (Jellinek and Kachi, 1984; Lede, 1986; Lede et al., 1987).

3.2. HYDROGEN FROM WATER RADIOLYSIS

Water can also be split by high-energy radiation. The overall reaction for water radiolysis is :

$$H_2O \rightarrow a\,H + b\,OH + c\,H_2 + d\,H_2O_2 + e\,OH^-(aq) + g\,H^+(aq)$$

Energy from nuclear sources is in the form of high energy photons, γ-rays, high-speed either charged particles of α-particles and β-radiation or high speed uncharged neutrons. The rate of energy transfer to the medium is called linear energy transfer (LET). The radiochemical yield, G value, is defined as the number of molecules converted per 100 eV absorbed energy, depends on the LET value of the radiation (Farhataziz and Rodgers, 1987). The G value of H_2 is greatly affected by the physical state of the medium and the chemical composition. The irradiation of water in the liquid state with γ-rays using ^{60}Co source gives a $G(H_2)$ yield of 0.45, in the vapor state at 110°C $G(H_2)$ is 5.2 (Danno, 1974). With ^{210}Po source a $G(H_2)$ value of 1.8 can be obtained. Hydrogen, is also obtained from the radiolysis of methane with a $G(H_2) = 6.4$. The radiolysis of higher aliphatic hydrocarbons produces hydrogen with G values of 6.6 for ethane, 8.2 for propane (Spinks and Woods, 1964).

3.3. HYDROGEN FROM WATER PHOTOLYSIS

Photolytic production of hydrogen is done with direct cleaving of water by solar energy. Direct photolytic cleavage of a water molecule initially produces H and OH radicals:

$$H_2O \rightarrow H + OH$$

The energy needed to split the H-O bond in water is 1.229 eV (Bull, 1988). Therefore a single photon with an energy of 2.458 eV is sufficient to split the water molecule. This energy can be principally obtained from solar radiation. The process must be made in stages using several light quanta and reactive intermediates should be separated.

Photolysis may also be carried out through photobiological processes. Photosynthetic organisms that use sunlight are utilized to convert water or carbon compounds into hydrogen in photobiological systems. The organisms that can be used for this purpose are the green algae, cyanobacteria and photosynthetic bacteria. Algae and cyanobacteria function to split water to hydrogen with an efficiency of 1-2 %. The photosynthetic bacteria can not split water, but they produce hydrogen from organic wastes with much higher efficiencies (Bull, 1988).

In photochemical systems synthesized molecules like porphyrin-quinones in a homogeneous solution are designed to imitate photosynthesis. Photochemical systems at present have low conversion efficiencies (Bull, 1988).

Photo electrochemical systems use semiconductor electrodes in contact with an electrolyte in aqueous solutions which absorb solar photons and produce electron-hole pairs. These pairs separate at the conductor liquid junction and drive chemical reactions at the surface. Hydrogen is reduced on one surface of a semiconductor plate and oxygen oxidized at the other surface of the plate in the presence of a catalyst. The major problem in producing hydrogen economically from photo electrochemical systems are to decrease corrosion of the semiconductor in the electrolyte, increase the photon conversion efficiency and reduce the cost of fabrication of the devices. The theoretical limit for conversion efficiency in photo electrochemical systems is 32 %, the practical conversion efficiency is 13 %, so there remains abundant opportunity for progress (Bull, 1988).

3.4. THERMOCHEMICAL CYCLES

Hydrogen is produced by chemical cleavage of water at temperatures lower than those used in thermolysis, in thermochemical methods through a series of cyclical chemical reactions which release hydrogen in a certain step. Temperatures around 1100-1200 K are considered maximum limit for performing thermochemical processes. This is due to the materials engineering problems. Thermodynamical considerations show that in the range of 300-1000 K at least three chemical reactions are necessary to dissociate water thermochemically. Model computations show that the theoretical efficiency of a thermochemical cycle decreases with the increasing number of reaction steps (Abraham and Schereiner, 1974). Although some 2000 to 3000 potential thermochemical cycles have been invented since the mid-1960 s only 20 to 30 remained applicable for the large scale hydrogen production after examining their practicability in terms of reaction and process technology. Some of the more extensively investigated thermochemical process cycles are sulphuric acid-iodine cycle, hybrid sulphuric acid cycle, hybrid sulphuric acid-hydrogen bromide cycle, calcium bromide-iron oxide cycle (UT-3) and iron chlorine cycle (Mark 9) (Engels et al., 1987; Winter and Nitsch, 1988; Yalçın, 1989).

Thermochemical cyclic processes can be classified according to the number of reaction steps or the maximum temperature. However, it has become accepted to group them in families according to the chemicals involved, e.g. iron-chlorine family (Fe, Cl_2, $FeCl_2$, $FeCl_3$), sulphur-halogen family (SO_2, halogen, X_2), copper-chlorine family (Cu, Cl_2, CuCl, $CuCl_2$). In order that the thermochemical cycle can be applicable the reactions should take place in the temperature range of 300-1000 K, the reaction rate should be adequately large, the chemistry and kinetics of the reactions taking place should be known (Schulten, 1978; Bowman, 1980).

If the heat losses and irreversible chemical reactions that are involved in practical processes are considered, the maximum efficiency attainable is only 50-80% of the theoretical value. However, the overall efficiency reached should still be clearly better than the total overall efficiency of water electrolysis which is 30%. Average efficiencies of thermochemical hydrogen production changes from 40% to 50% for different processes. More efficient thermochemical methods are not probable to become economic due to movement of large mass of materials in the chemical reactions to produce relatively small amounts of hydrogen. Various chemicals used in the thermochemical processes may not be recycled and because of toxic nature of some of them, they may not be environmentally unacceptable for large scale applications. High temperature thermal energy storage and suitable interface installation and special construction materials are needed because of the high temperatures attained and corrosive chemicals used in the processes (Plass et al. 1988; National Hydrogen Association, 1991).

4. Hydrogen from Unconventional Sources

4.1. HYDROGEN FROM HYDROGEN SULFIDE

A possible source of hydrogen is hydrogen sulfide which is recovered in large amounts from desulfurizing processes in synthesis gas and natural gas production and in refineries. Hydrogen sulfide is a fairly abundant resource whose potential is not being fully utilized by the industry (Zaman and Chakma, 1995). Hydrogen sulfide is converted to elemental sulfur by means of the Claus process or by other oxidizing processes. The recovery of hydrogen and sulfur from waste hydrogen sulfide is an interesting alternative to the manufacture of sulfur alone. Efforts to produce both hydrogen and sulfur from hydrogen sulfide has been made in contemporary years through a different type of technologies. These include thermal, thermochemical, electrochemical, photochemical and plasmochemical methods.

The decomposition reaction of hydrogen sulfide is endothermic. At temperatures less than 1800 K the equilibrium state is thermodynamically unsuitable with respect to hydrogen formation (Bartlett, 1979).

$$2\,H_2S \rightarrow 2\,H_2 + 1/4\ S_8\,; \qquad \Delta H_{298\,K} = 84.90\ kJ\,/\,mol$$

The reaction occurs effortlessly by the use of platinum-cobalt catalysts at 1000°C (Raymond, 1974), molybdenum disulfide or tungsten disulfide catalysts at 800°C (Kotera et al. 1976) or other transition metal sulfides catalysts supported on alumina at 500-800°C (Chivers et al. 1980).

The hydrogen yield can be extensively increased by vacating either sulfur or hydrogen from the reaction medium, thus equilibrium shifts to the benefit of decomposition products. Different methods have been utilized for the separation of hydrogen from hot hydrogen sulfide. For this purpose membranes made of zeolites, polymers, metals, glass and ceramic oxides are used and pressure-swing adsorption and thermal diffusion methods can be applied (Chivers and Lau, 1987).

4.2. HYDROGEN FROM CONVERSION OF BIOMASS

Hydrogen can be obtained from biomass with an inexpensive method. Hydrogen production from biomass would embrace accumulation, transportation and preparation of the biomass for pyrolysis/gasification process. The preparation step of biomass conversion involves heating of the biomass/water slurry to high temperature under pressure in a reactor. This heating treatment decomposes and partially oxidizes the biomass, producing a gas product abundant in CO_2, CO, nitrogen, methane and hydrogen. Mineral matter is removed from the bottom of the reactor. The gas stream going to a high temperature shift reactor is converted to a product composed of hydrogen and minor amounts of carbon oxides is purified by a pressure swing adsorption unit to produce high purity hydrogen. The system is similar to a coal gasification plant with the exception of the unit for the pretreatment of the biomass and the design of the reactor. The processing facility is larger than that of comparably sized coal gasification plant because of the lower calorific value per unit mass of biomass as compared to coal. This might result in higher capital costs (National Hydrogen Association, 1991).

5. References

Abraham, B.M. and Schreiner, F. (1974), Ind. Eng. Fundam., **13**, 30.

Barbir, F. and Veziroğlu, T.N., (1992), 'Hydrogen Energy System and Hydrogen Production Methods', In, Clean Utilization of Coal, Y.Yürüm (Ed.), NATO ASI Series C 370, Kluwer Academic Publishers, Dordrecht, p. 277.

Bartlett, R.W. (1979), Preliminary Evaluation of Process for Recovering Hydrogen from Hydrogen Sulfide, SRI International, File No. 79-447, p. 55.

Baykara, S.Z. and Bilgen, E. (1988) 'Solar Hydrogen by Hybrid Process of Water Thermolysis and Electrolysis', in T.N. Veziroglu and A.N. Protsenko (Eds.) Hydrogen Energy Progress VII, **2**, 809-818.

Baykara, S.Z. and Bilgen, E. (1989), 'An Overall Assessment of Hydrogen Production by Solar Water Thermolysis', Int. J. Hydrogen Energy, **14**, 881-889.

Bernard, G. and Vidal, J. (1984), Petrole et Techniques, **307**, 28-38.

Bertling, H. and Nashan, G. (1981), Erdöl Kohle Erdgas Petrochem. **34**, 397-401.

Bowman, M.G. (1980), 'Interfacing Primary Heat Sources and Cycles for Thermochemical Hydrogen Production', 3rd World Hydrogen Energy Conference, Tokyo.

Bridger, G.W. (1972), Chem. Process Eng. (London), **53**, 38.

Bull, S.R. (1988), 'Hydrogen Production by Photoprocesses', Proc. Int. Renewable Energy Conf., Honolulu, Hawaii, p. 413-426.

Catalyst Handbook ICI (1972), Wolf Scientific Books, London.

Chivers, T., Hyne, J.B. and Lau, C. (1980), Int. J. Hydrogen Energy, **5**, 499.

Chivers, T. and Lau, C. (1987), Int. J. Hydrogen Energy, **12**, 561.

Clauson, F.A. and Marion, C.P. (1980), AIChE-IMIQ Energy Meeting, Acapulco.

Dalluge, G. and Polster, P. (1981), Chem. Tech., **33**, 636-640.

Danno, A. (1974), CEER Chem. Econ. Eng. Rev. 7.

Davis, G.R. (1990), 'Energy for Planet Earth', Scientific American, **263**, 55-62.

Engels, H., et al. (1987), 'Thermochemical Hydrogen Production', Int. J. Hydrogen Energy, **12**, 291-295.

Farhataziz, A. and Rodgers, M.A. (1987), Radiation Chemistry, VCH Verlagsgesllschaft, Weinheim.

Hüttinger, K.J. (1988), 'Kinetics of Coal Gasification', In, New Trends in Coal Science, Y.Yürüm (Ed.), NATO ASI Series C 244, Kluwer Academic Publishers, Dordrecht, p.433.

Jellinek, H.H.G. and Kachi, H. (1984), Int. J. Hydrogen Energy, **9**, 677.

Kotera, Y., Todo, N. and Fukuda, K. (1976), US 3 962 409.

Meckel, J.F. and Flockenhaus, C. (1981), Chem. Anlagen Verfahren, 122-128.

Minet, R.G. and Desai, K. (1983), J. Hydrogen Energy, **8**, 285.

Moore, R.B. (1983), J. Hydrogen Energy, **8**, 905.

Lede, J., Lapique, F. and Villermaux, J. (1983), Int. J. Hydrogen Energy, **8**, 675.

Lede, J. (1986), Int. Chem. Eng. **26**, 647.

Lede, J., Villermaux, Ouzane, R., Hossain, M.A., Ouahes, R. (1987), Int. J. Hydrogen Energy, **12**, 3.

National Hydrogen Association (1991), 'The Hydrogen Technology Assessment', Phase I, A Report for The NASA, Washington, D.C.

Ohta, T. (Ed.), (1978), Solar Hydrogen Energy System, Pergamon Press, New York.

Plass, Jr. H.J., Barbir, F. and Veziroğlu, T.N. (1991), 'Hydrogen Systems Application Analysis', Proc. DOE/SERI Hydrogen Program Review Meeting, Washington, D.C.

Raymond, M.E.D. (1974), can. Sulfur Symp. Calgary, Canada.

Rostrup-Nielsen, J. (1977), Chem. Eng. Prog., **73**, 87-92.

Schulten, R. (1978), 'Nuclear Energy as a Primary Energy Source for Hydrogen Production', 2nd World Hydrogen Conference, Zürich.

Seider, W.D., Gautam, R. andWhite, C.W. (1980), 'Computation of Phase and Chemical Equilibrium, A Review', ACS Symp. Ser., **124**, 115-134.

Spinks, J.W.T. and Woods, R.J. (1964), An Introduction to Radiochemistry, Wiley and Sons, New York.

Steinberg, M. and Cheng, H. C. (1988), 'Modern and Prospective Technologies for Hydrogen Production from Fossil Fuels', in T.N. Veziroglu and A.N. Protsenko (Eds.), Hydrogen Energy Progress VII, Pergamon Press, Oxford, **2**, 699-740.

Stull, D.R., Westrum, E.F.,and Sinke, G.C. (1969), The Chemical Thermodynamics of Organic Compounds, Wiley, New York, p. 631.

Tarman, P.B. and Biljetina, R. (1978), The Steam Iron Process for Hydrogen Production, 85th National Meeting of the AIChE, Philedelphia.

van Krevelen, D. W. (1993), Coal, Third Edition, Elsevier, Amsterdam, pp. 741.

Winter, C.-J. and Nitsch, J. (1988), 'Hydrogen as an Energy Carrier', Springer-Verlag, Berlin, Heidelberg.

Yalçın, S. (1989), 'A Review of Nuclear Hydrogen Production', Int. J. Hydrogen Energy, **14**, 551-561.

Zaman, J. and Chakma, A. (1995), Fuel Proc. Technol., **41**, 159.

PROSPECTIVE TECHNOLOGIES FOR HYDROGEN PRODUCTION

M. STEINBERG
Department of Advanced Technology
Brookhaven National Laboratory
Upton, NY 11973
USA

ABSTRACT. A study is presented assessing the technology and economics of hydrogen production by advanced processes. The advanced processes include 1) high temperature electrolysis of steam, 2) coal gasification and electrochemical shift, 3) integrated coal gasification and high temperature electrolysis, 4) thermal cracking of natural gas and 5) the HYDROCARB thermal conversion of coal. High energy radiation, plasma and solar photovoltaic-water electrolysis production of H_2 are discussed. The advanced high temperature electrochemical systems suffer from high electrical energy and capital cost requirements. The thermal cracking of methane is potentially the lowest cost process for hydrogen production. This is followed closely by the HYDROCARB coal cracking process. To reach full potential, the thermal cracking processes depend on taking credit for the clean carbon fuel by-product. As the cost of oil and gas inevitably increases in the next several decades, emphasis will be placed on processes making use of the worlds reserve of coal.

1.Introduction

The unconventional advanced systems presented include 1) High Temperature Electrolysis (HTE) of Steam, 2) Electroconductive Membrane Process for Production of Hydrogen from Gasified Coal, 3) Combined Coal Gasification with High Temperature Electrolysis (CG-HTE), 4) Thermal Cracking of Natural Gas and 5) Thermal Cracking of coal (HYDROCARB). Other advanced processes including high temperature plasma, photolytic and radiation systems are also briefly discussed.

The processes presented were all assessed at a hydrogen production capacity of 100×10^6 SCF/D. This capacity is large enough to take advantage of economy-of-scale and small enough to still be of use for synthetic fuel manufacture. The technologies and capital and operating cost estimates are compared including by-product credit values and some conclusions are reached.

2. Hydrogen Production by Advanced Processes

2.1. HIGH TEMPERATURE ELECTROLYSIS OF STEAM

A significant improvement in the electrolytic production of hydrogen can be achieved by the high temperature electrolysis (HTE) of steam. The low temperature water electrolysis process for hydrogen production as described previously under conventional processes depends on the electrolytic conduction of an aqueous media such as alkaline or acid

Y. Yürüm (ed.), Hydrogen Energy System, 31–44.

solutions. Most of the energy of splitting water comes from the electrical energy fed to the cell. However, by raising the temperature of the cell an increasing amount of the energy for decomposing water can be thermal energy according to the second law of thermodynamics :

$$\Delta G = \Delta H - T\Delta S, \text{ and}$$
$$\Delta G = nFe$$

where

ΔH is the enthalpy of water decomposition
ΔG is the free energy of water decomposition
ΔS is the entropy of water decomposition
T is the absolute temperature
e is the voltage or energy of the cell
n is the number of electron equivalents in producing a molecule of H_2.

The electronic equations for HTE at each electrode are as follows:

Cathodic Reaction	$2\,H_2O + 4\,e^- = 2\,H_2\,(g) + 2\,O_2^=$
Anodic Reaction	$2\,O_2^= = O_2\,(g) + 4e^-$

At temperatures where water exists as steam the conduction must occur by ionic means rather than by electrolytic as in aqueous condensed systems. Ionic conductors have been developed by a number of research groups including Doenitz et al. [1] and Westinghouse Electric Corporation [2]. They consist of a porous metal oxide ceramic, usually zirconium oxide (ZrO_2) stabilized with yttrium oxide (Y_2O_3). Electrodes are placed against a thin slab of gastight ZrO_2 - (Y_2O_3). Steam is supplied on one side of the ionic conductor. On applying a direct current electrical field, hydrogen forms at the cathode while oxygen ions migrate through the ionic conductor and from oxygen molecules at the anode. The ionic conductors can be arrayed in various geometrical configurations (such as in tubular from) to produce a single cell which can be placed either in parallel or series circuitry to form a battery. These in turn can be hooked up in a bank of batteries to form a large scale reactor.

Several HTE process designs have been developed based on coal as the primary energy source [3]. One lower cost system shown in Figure 1 feeds electricity and steam from a coal-fired steam power station and coal-fired steam heater. Saturated steam from a power plant is preheated to 1100°C in ceramic conductor tubes by a coal-fired heater before being fed to the HTE plant. The steam is electrolyzed with a temperature drop of 200°C. The process can be designed to convert 50% to 90% of the steam to hydrogen and oxygen with an overall thermal efficiency of 35 to 38% based on 35% electrical power generation from coal. For increased hydrogen productivity of the cell, current densities can vary between 3000 to 5000 amps/M^2 with cell potentials varying between 1 and 1.6 volts.

A distribution of the hydrogen production cost is as follows:

	% of Production Cost:
Feedstock (includes fuels)	66.2
Capital Cost	17.6
O&M Cost	16.2
	100.0

It is noted that the feedstock cost which is primarily electricity strongly dominates the cost of H_2 production and is therefore very sensitive to the cost of electricity. Furthermore, the by-product oxygen can reduce the production cost by 16.6%.

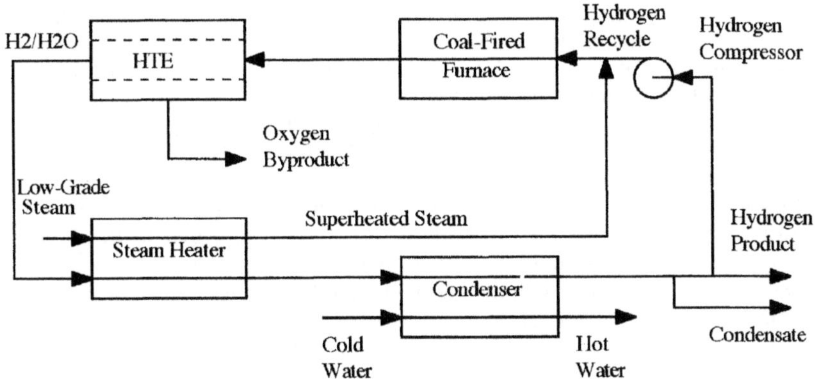

Figure 1. High temperature electrolysis (HTE) of steam.

2.2. ELECTROCONDUCTIVE MEMBRANE PROCESS FOR THE PRODUCTION OF HYDROGEN FROM GASIFIED COAL

In the realm of advanced electrochemical techniques for hydrogen production there exists a combination process wherein gasified coal is electrochemically treated to produce a highly concentrated hydrogen gas and by-product steam [4].

The gasification of coal is generally represented by the following reaction

$$C + H_2O = CO + H_2$$

The second step in the process consists of cleaning and shifting the CO to form hydrogen and CO_2 by reacting with steam. This step is conducted in a high temperature electrolysis cell across a gas impervious mixed ionic-conducting ceramic membrane. The overall reaction for converting the CO is the chemical shift reaction,

$$CO + H_2O = CO_2 + H_2$$

However, unlike the conventional water gas shift reaction which is carried out thermocatalytically, it is unnecessary to remove sulphuric impurities to a high degree prior to shifting and to remove CO_2 after the shift reaction. This results in a savings in process cost compared to the conventional process.

The core technology for this advanced shifting process is the ability of certain ceramic materials, e.g., doped stabilized zirconia, to conduct oxygen ions at high temperature while simultaneously being electronic conductors. On one side of the membrane the following reaction occurs (cathodic)

$$H_2O + 2e^- = H_2 + O^+$$

and on the other side (anodic)

$$CO + O^= = CO_2 + 2e^-$$

The overall reaction is the shift reaction given above. The oxide membrane thickness is about 20 microns. When steam passes on one side of the membrane it will be reduced at reaction sites along the membrane by the addition of electrons from the membrane resulting in the production of hydrogen molecules and the release of oxygen ions to the membrane. On the opposite side of the membrane, at reaction sites, carbon monoxide will be oxidized by reaction with the oxygen ions, which have been transported across the membrane to from carbon dioxide and release electrons to the membrane. The oxygen ions move across the membrane due to the large difference in oxygen activity on each side of the membrane. This process will continue as long as the flow of the separate reactants continues maintaining the oxygen activity gradient across the membrane.

Typically this process would be carried out at approximately 1000°C and the partial pressure of oxygen in steam would be about 10^{-8} atm. Compared to the corresponding partial pressure of oxygen in carbon monoxide at this temperature, which is about 10^{-18} atm, this ten orders of magnitude gradient would result in a theoretical voltage across the membrane of about 630 mV. However, this voltage would be reduced by the resistance of the material to ionic and electronic direct current flow, as dissipated heat (IR losses). Further reduction in the voltage across the film would result from so-called irreversible polarization losses due to oxygen adsorption and desorption reactions on the surfaces. Also in a actual reactor, where the thin membrane would be supported on a porous substrate, there would be diffusional losses through the substrate, associated with getting reactants and products to and away from electrolyte surface reaction sites.

The technology for the high temperature electrolytic membrane reactor is based on the development work for the high temperature electrolysis technology. Since the free energy for the above shift reaction is exothermic, an impressed electrical potential is not required to drive the reaction. Thus no electrical power is heeded for this process. A preliminary process design and economic estimate has been attempted as follows.

A simple flow diagram of the coal-to-hydrogen plant is shown in Figure 2, consisting of three major plant sections: The Fuel Gas Generation Section; the Hydrogen Production Section and the Steam Generation Section.

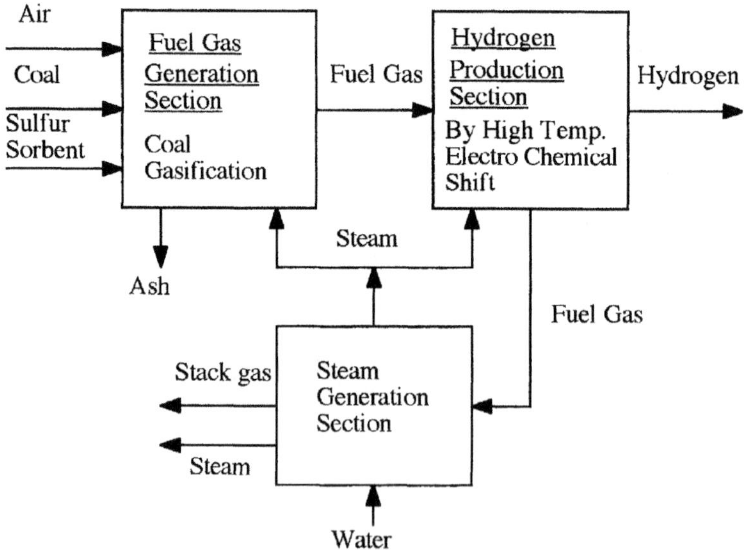

Figure 2. Simplified diagram for coal to hydrogen by gasification and electrochemical shift.

The Fuel Gas Generation Section consists of the following major components and subsystem for a 100 x 10^6 SCF/D hydrogen plant,

Coal receiving, storage, preparation, handling and feeding
Dolomite receiving, storage, preparation, handling and feeding
Coal gasifier modules with in-bed desulfurization and two stages of recycle cyclones
Air blowers
Hot-gas particulate removal using a high-temperature ceramic bag house
Ash transport, processing (calcium sulfide oxidation) and disposal
Hot-gas piping

The Hydrogen Production Section consists of:

Hydrogen Production reactor
Hydrogen dryer system

The Steam Generation Section consists of:

Fuel gas combustion system
Heat recovery steam generation system

The selected process design operates at near atmospheric pressure using air blown fluidized bed gasifier technology and incorporates desulfurization with limestone or dolomite. The gasifier operates in the range of 870 to 1040°C. The process performance is sensitive to the selected gasifier pressure and temperature, the gasifier type and the oxidant type (air or oxygen).

The process considered here is designed so that steam generated from the drying of the hydrogen stream and from the utilization of the lean fuel gas exhausted from the hydrogen production reactor is sufficient to provide all of the plant steam requirements and air compression requirements while minimizing steam generated for export which is substantial. This selection is based on the commodity value of hydrogen (about $8.00/MSCF) being higher than the equivalent commodity value of industrial steam or electricity (at about $5/1000 lb or $ 0.039 per kWh). The plant coal feed rate is about 3800 tons per day.

Except for the high temperature electrolytic membrane, most of the plant components are commercial or near commercial items. The electrolytic membrane reactor is conceived to be of a tube-in-shell design and is based largely on the extensive development and commercialization work performed by Westinghouse on solid electrolyte fuel cells. The tubes are taken as 1.3 cm diameter by 100 cm long, with a 0.1 cm wall thickness based on presently established extrusion and coating technology. The assumed hydrogen production rate per unit o tube area is 33 ml / min-cm^2, and the evaluation depends on how successful the technology development will be. The reactor costing is based on about 150 000 tubes, each weighing about 390 gms. With a stabilized zirconia price of $15/Kg, the resulting tube material cost is about $880,000. Based on solid electrolyte fuel cell experience, the manufactured tube cost of will be about $2 640 000. The total reactor cost for the installation is estimated to be less than $20 000 000, which is a relatively small portion of the total plant cost.

The plant consumes the following raw materials and power :

coal : 3800 tons/day
dolomite : 1050 tons /day
water : 4200 tons / day
auxiliary power : 5 000 kW

The plant produces for export about 1 500 000 lb/hr steam, with the pressure and temperature depending on the end use.

Costs are reported for a site located in western Pennsylvania, with an expected accuracy of 40%. A contingency of 15% is included on the process investment. Interest during construction is not included.

The distribution of production costs is shown as follows.

	% of Production Cost
Feedstock (includes fuels)	33.7
Capital Cost	43.0
O&M Cost	23.3
	100.0

The production cost is very sensitive to the gasifier type. By allowing credit for by-product steam at \$5.00 / 10^3 lb, the hydrogen product cost can be reduced by as much as 40% .

2.3. COMBINED COAL GASIFICATION WITH HIGH TEMPERATURE ELECTROLYSIS (CG-HTE)

An advanced hybrid coal gasification-HTE process is useful because the two process systems complement each other. The coal gasification system produces synthesis gas (CO and H_2) and steam at high temperature which can be used in the HTE. The HTE produces H_2 and O_2 from the steam and the O_2 produced in the HTE, can in turn be used in the gasification of coal. A simplified flow sheet of the process is shown in Figure 3. An analysis of the system based on the contribution to hydrogen production from each section of the process can be obtained as follows.

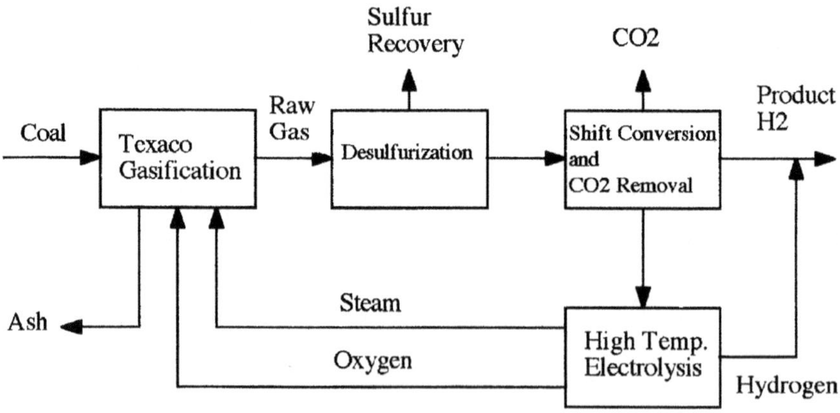

Figure 3. Simplified diagram for coal gasification integrated with high temperature electrolysis (CG-HTE).

In the gasification unit the reaction stoichiometry and energetics are :

$$C + H_2O = CO + H_2 \qquad \text{Endothermic, } \Delta H = +42 \text{ Kcal /mol}$$
$$0.5\,C + 0.5\,O_2 = 0.5\,CO_2 \qquad \text{Exothermic, } \Delta H = -47 \text{ Kcal /mol}$$

and for the shift reaction :

$$CO + H_2O = CO_2 + H_2 \quad \text{Neutral,} \qquad \Delta H = 0 \text{ Kcal / mol}$$

overall reaction: $1.5\,C + 2H_2O = 1.5\,CO_2 + 2\,H_2$

The oxidation of 0.5 mole of carbon (coal) produces the necessary energy to sustain the steam gasification of carbon.

In the HTE unit the reaction is

$$H_20 = H_2 + 1/2\,O_2 \qquad \text{Endothermic } \Delta H = +68 \text{ Kcal / mol}$$

Two moles of H_2 are produced in the gasifier for every mole of H_2 produced in the HTE. Therefore, for a combined CG-HTE process two thirds of the hydrogen in the steam electrolyzer. This production allocation is utilized in making the cost estimates. Data from the earlier estimates in the conventional Texaco gasification process and the advanced HTE process were scaled to make the estimates.

The distribution of production cost is as follows:

	% of Production Cost
Coal Feedstock (includes fuels)	39.2
Capital Cost	40.7
O&M Cost	20.1
	100.0

It is noted that the feedstock cost has been significantly reduced due to the fact that a large fraction of hydrogen is produced from the coal gasification and thus a lower fraction of electrical energy requirements for HTE. On the other hand, capital investment has gone up. The total production cost, \$4.43 / 10^3 SCF for the combined GC-HTE plant is lower than for the HTE plant alone, \$5.07 / 10^3 SCF before by-product credit is taken. however, by taking by-product oxygen credit for the HTE plant, the HTE plant alone becomes more economical than the combined plant which has no by-product credit.

2.4. THE THERMAL CRACKING OF NATURAL GAS

The thermal decomposition of natural gas ($CH_4 = C + 2\,H_2$) has been practiced for many years for the production of carbon black for rubber tire vulcanization, for pigment and for the printing industry [5]. In the industrial process, the hydrogen produced is used to provide part of the thermal energy in the process. The process is practiced batchwise with the use of tandem furnaces at near atmospheric pressure. A methane-air flame is used to heat up firebrick to temperatures in the order of 1400^oC. The air is then turned off and the methane alone decomposes on the hot firebrick until the temperature drops to about 800^oC. The micron sized carbon particulates are collected in the effluent gas stream in bag filters. The methane-hydrogen effluent gas is then used to heat up a second furnace while the methane decomposition is continuing in the first furnace. Then the flow of gas is reversed so that while the first furnace is producing carbon black , the second furnace is being heated up.

Attempts have also been made to thermally crack natural gas in a continuous fixed bed reactor [6]. It is noted that catalysts were used to increase the rate of decomposition and even carbon itself might act as a catalyst. It appears that an efficient continuous process should be possible of development for production of hydrogen with the carbon as a by-product. The decomposition reaction is endothermic by 18 Kcal/mol so that a minimum of 9% of the heat of combustion of methane (212 Kcal/mol) is needed to drive the process. A preliminary process design and economics estimate of a continuous process has been made [7,8]. Figure 4 shows a flow sheet of the continuous process. About 60% of the total Btu output of the plant can be recovered as H_2 and about 40 % as carbon. Thus the carbon can be taken as a significant by-product to reduce the cost of the hydrogen. Carbon black is a clean ashless, sulphur-less premium fuel which can command a higher value, however, credit has been taken conservatively at the same cost as the raw feedstock coal. Even though this is a new continuous process that has not been practiced on a commercial scale, the same economic rules as the conventional system have been applied here. These estimates are probably not better than about ±40 %.

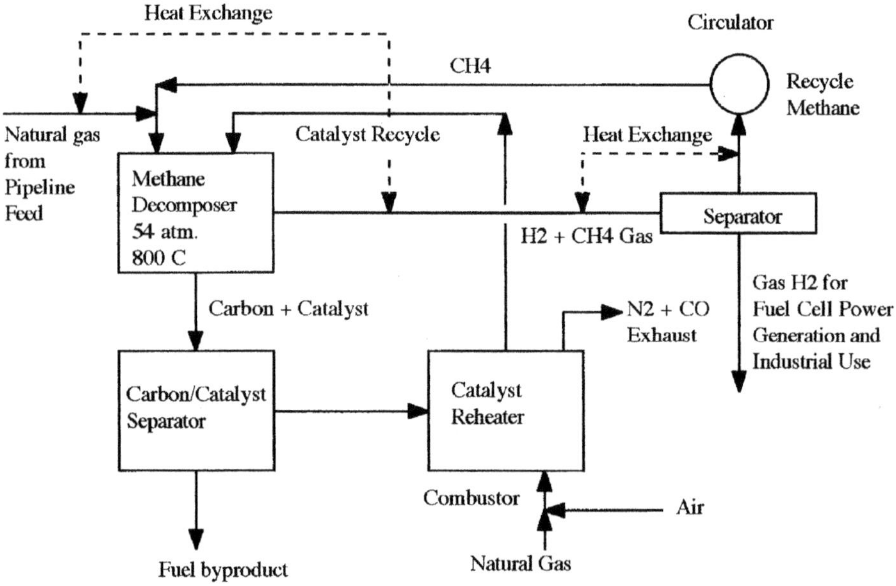

Figure 4. Simplified diagram for hydrogen production by methane cracking process.

The distribution of the production cost before credit for the by-product clean carbon fuel by-product is as follows:

	% of Production Cost
Feedstock (includes fuels)	76.2
Capital Cost	12.8
O&M Cost	11.0
	100.0

Even with the uncertainties, the net cost of hydrogen for the thermal decomposition of hydrogen comes out to be the lowest in this study at \$1.64 /10^3 SCF or \$5.10/10^6 Btu after by-product credit is taken . The by-product credit in the from of carbon as a fuel amounts to only 28% of the production cost and is not heavily dependent. However, as indicated in the above table, the cost is highly dependent an the natural gas feedstock cost. At present, the price of natural gas is low because of the oil and gas glut. As the supply becomes more restricted, the cost of hydrogen by thermal decomposition of natural gas will increase relatively more rapidly than by other processes, but by allowing by-product credit, it should always be competitive with other natural-gas-based processes such as steam-gas reforming. This analysis indicates that if carbon-black can be fully used as a fuel as claimed above, and advanced continuous thermal process should be explored for the decomposition of natural gas to produce hydrogen for near-term and future use.

2.5. THE HYDROCARB THERMAL CONVERSION OF COAL

An advanced process (called HYDROCARB) has been conceived for the production of carbon black from carbonaceous raw materials, the most abundant of which is coal [9,10]. A co-product of this process is a hydrogen-rich or a methane-rich gas. The process consists of the deep hydrogenation of the carbon containing raw material (coal) to produce mainly gaseous methane together with smaller amounts of CO, CO_2 and water. After removing water the methane-rich process stream, is sent to a thermal cracker in which the methane is decomposed to carbon black particulate and hydrogen.

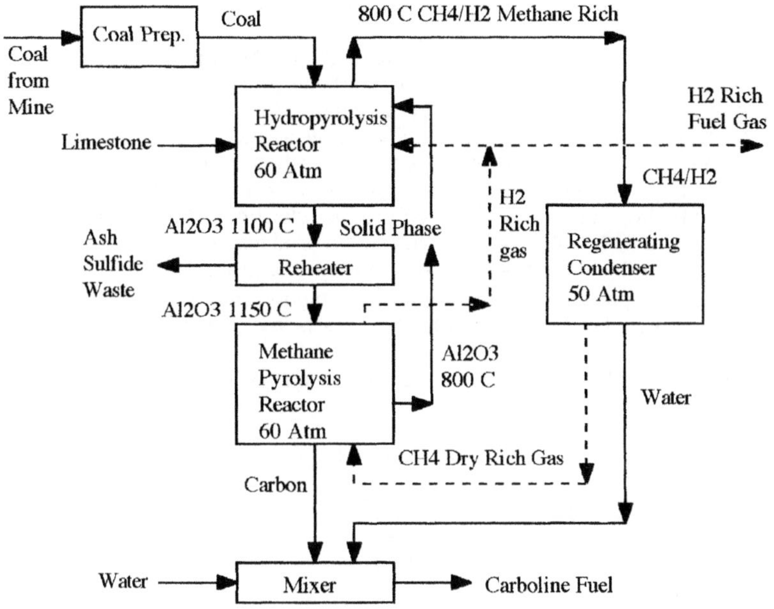

Figure 5. Simplified diagram for hydrogen production by HYDROCARB coal cracking process.

The hydrogen-rich gas is recycled back to the coal hydropyrolyzer. Depending on the market, the excess hydrogen-rich gas or methane-rich gas is recovered as a co-product and is therefore a source of hydrogen. A flowsheet of the conceptual process is shown in Figure 5.

The reaction taking place in the hydropyrolyzer operating on a bituminous coal can be expressed as follows:

$$CH_{0.8}O_{0.08}S_{0.016}\,N_{0.015} + 2\,H_2 = CH_4 + 0.08\;H_2O + 0.32\;H_2 + 0.008\;N_2$$

The sulphur is removed by addition of limestone to form CaS which is oxidized to $CaSO_4$ for disposal. The ash remains behind. The methane, together with the equilibrium amounts of CO, CO_2, and H_2O, is sent to a regenerative condenser to remove the oxygen impurities as water and then fed to a methane pyrolysis reactor at elevated temperatures. The methane is decomposed to carbon black and hydrogen according to the following reaction which has been discussed in the previous section.

$$CH_4 = C + 2\,H_2$$

The hydrogen-rich gas is then sent back to the hydropyrolysis reactor to begin the cycle once again. In effect, the process cracks coal into its elements, C, H_2, and oxygen as H_2O, with the assistance of excess hydrogen as a transfer agent according to the following overall reaction.

$$\begin{aligned} CH_{0.8}O_{0.08}S_{0.016}\,N_{0.015} &= C + 0.32\,H_2 + 0.08\,H_2O + 0.008\;N_2 \\ + \text{ash} + CaCO_3 &\quad + \text{ash} + 0.016\,CaS \end{aligned}$$

The hydropyrolysis of coal is an exothermic reaction while the methane decomposition reaction is endothermic. Process design and mass and energy balance estimates indicate that the process can be made 90% mass and energy efficient. Moving beds of granular Al_2O_3 are used in the two reactors to transfer the heat of reaction to the process gas. In one design 79.8% of the energy comes out of the process as carbon black energy and 20.2% as hydrogen energy. Thus the product from this plant is carbon black as a clean fuel for the energy market and hydrogen as a significant co-product clarified. The carbon can be burned as fuel or can be mixed with water, oil, or kerosene to produce liquid fuel mixes.

Cost estimates for the HYDROCARB process have been made [10]. These have been interpolated in terms of producing hydrogen at a production rate of 100×10^6 cu ft/day and taking the carbon as a by-product fuel with cost factors used in the conventional process mentioned earlier.

The distribution of the production cost before credit for the by-product clean carbon fuel is as follows:

	% of production Cost
Coal Feedstock (includes fuels)	50.5
Capital Cost	30.4
O&M Cost	19.1
	100.0

It is noted that with by-product credit, the production cost of hydrogen amounts to $\$1.78/10^3$ or $\$5.52/10^6$ Btu which is second to the lowest cost by-product carbon which accounts for 69% of the production cost of hydrogen. Development of this process must then depend on opening up a market for a clean carbon fuel. The significant advantage of the HYDROCARB process is that it makes use of one of the world's largest energy resources and that is coal. Coal in the near- and far-term future will always be available at a reasonable cost and thus the clean carbon and hydrogen co-products will be highly competitive with all other coal-based processes for hydrogen production such as coal gasification processes which are 1.74 to 2.40 times higher than the HYDROCARB process.

3. Other Advanced Processes for Hydrogen Production

3.1. PLASMA, SOLAR AND RADIATION PROCESSES

Several high temperature and high energy processes have been suggested in the past and are listed as follows [11]: a) Plasma-arc process, b) Photolytic laser process, c) High energy radiation process.

a) The plasma-arc essentially hearts water to high temperature by means of an electric field to an extreme temperature reading up to 5000°C. The process is equilibrium limited in that the water is cracked to seven radical and molecular components H, H_2, O, O_2, OH, HO_2 and H_2O. Up to about 50% concentration by volume of H and H_2 components is possible. In order to stabilize the hydrogen components, so that no recombination with the oxygen constituents takes place a very rapid quench of the plasma gases must take place. This can be accomplished with direct water injection or even with a cryogenic liquid. Because of this highly irreversible quench process the process, is very wasteful of energy. The requirement of electrical energy for the arc taken together with a wasteful energy system makes the process extremely expensive.

b) The photolytic process which uses light absorption in a mercury catalyzed water vapor system at a wavelength of 3060 Å is a low efficiency system in that energy must be converted from thermal to electrical to photolytic radiant energy which is then transmitted to mercury which then transmits it to water. The overall efficiency is usually less than 10%. Lasers in recent years which can emit more intense photon radiation in specific wavelengths, improves efficiencies somewhat but even these devices cannot indicate an overall efficiency greater than 10%.

c) With the advent of nuclear energy there was much effort over several decades in developing high energy radiation processes for the synthesis and production of chemicals. One system investigated was the radiation decomposition of waste. γ, β and fission fragment radiation were used. The highest yield obtained was with fission fragment radiation directly in a nuclear reactor however, the energy efficiency obtained for the process, never exceeded about 18%. Furthermore the separation of highly radioactive particulates and gases as well as the separation of hydrogen from oxygen are severe process problems.

3.2. SOLAR PHOTOVOLTAIC WATER ELECTROLYSIS PROCESS

There appears to be a growing interest in directly generating electrical energy using solar energy by means of photovoltaic (PV) devices. New solar cell materials such as amorphous silicon have reached solar to electrical energy conversion efficiencies of 12% for large area laboratory modules. Manufacturing plants are presently expected to produce cells for a cost of $1/peak watt. By directing this PV cell output to water electrolysis at 84% efficiency, it is projected that in the time frame of the next 15 or 20 years, it will be possible to produce PV cells for as low as $ 0.20 to $0.40/peak watt and to generate hydrogen in solar intensive areas such as in the Southwestern part of the U.S. for \$3 to \$5/10^3 ft^3 at capacities of 0.5 million cubic feet/day [12].

4. Comparative Assessment Discussion

Comparing the ranking of the advanced processes indicates that the methane cracking (decomposition) process has the lowest gross and net H_2 production cost. In this case, although the H_2 efficiency is not very high, the capital investment is very low which compensates for the lower efficiency. The by-product credit is significant and amounts to about 28% of the total cost.

The second highest ranking non-conventional process is the Hydrocarb Coal Cracking Process. In this case although the capital cost in seven times the methane cracking process and the H_2 efficiency is low (only 17.3%), the by product carbon black credit is so high, up to 70% of the total cost, that the net production cost is only 8.8 % higher than the methane cracking process. The Westinghouse coal gasification with electrochemical shift is the next increased production cost process exceeding Hydrocarb by another 50%, due mainly to increased capital, and O&M cost. The by-product steam credit is very significant amounting to 40% of the production cost. The next ranking process is the high temperature electrolysis followed by the hybrid coal gasification and/HTR process, the latter giving the highest net product cost which is due primarily to the fact that there is not by product credit. It should also be noted that the hydrocarb process yields the highest thermal efficiency of 91.7%. The thermal cracking processes are much lower in cost than the advanced high temperature electrolysis processes.

It should be noted that the above comparisons and conclusions hold for feedstock costs as of mid-1987. Generally it is projected that oil and gas will be available at the present low cost of about \$3/MM BTU in the U.S. for the next 10 to 20 years after which these fossil fuels will increase in cost again. However, coal cost, mainly because of the large U.S. indigenous natural resource, should remain reasonably stable. This means that natural gas and oil based processes can be very sensitive to feedstock cost increases and that the total and net cost can catch up or even exceed the coal cost. The electrolytic process cost should remain fairly stable, increasing slowly because electrical power cost based on coal and nuclear will tend to maintain lever electrical power costs. The conclusion can be reached that for the future, effort should be directed toward supporting R&D for the processes based on coal feedstock with emphasis on the thermal processes because they show the lowest hydrogen production cost after by-product credit. Thus the carbon black processes should strongly be considered for support.

Another point which should be stressed is that hydrogen production is usually integrated with production of another end-product. For example, for synthetic fuel production from coal, synthesis gas which contains hydrogen is used to produce substitute natural gas (SNG), methanol and substitute gasolines. In ammonia production the hydrogen production process is integrated with the ammonia plant. Thus the integration of hydrogen with end-use production processes, yield some economy of an integral operation. The stand alone process system is of value where hydrogen alone is used as a fuel for transportation, automotive, aircraft and rocket engines. The purpose of using hydrogen as fuel in a so-called "hydrogen economy" is its clean burning qualities, largely eliminating acid rain, particulates emission, and CO_2 emissions which lead to the greenhouse effect. The hydrogen economy is usually related to the use of a non-fossil solar and nuclear energy source, eliminating the emissions of CO_2. However, there is a possibility of basing a pure hydrogen economy on fossil fuel when applying the processes of methane cracking and HYDROCARB coal cracking. It is only in these cases that carbon can be extracted and stored, while using only the hydrogen product as a fuel [13]. The gross cost for the HYDROCARB process, although high, is nevertheless not that much more than conventional coal gasification processes and less than the conventional water electrolysis process.

As noted in each of the sections above, hydrogen by the exotic process of water splitting, lasers and plasma are in principal much higher in cost than the conventional and the advanced processes.

Hydrogen from by-product chemical and electrochemical processes are not economic because of the limited available production capacities.

Much government and industrial effort has already been expended on high temperature electrochemical and water splitting thermal processes. However, no significant reductions in hydrogen costs have resulted compared to conventional processes. On the other hand very little effort has been put into thermal cracking processes. It has been shown in this study that the thermal cracking processes have greater potential in reducing hydrogen costs especially with a market for the by–product carbon. It is recommended that development effort be expended to explore and advance the concept of thermally cracking fossil fuels for hydrogen production.

5. Conclusions

1. Of the advanced processes, the thermal cracking of methane and coal systems are worthwhile pursuing because of the large economic potential compared to the advanced electrochemical and conventional systems.

2. The advanced thermal process for hydrogen production from coal should be developed as an integrated system with a market for by-product carbon and the use for hydrogen as a fuel.

3. The high temperature electrochemical processes suffer from high energy and high capital cost.

4. Thermochemical and high energy water splitting systems are inherently less efficient and more costly than the conventional and advanced electrochemical and thermal cracking processes. The solar photovoltaic water electrolysis process has the potential of becoming a viable hydrogen producer in solar intensive areas of the world, sometime beyond the turn of the century.

5. It is recommended that research and development effort be expended in the development of the concept of thermal cracking of fossil fuels for clean carbon and hydrogen fuel production.

6. References

[1] Doenitz, W.R., Schniedberger, R., Steinheid, E. and Strucker. R. (1980) 'Hydrogen production by high temperature electrolysis of water vapor', Int.J. Hydrogen Energy, **5**, 55.

[2] Westinghouse Electric Corporation (1983) 'High temperature water vapor electrolysis using solid electrolysis cells', Technical Progress Report No.1: BNL Contract No. 585847-S, Pittsburg, PA.

[3] Lipa, M.A. and Borhan, A. (1983) 'High Temperature steam electrolysis: technical and economic evaluation of alternative process designs, BNL 51798, Brookhaven National Laboratory, Upton, N.Y.

[4] Brown, T.T., Newby, R.A. and Keirns, D. (1987) 'Letter communication, Westinghouse Electric Corp.

[5] Donnet, J.B. (1976) Carbon Black, Marcel-Dekker, N.Y., pp 16-18.

[6] Pohleny, J.B., and Scott, N.H. (1966) 'Method for hydrogen production by catalytic decomposition of a gaseous hydrocarbon stream' U.S. Patent 3,284,161 assigned to Universal Oil Products Co.

[7] Grohse, E. W. and Steinberg, M. (1987) 'A process to convert remote gas to a fungible fuel with capability for EOR', BNL 40424, Brookhaven National Laboratory, Upton, N.Y.

[8] Steinberg, M. (1985) 'A proposal to develop a process for converting natural gas (methane) to premium liquid and gaseous fuel', BNL Proposal, Brookhaven National Laboratory, Upton, N.Y.

[9] Grohse, E. W. and Steinberg, M. (1987) 'Economical clean carbon and gaseous fuels from coal and other carbonaceous materials', BNL Report 40485, Brookhaven National Laboratory, Upton, N.Y.

[10] Steinberg, M. (1987) 'The HYDROCARB Process-Conversion of carbonaceous materials to clean carbon and gaseous', BNL Report 40731, Brookhaven National Laboratory, Upton, N.Y.

[11] Steinberg, M. Beller, M. and Powell, S.R. (1974) 'A survey of applications of fusion power technology to the chemical and material processing industry', BNL Report 18866, Brookhaven National Laboratory, Upton, N.Y.

[12] Ogden, J. M. and Williams, R.H.(1988) 'Hydrogen and the revolution in amorphous silicon solar cell technology', Princeton University, Princeton, NJ.

[13] Steinberg, M. and Cheng, H. C. (1987) 'Advanced Technologies for Reduced CO_2 Emissions', BNL Report 40730, Brookhaven National Laboratory, Upton, N.Y.

ELECTROCHEMICAL AND PHOTOELECTROCHEMICAL HYDROGEN PRODUCTION

A. YILDIZ and K. PEKMEZ
Department of Chemistry
Hacettepe University
Ankara 06532
Turkey

ABSTRACT. Basic principles of the conventional water electrolysis will be presented. New developments including electrolysis with ion exchange membranes, water vapor electrolysis will be discussed. Electrochemical processes yielding hydrogen as by - product will be given. Electrochemical and photophysical details related to the photoelectrochemical hydrogen generation using semiconducting materials will also be included.

Feasibility of using electrolysis for the production of hydrogen depends largely on the availability of cheap electricity from hydropower stations. For this reason only about five percent of the world hydrogen production is accomplished by means of electrolysis.

1. Conventional electrolysis cells

The following electrode reactions take place at the electrodes of an electrolysis cell filled with a suitable electrolyte (aqueous solutions of KOH or NaOH or NaCl or electrolytes which are immobilized in polymer matrices) upon the application of a potential.

$$2\,H_2O\,(l) + 2\,e^- \longrightarrow H_2\,(g) + 2\,OH^-\,(aq) \qquad \text{Cathode reaction}$$

$$2\,OH^-\,(aq) \longrightarrow 1/2\,O_2\,(g) + H_2O\,(l) \qquad \text{Anode reaction}$$

$$H_2O\,(l) \longrightarrow H_2\,(g) + 1/2\,O_2\,(g) \qquad \text{Cell reaction}$$

From the thermodynamic data the reversible decomposition potential (E_{rev}) at standard conditions can be calculated to be 1.228 V using the following relationship.

$$w = \Delta G = n \cdot F \cdot E_{rev}$$

Y. Yürüm (ed.), Hydrogen Energy System, 45–52.

ΔG is lower than ΔH by an amount that is equal to $T\Delta S$. In other words addition of heat, $Q = T\Delta S$, is also essential for the operation of an electrolysis cell besides electrical energy. If the heat energy is to be supplied in form of electrical energy the thermal potential of 0.252 V (at standard conditions) must be added to the E_{rev} value according to the following relationship.

$$E_q = \frac{T\Delta S}{n \cdot F}$$

This means that the theoretical decomposition potential of water at standard conditions is 1.480 V. It must also be added that the irreversible processes occurring on the anode and cathode and the electrical resistance of the cell cause the actual decomposition potential of water, E_{act}, to be always higher than 1.480 V (at standard conditions). The irreversible processes on the electrodes require the application of overpotentials. Some specific precautions, such as placing suitable catalysts to the surfaces, must be taken to reduce these overpotentials. The electrical resistance value of the cell depend on the conductivity of the electrolyte, on the ionic permeability of the gas-thight diaphragm which separates anodic and cathodic regions, on the distance of the anode from the cathode and on the current density used. Common electrolytes are better conductors at 70 - 90°C, so the electrolysis cells are operated at these temperatures. Diaphragms are made of either polymers of polysulphone type or of ceramic material such as barium titanate. In some applications the electrodes are directly placed on the surface of the diaphragm. This way the space between the electrodes can be reduced and voltage drop and heat loss are minimized. The cathode material is generally steel. The overvoltage due to hydrogen evolution is lowered by coating the cathode material with a catalyst such as platinum black. The anode material is nickel or nickel coated steel. The oxygen evolution overvoltage at the anode can be lowered by using lanthanum containing perowskites or nickel - cobalt oxides. The cell is made of a corrosion resistant material such as carbon steel. Since the energy losses in the electrolyte are converted to heat the cell must be cooled by water. Water has to be added to the cell because of the water consumed by

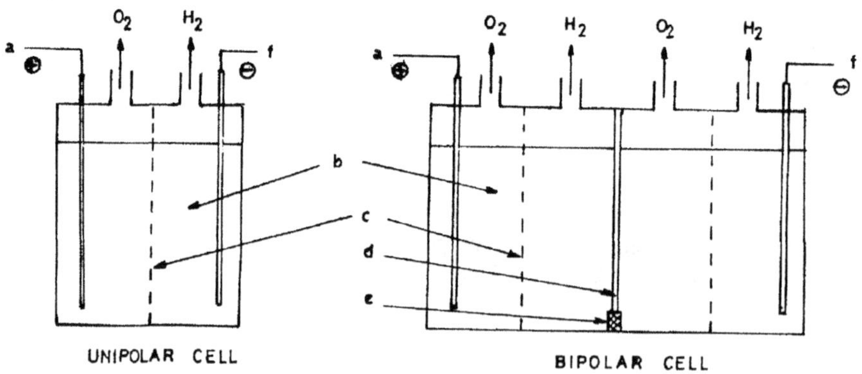

Figure 1. Schematic illustration of electrolysis cell a) anode b) electrolyte c) diaphragm d) bipolar electrode e) insulator f) cathode

electrolysis in the cell. The cells are either unipolar or bipolar. There is metallic separator between two cells in a bipolar cell and these separators are connected in series. The metallic separators which are made of nickel or nickel coated steel serve as cathodes in one of the cells and as anodes in the neighboring cells. Maintainance or renewal of a single unit in a bipolar cell is difficult in that the whole electrolysis process must be interrupted. Since the cells are connected in parallel maintainance or renewal of a single cell does not necessarily require the interruption of the whole process in unipolar cells. Figure 1 illustrates a typical unipolar and a bipolar water electrolysis cell.

The kinetic inhibition of the anodic and the cathodic processes and the voltage drop due to the electrical resistance of the cell itself cause the efficiency of the electrolysis to drop. The efficiency of the electrolysis, η, is defined as

$$\eta = \frac{E_{rev} + E_q}{E_{act}}$$

Typical cell voltages in conventional electrolysis plants vary between 1.85 to 2.05 V. This means that the efficiencies vary between 72 to 80 per cent. The current densities under typical operating condition are between 2 to 3 A / m^2. 4 - 5 kwh of power is consumed per m^3 of hydrogen production (at STP) in a typical water electrolysis unit. Electrolysis of water at increased pressures in the range between 0.6 - 20 MPa is also possible. Construction and sealing of the pressure electrolysis plants bring additional complications. Hydrogen compression is achieved in pressure electrolysis plants without significant power consumption.

2. Electrolysis using solid polymers

Thin film (0.25 mm) of a proton conducting ion exchange material such as Nafion is used in place of a liquid electrolyte as a diaphragm in these electrolyzers. Electrical charge is transported by means of hydrated hydrogen cations of the sulphonic acid group of this polymer. Cathode material is graphite and anode material is either sintered Ni or Ti. The catalysts are Pt metal on cathode and RuO_2 on anode. In a solid polymer electrolysis (SPE) unit water is added at the anodic side and serves as cooling material as well. The volume of the electrolysis unit is reduced extremely due to the absence of the KOH electrolyte system. This way extremely high current density values (up to 20 000 A / m^2) can be achieved. Principles of an SPE electrolysis unit is shown in Figure 2.

3. High temperature electrolysis of water

The total energy requirement of water decomposition is almost temperature independent. E_q value increases with increasing temperature. This means that the reversible portion of the energy, E_{rev}, which must be supplied in the form of electrical energy decreases with increasing temperature. If the thermal energy part ($\Delta H = T\Delta S$) of the total energy required for electrolytic water decomposition is supplied from thermal energy sources at high temperatures, the decomposition potential is lowered significantly. Operating the cell at high temperatures has an additional advantage of decreasing the overvoltages on the electrodes which result from the kinetic inhibition of hydrogen and oxygen evolution.

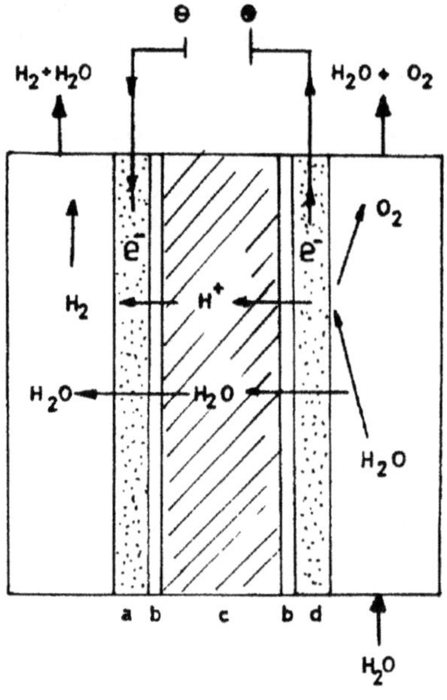

Figure 2. Electrolysis unit using solid polymer electrolyte a) porous cathode b) catalyst c) membrane, solid electrolyte d) porous anode

Doped ZrO_2 or some other rare - earth oxides which are called lattice defect conductors are used as electrolyte in high temperature electrolysis cells. These materials become oxygen ion conductor at temperatures above 800°C. 0.5 mm thin membrane made of these materials are used in a cell to reduce the electrical resistance. Porous electrodes are placed on the surface of this membrane. The cathode material is ZrO_2. Ni is placed on the cathode as hydrogen evolution catalyst. Mixed oxides with perowskite structures such as $LaNiO_3$ or $LaMnO_3$. Water vapor is fed to the cathode. The cell reactions are;

$$H_2O\,(g) + 2\,e^- \longrightarrow O^{2-} + H_2\,(g) \qquad \text{Cathode reaction}$$

$$O^{2-} \longrightarrow 1/2\,O_2\,(g) + 2\,e^- \qquad \text{Anode reaction}$$

$$H_2O\,(g) \longrightarrow H_2\,(g) + 1/2\,O_2\,(g) \qquad \text{Cell reaction}$$

About 1.3 V is applied between the electrodes and 4 A / m^2 current densities are obtained in a typical water vapor electrolyzers operating at 900°C and 3 kwh of power is consumed per m^3 of hydrogen produced at STP. Operating principles of a high temperature electrolysis cell is illustrated in Figure 3.

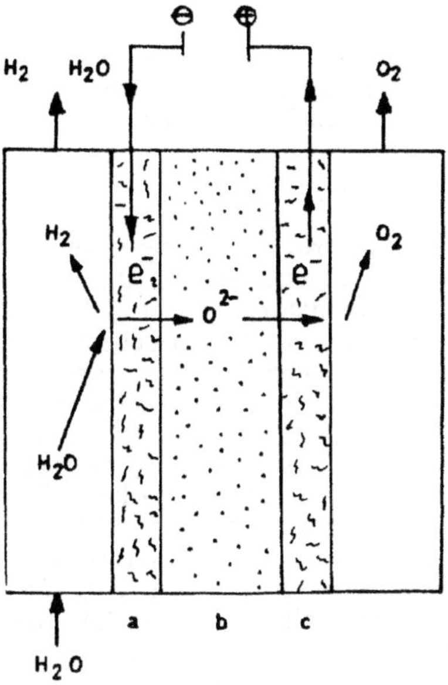

Figure 3. High temperature electrolysis cell using ZrO_2 solid electrolyte a) cathode b) electrolyte c) anode

4. Electrolysis of sodium chloride

In chlor - alkali process high purity hydrogen is obtained as by - product at the cathode. The overall cell reaction in so - called membrane process which produces chlorine at the anode is as follows

$$NaCl\,(aq) + H_2O\,(l) \longrightarrow 1/2\,H_2\,(g) + 1/2\,Cl_2\,(g) + NaOH\,(aq)$$

Sodium hydroxide is also another main product. The cathode material is stainless steel or nickel and the anode material is titanium. Diaphragm of the electrolysis cell is made of polysulphones. Electrolysis of hydrochloric acid for waste disposal of aqueous acidic solutions also produces pure molecular hydrogen as by - product at the cathode.

5. Photoelectrochemical water cleavage

Cleavage of water using semiconducting electrode materials or semiconducting microheterogeneous particles is also possible. The photoelectrochemical process is based on the band - gap absorption of photons by the semiconductor upon which the electrons are elevated from the valence band to the conducting band creating electron - hole pair. Following this charge separation oxidation of water molecules at the surface of the n - type semiconducting anode takes places

$$H_2O\ (l) \longrightarrow 1/2\ O_2\ (g) + 2\ H^+\ (aq) + 2\ e^-$$

The cathode of the photoelectrochemical cell is made of a metal onto which hydrated protons are transported. Transfer of electrons from the Fermi level of the metal causes the reduction of hydrated protons producing molecular hydrogen.

$$2\ H^+\ (aq) + 2\ e^- \longrightarrow H_2\ (g)$$

In order to achieve the cleavage of water the band gap energy of the semiconductor must be greater than the decomposition potential of water and the edge of the conduction band must be above the redox potential for the oxidation of water. $SrTiO_3$ anodes are usually used as n - type semiconducting surfaces in photoelectrochemical cells. This stable material has a band gap energy of 3.2 eV which permits the absorption of only ultraviolet radiation. Figure 4 illustrates the basic principles of photoelectrolysis of water using strontium titanate anode. In order to harvest the visible portion of the light semiconducting materials with much smaller band gap energies, such as CdS are used. Photolytic corrosion of photoactive semiconductors like CdS bring a further complication. Wide band gap semiconductors are more stable although they suffer from intrinsically inefficient use of the solar spectrum. Photoresponse of these stable materials can be shifted into the visible region of the solar spectrum using chromophores called sensitizers. One approach is to attach the anionic dyes to the ionic structure of an electrically conducting material such as polypyrrole which is coated on the surface of the wide band gap semiconductor. One other advantage of using polypyrrole coatings at the semiconductor surface is that an improved charge separation is obtained. Reduced photocorrosion is also observed when polypyrrole is used as coating on narrow band gap semiconductors such as CdS.

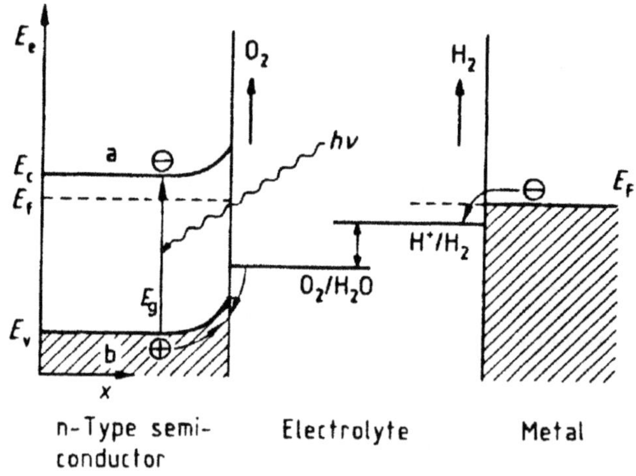

Figure 4. Photoelectrolysis cell using n - type semiconductor anode ($SrTiO_3$) and a metallic cathode a) conduction band b) valence band Ef: Fermi energy Eg: Energy gap between valence and conduction band

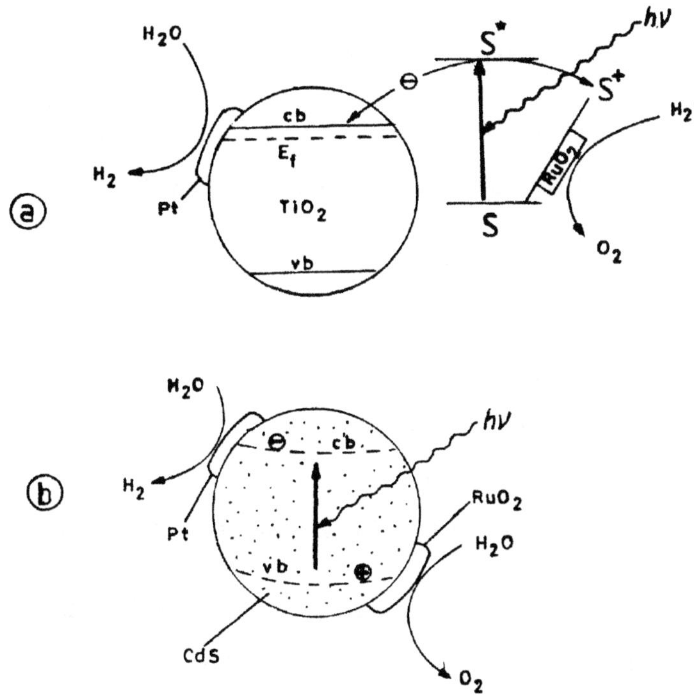

Figure 5. Catalytic photodecomposition of water using n - type semiconducting powders a)with b) without sensitization. vb: valence band, cb: conduction band, E_f: Fermi energy

Photocleavage of water is also achieved using microheterogenesus semiconducting dispersions of titanium oxide or cadmium sulphide. Pt and RuO_2 catalysts are deposited on these particles whose diameters are 100 - 500 nm to facilitate to hydrogen and oxygen evolution. Oxygen remains attached to the surface as peroxide while hydrogen is continuously evolved from the irradiated slurry. The same sensitization and / or photocorrosion protection methods that are discussed above are also applicable to the semiconducting dispersions. Figure 5 shows the schematic diagram photoelectrochemical dispersion systems with and without sensitization.

6. References

Bard, A. J. and Faulkner, L., 1980, 'Electrochemical Methods' John Wiley and Sons, New York

Carpetis, C., 1981, 'Dtsch. Forsch. Versuchsanst. Luft Raumfahrt Nachr.' **34**, 15

Casper, M. S., 'Chemical Technol.' Review No 102, Noyes Data Corp. (Park Ridge, N. J.)

Duhong, D., Serpone, N. and Gratzel, M., 1984, 'Helv. Chim. Acta' **67**, 1012

Gerischer, H. and Mindt, W., 1968, 'Electrochim. Acta' **13**, 1329

Tomkiewicz, M. and Fay, H., 1979, 'Appl. Phys.' **18**, 1

Ullmann's Encylopedia of Industrial Chemistry, 1985, **A 13**, 333

Wendt, H.,1984, 'Chem. Ing. Tech.' **56**, No 4, 265

Wendt, H., Winter, C. J. and Nitsch J., 1986, 'Wasserstoff als Energietager' Springer Verlag, Berlin, 179

Yıldız, A., Sobezynstki, A., Bard, A. J., Campion, A., Fox, M. A., Mallouk, T. E., Webber, S. E. and White J. M., 1989, 'Sensitized polypyrrole - coated semiconducting powders as materials in photosystems for hydrogen generation' Langmuir, **5**, 148

TECHNOLOGIES FOR REDUCING CARBON DIOXIDE EMISSIONS FROM FOSSIL FUEL FIRED INSTALLATIONS

M. STEINBERG
Department of Advanced Technology
Brookhaven National Laboratory
Upton, New York 11973
U.S.A.

ABSTRACT. All mitigation scenarios proposed to date either tend so slow down the rate of atmospheric CO_2 emissions or level out the CO_2 concentration in the atmosphere. A unique system has been devised which offers a method of actually reducing the CO_2 concentration in the atmophere while still generating energy from fossil fuel. The only economically and technically reasonable method for removal of CO_2 from the atmosphere is by the process of solar photosynthesis which extracts carbon from atmospheric CO_2 by formation of biomass, e.g., lignocellulose. The biomass is then thermochemically converted by the HYDROCARB technology with fossil fuel, gas oil or coal to produce carbon black and methanol. The carbon is returned to the earth, for longterm storage, while the methanol is used as fuel. In this manner, methane gas and oil coprocessed with biomass results in a net removal of CO_2 from the atmosphere of about 78 lb CO_2 per million Btu of methanol generated energy. There is also large energy enhancement in utilizing gas and oil for producing methanol compared to conventional methanol processes using these fuels. Coprocessing with bituminous coal results in no net CO_2 emitted or removed per unit of methanol energy generated. The process chemistry system is described and quantified in this paper.

1. Introduction

Much attention is being given currently to understanding global climate change and the effect of increasing concentrations of atmospheric CO_2 due to fossil fuel combustion. Studies have been conducted for the Carbon Dioxide Research Division (CDRD) of the U.S. Department of Energy relating to possible mitigation technologies which would tend to reduce CO_2 emissions in the future.

2. Fossil Fuel Substitution

Natural gas produces the least amount for CO_2 unit of energy generated, followed by petroleum. Coal produces the greatest amount of CO_2 per unit of energy released as shown in Table 1 (Steinberg 1983 a, b). Natural gas can generate about twice as much energy per unit of CO_2 as coal. The nearterm approach to the reduction of CO_2 emissions is to burn more methane and less coal. However, the projected recoverable reserves for each of these fuels indicate a natural gas reserve lasting only decades while

Y. Yürüm (ed.), Hydrogen Energy System, 53–68.

coal can last hundreds of years. The theory of abiogenic origin of natural gas which projects much larger reserves of deep natural gas has not been proven and if indeed these deposits do exist, they may be too expensive to recover.

Table 1 also indicates CO_2 emissions from coal based synthetic hydrocarbon fuel processes which are even larger than emissions from natural fuels due basically, to the inefficiencies of the coal conversion processes.

TABLE 1. CO_2 Generation from Various Fuels

Fuel Type	Chemical Formula	Value BTU/lb	CO Generated $lbsCO_2/lb$ fuel	CO Generated lbs CO_2 /MMBTU	Energy Generated Kw H(E)/lbCO_2
Natural					
Bituminous coal	$CH_{0.8}O_{0.1}$	12 700	2.59	215	0.56
Fuel oil and gasoline	$(CH_2)N$	19 600	3.14	160	0.70
Biomass(wood)	$(CH_2O)N$	8 000	1.47	180	0.62
Natural gas	CH_4	24 000	2.75	110	1.01
Synthetic					
SNG from coal gas	CH_4	24 000	7.90	330	0.34
Liquid HC from coal liq.	$(CH_2)N$	20 000	6.30	310	0.36
Hydrogen from Natural gas	H_2	61 000	7.00	110	1.01
Hydrogen from Coal gas	H_2	61 000	16.50	270	0.41
Hydrogen from non-fossil energy (Hydro, solar, nuclear)	H_2	61 000	0	0	-

3. Nonfossil Solar and Nuclear Energy

Solar energy and nuclear energy substitutes for our energy needs are ideal for mitigating the CO_2 problem since these sources do not generate CO_2. A "hydrogen economy" has been suggested based on solar and nuclear energy. However solar energy is a diffuse form of energy and, for generating large quantities of heat or electrical energy, large land areas are required. It is estimated that due to economic and technical restraints it may not be possible to generate more than about 20% of our energy needs by solar energy in the next fifty years.

Nuclear energy is a good form of energy to mitigate the CO_2 problem. However, it suffers from unfinished technology and safety risks including (1) closing the fuel cycle by finding a final resting place for long-lived fossion wastes. (2) converting or breeding fossil fuel from fertile natural uranium and thorium fuel, (3) improved safety of reactors, to reduce the risk of such accidents as those which occurrred at Three Mile Island and Chernobly and (4) disposal of excessed nuclear weapons material.

4. Removal, Recovery and Disposal of CO_2

A systems study for removal, recovery, and disposal of carbon (Steinberg and Cheng 1984) dioxide from fossil fuel power plants in the U.S. has been made. The U.S. contributes about 30% of the world's total man-made CO_2 emissions. Utilities which consitute large central sources of CO_2 account for about 30% of the total U.S. emissions. Estimates indicate that a reduction in the annual incremental atmospheric CO_2 content of up to 10% would be achieved by application of CO_2 control systems to the U.S. utility industry.

The projected removal and recovery system employed at each power plant site is based on an absorption/stripping system using an improved stack gas scrubbing solvent. The recovered CO_2 is liquefied for transmission to the ultimate disposal site. The CO_2 control process is integrated with the power plant operation in that low back-pressure steam from the power-generating turbines is used to regenerate the solvent. In this manner, for 90% removal of CO_2, the power plant efficiency is reduced by only 3 percentage points from conventional plant efficiency of 38% to a value of 35% with the control system.

Three methods of ultimate disposal are possible: (1) injection into 300-M-deep ocean below the thermocline, located approximately 100 miles off the coastal shores of the U.S. No CO_2 exists in the waters below the thermocline and it thus constitutes an unlimited sink or storage site for CO_2; (2) storage in depleted oil and gas wells and (3) storage in excavated salt domes. The U.S. Department of Energy, Federal Regional System, which divides the U.S. electrical generating plant capacities into ten separate geographical areas, was utilized to design the disposal system. Ocean disposal was used for the coastal regions as shown in Figure 1. CO_2 is disposed of by dissolution below the thermocline or by deep pumping of the liquid CO_2 to 3000 M where its density is greater than that of sea water at that depth and the liquid CO_2 can sink to the bottom of the ocean floor and remains there. Lower capacity depleted gas, oil, and excavated salt wells can be employed for the inland regions. Small (6 in. dia.) pipelines carry the recovered liquefied CO_2 from each power plant to collection centers in each region, and from there the CO_2 is transmitted in larger lines (36 in. dia.) to the ultimate ocean disposal site. The operating and investment costs for the entire control system were estimated. Approximately 16% of the power plant capacity is required to drive the entire control system in those regions where coal is mainly consumed. whereas only about 11% is required where gas is predominantly used. Over 95% of the power used by the control system is needed for the removal, and liquefaction of CO_2 at the power plant sites, and the remaining 5% is used for pumping to ultimate disposal. The overall capital investment required for the control system ranges from 70% to 150% of the existing capital investment in the power plants, depending on the region. Thus, on the average, the CO_2 control system could double the investment cost for power generation. Depending on the region of the country, the productions cost of electricity with the control system would increase by 56% to 100% over the cost without the control system. Thus, the average cost of electricity would increase by 75%.

It is estimated that the worldwide application of a CO_2 control system could reduce the annual incremanetal CO_2 concentrotion only up to about 30%. However further studies are needed concerning the CO_2 concentration distribution for various ocean depths to determine which depth is most appropriate for ocean disposal of CO_2 and most important, it is absolutely necessary to study the effects of CO_2 injection the ocean ecology.

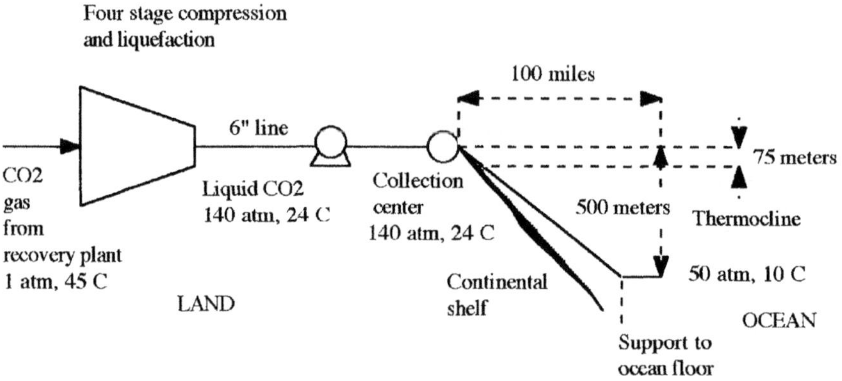

Figure 1. Schematic diagram for ocean disposal system.

5. Markets for CO_2

Markets for CO_2 utilization as in such products as plastics, refrigeration, carbonated beverages and enhanced oil recovery are too small to absorb the large quantities of CO_2 generated by the energy economy, Assuming the problems of nuclear energy can be solved, then the recovered CO_2 from fossil fuel plants can be reduced by hydrogen generated with the use of nuclear energy to produce such widely useful high volume hydrocarbon fuels as gasoline and methane (Steinberg, 1983 a, b).

6. Reforestation

Another approach to keeping the atmospheric CO_2 in balance while utilizing fossil fuel is to reforestate (Dang and Steinberg, 1980). This requires planting enough trees to absorb CO_2 at a rate which would balance the emission of CO_2 from fossil fuel burning power plants. The trees at the end of life could then be cut and buried. The land mass areas required to compensate for fuel power production are very large and besides not enough is yet known about the effect of CO_2 concentration on growth rate.

7. Improved Energy Technology

A more viable nearer term study has been made which explores the potential for more effective use of energy through improvements in energy technology and projects the consequent reduction in future buildup of atmospheric CO_2 which would mitigate the expected climatic and environmental changes (Steinberg, 1983 a,b). The study was done from the perspective of the perspective of the physical sciences rather than the social and political sciences. It focused on how energy is consumed for a specific energy service rather than on technical tradeoffs such as the use of mass transit instead of automobiles.

The efficiencies of end-use technologies vary widely among different countries. The evolution of these efficiencies, and differences among countries, indicate the magnitude of the potential for global energy savings through development and implementation of improved end-use technologies. In addition, the continuing trend toward a more electrified future world, along with the high energy-intensiveness of electricity generation, demonstrates the importance of electricity-generation technologies to future fuel consumption. The study was therefore performed to investigate the effects on global energy consumption and CO_2 emissions of end-use and electricity-generation technologies, with emphasis on efficiency improvement.

The energy technologies were first grouped by evolutionary stage as (1) now in use. (2) emerging, and (3) advanced or "far-out". The energy-efficiency levels associated with each technology were evaluated in terms of (1) present, (2) achievable, and (3) theoretical maximum values. In screening for global applicability, the improved technologies were considered with regard to the potential energy savings achievable by their introduction into three end-use sectors: (1) residential and commercial, (2) industrial, and (3) transportation.

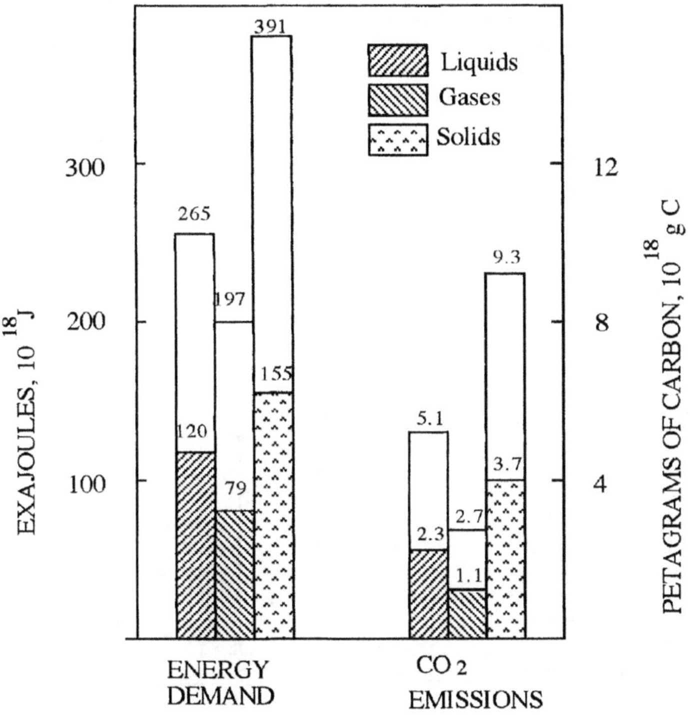

Figure 2. Global energy demand and CO_2 emissions by fuel type with and without technology improvements for year 2050. The difference between the two values in each column represents the effects of technology improvements on energy demand and CO_2 emissions.

The method for deriving the benefit in terms of the reduction in CO_2 emission resulting from a technology improvement is to compare the global energy demand and associated CO_2 emission with and without the improvement. The projections were made for the year 2050 with 1975 technology data used as a base. The global energy demand projections by fuel type were obtained from the Edmonds and Reilly energy-economic scenario by using the Institute for Energy Analysis and Energy-CO_2 model. The possible energy savings due to improved end-use technologies were estimated first in terms of efficiency improvements applicable to specific energy services for each end-use sector. Then the use of the electricity-generating technologies was assumed to determine the additional savings in fuel. This approach allows estimation of the possible minimum fuel requirements for each energy end-use service, and it therefore allows determination of the potential savings of fossil fuels and the associated reduction of CO_2 emissions which can be expected from technology improvements that might be adopted globally.

Consideration of the effects of end-use technology improvement on future energy consumption requires detailed data on energy demand by energy service for each fuel type and end-use sector. These date are not available, except for the U.S. in the mid-1970's. Furthermore, data for energy demand by end-use sector for each fuel type are not available for developing countries. In this study, the detailed energy demand data by service for each fuel type and end-use sector in the U.S. in 2050 were estimated from the allocation of energy needs for services in the mid-1970's by assuming the patterns of energy use to be constant.

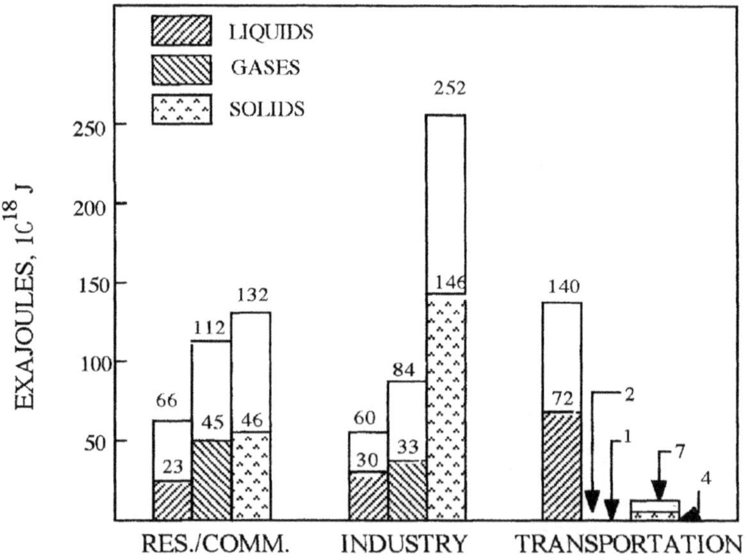

Figure 3. Global sectoral energy demand by fuel type with and without technology improvements for year 2050. The difference between the two values in each column represents the fuel savings due to technology improvements.

It was thus possible to apply the high-efficiency end-use technologies to each energy service to calculate the energy needs and savings for each fuel type and sector in the U.S. in 2050. Since the industrially developed regions such as Canada, Western Europe, and Japan have technology bases and economic-growth pattern trends similar to those in the U.S., the percentage fuel savings (i.e., technology improvement factors) calculated for the U.S. for each fuel type and end-use sector an be used to obtain the fuel savings in these developed regions.

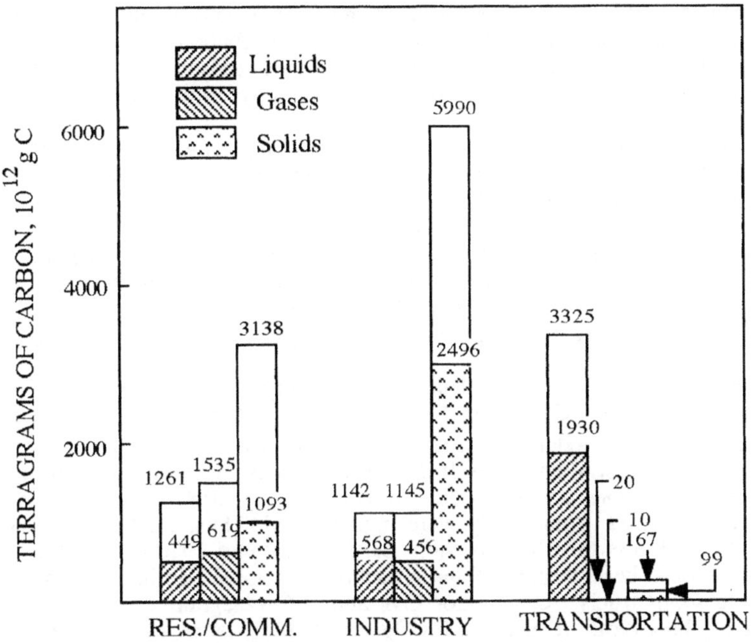

Figure 4. Global sectoral CO_2 emissions by fuel type with and without technology improvement for Year 2050. The difference between the two values in each column represents the reduction in CO_2 emissions due to technology improvements.

Only aggregate energy-demand data by fuel type for end-use are available for the developing regions including the U.S.S.R., Eastern Europe, China Africa, etc., Economic activities are by far the predominant factor in determining the way energy is consumed. Therefore, the way industrial countries use energy in the future should help to set the long-term economic prospects and thus the energy use for the developing regions. With differences in available resources and stages of industrialization, as well as other factors taken into account, the patterns of future energy consumption in the developing regions were approximated by those of the pertinent developed regions for various future times. It was noted that a smaller industrial base and higher future growth rate in the many developing countries provide substantial opportunities for introducing energy-efficient technologies at an earlier stage of development. Therefore, the effects of technology on future energy demand in the developing regions can be estimated by applying the relevant technology-improvement factors calculated for the developed regions. Another method of estimating the energy demand and savings for the developing regions was also used. Instead of estimating the pattern of future energy consumption in the developing regions, the aggregate percentage of fuel savings for each fuel type calculated for the industrially developed regions was applied directly to the corresponding aggregate fuel demand in the developing regions. The differences between the values of energy demand and savings found by these two approaches were negligible.

To account for the effects of improvements in electricity-generation technology, the electricity demands and the savings resulting from improvements in electricity utilization were assigned to each primary fuel according to its contribution to electricity demand without technology improvement. The resultant fossil-fuel needs for electricity generation and the corresponding fuel savings due to technology improvements were then calculated.

Estimates obtained with the foregoing methodology indicated that the more efficient end-use and electricity-generation technologies would lead to global savings in the year 2050 of about 145 exajoules (10^{18} joules) of liquid fuel, 118 exajoules of gaseous fuel, and 236 exajoules of solid fuel that is overall fossil fuel savings of about 500 exajoules. These savings amount to a 58% reduction in the demand for fossil fuel in 2050 compared with the demand based on the technologies and energy efficiencies of the mid-1970's. The actual fuel savings should be even larger when account is taken of the associated additional energy that would have been needed for producing and transporting the fossil fuels that were not used.

The amount of CO_2 emitted is calculated by multiplying the amount of a fossil fuel consumed by the amount of CO_2 emitted per unit quantity of that fuel. Such calculations showed that an overall reduction of 10 petagrams (10^{15} g) of carbon emission (i.e., reduction of carbon emission from 17 petagrams without improvement to 7 petagrams with improvement) could be achieved in 2050 if the technology improvements discussed in this study were implemented. Thus, the CO_2 emissions in 2050 could be 59% less than the emissions produced by mid-1970's technology. Solid fuel accounts for the largest share of the reduction in emitted carbon, 5.6 petagrams; liquid and gaseous fuels account for 2.8 and 1.6 petagrams, respectively. The distribution of the overall reduction of CO_2 emission in terms of end-use sector is 3.8 petagrams of carbon from the residential and commercial sector, 4.7 petagrams from the industrial sector and 1.5 petagrams from the transportation sector. These data are shown graphically in Figures 2, 3 and 4.

With the savings of electricity generated from non-fossil energy (i.e., nuclear, solar, hydro) due to technology improvements, consideration was also given to the potential for even greater fossil-fuel savings through replacement of fossil fuels by non-fossil energy froms for electricity generation. It is estimated that such replacement would produce and additional savings of 13 exajoules of liquid, 14 exajoules of gaseous fuels, and 56 exajoules of solid fuels. This result is in line with the forecast of an increasing role of coal and a shrinking role of liquid and gaseous fuels for electricity production. Combining the improvements of end-use and electricity-generation technologies with the replacement of fossil fuels for electricity generation indicates on overall fossil-fuel saving of 580 exajoules in 2050, which is 68% of the fuel demand for the 1975 base case. The associated overall reduction of CO_2 emissions would be 11.8 petagrams of carbon or 68%. If the current trend toward increased electrification continues, nonfossil fuel electricity generation will probably have a significant impact on the future structure of energy use.

An order-of magnitude assessment of the costs of implementing improved energy technology was also made, assuming instantaneous or overnight construction and mature technologies. The total capital costs for implementing the improved technologies for 2050 were estimated at about 6 trillion 1980 U.S. dollars and 24 trillion for the world. The return on investment (ROI) calculated purely on the basis of fuel savings alone was 8% for the U.S. and 10% for the world. The industrial sector is more effective than the others in reducing CO_2 emissions, yielding a 50% overall reduction in CO_2 emissions, requiring lower capital investment for technology implemention (6% of the overall capital requirement) and showing a higher return on investment (40%).

Whether these estimated energy savings and the associated reduction in CO_2 emissions will in fact be achieved, and to what extent, is difficult to predict because of uncertainties related to future economic growth, industrial investment activity, and infrastructure changes. In general , the energy-efficient technologies discussed in this study require substantial capital investment either for changes in existing stocks or for construction of new stocks employing them. The rate and structure of economic growth will certainly affect the rate of introduction of the new and improved technologies. The turnover of capital stock, and thus of energy technology, is a function of investment activity, just as the choice among different new technologies is a function of the perceived relative cost assumed to prevail over the expected economic lifetime of the technologies. Thus, the choice of future technologies and efficiencies is a function of costs and the related socio-political structure which determine the choice of sectoral energy services and processes as well as the supply of energy sources. however, the large margins of energy efficiencies between present and future improved end-use technologies indicate the available potential for major longterm reductions in energy consumption through technological improvement.

The calculated energy savings and CO_2 reductions may represent an upper limit; those actually achievable in practice could be somewhat lower. Nevertheless, the fact that improved technologies are sufficiently well known for use today plus the finding that the potential for energy savings and CO_2 reduction is very large indicate the great benefit of technology improvement in determining future energy consumption and CO_2 emissions. The results of this study strongly suggest the need for a deeper investigation into the technical, economic, and socio-political uncertainties of achieving the possible technological improvements in global energy production and use. Compared with other suggested ways of coping with the possible global "greenhouse effect", the pursuit of improvements in energy technology appears to be, by far the most realistic and cost effective. Furthermore by applying improved energy technology, the world stands to realize an economic return on the required capital investment, which is an incentive that cannot be ignored and is in itself a worthwhile goal. Studies of this type should be pursued principally in gathering much improved information on the demographics and potential of applying improved technologies in the major and potentially major CO_2 emitting countries of the world.

8. Hydrogen Economy Based on Fossil Fuel

A very advanced technology based on fossil can be cited. This is the possibility of generating and utilizing hydrogen from fossil fuel sources without generating any CO_2 whatsoever. A two-step process has been devised which consists of (1) hydrogenating coal (or any other fossil fuel) to produce methane followed by (2) thermal decomposition of methane to pure carbon and hydrogen (Steinberg, 1987). The steps in the process which we call HYDROCARB are shown in Table 2.

TABLE 2. HYDROCARB (CARBEX) Process for Production of Ash-free and Sulphur-free Clean Carbon Fuel (Carboline) and Fuel Gas from Coal

Two basic steps

I. *Hydrogasification* of coal with hydrogen to methane.

Bituminous coal + hydrogen → methane + water + hydrogen

$CH_{0.8}O_{0.08}S_{0.016}N_{0.015} + 2\,H_2 \rightarrow CH_4 + 0.08\,H_2O + 0.32\,H_2 + 0.008\,N_2$
+ Ash+ Limestone ($CaCO_3$) + Ash + 0.016 CaS

Reaction is ~ - 18 Kcal/mol *Exothermic.*

II. *Thermal decomposition* of Methane

Methane → Carbon black + Hydrogen

$CH_4 \rightarrow C + 2\,H_2$

Reaction is ~ + 18 Kcal/mol *Endothermic*

Hydrogen is recycled. Excess hydrogen is a byproduct and is used as fuel. and carbon black is slurried with water to produce CARBOLINE CWM fuel product.

Overall Coal Cracking Reaction

$CH_{0.8}O_{0.08}S_{0.016}N_{0.015} = C + 0.32\,H_2 + 0.08\,H_2O + 0.008\,N_2$
+ Ash + $CaCO_3$ + Ash + 0.016 CaS

By the principle of recycling, the process can be made highly energy and mass efficient and thus economical. The net effect is a cracking of the fossil fuel to its elements carbon and hydrogen while the oxygen and sulphur impurities are disposed of as water and sulphated ash. The amount of energy in the fossil fuels which can be converted to hydrogen energy is shown in Table 3. From natural gas, 56% of the energy content results in hydrogen generation. This reduces to a net of 24% when utilizing coal to generate hydrogen in this fashion. This is not inconsequential, considering that the so-called "hydrogen economy" based either on solar or nuclear energy must generate electricity first at approximately 30% efficiency and then utilizing the electrical energy in an electrolytic cell to decompose water to hydrogen and oxygen which is, at best, 80% efficient, results in an overall efficiency of 24% and this is about the same as generating hydrogen from coal by the above described process. It should be noted that the bulk of the energy is in the clean carbon produced which can be placed into "monitored retrievable storage". Should it be found, in the future, that the CO_2 greenhouse effect is not taking place, then this clean carbon can be readily removed from storage and used as a clean source of energy. The HYDROCARB process allows a "hydrogen economy" to be based on fossil fuel energy.

TABLE 3. Generating energy from fossil fuel without CO_2 emission.
The hydrogen economy based on fossil energy.

Principle: * Extract hydrogen from fossil fuels
* Use hydrogen only as fuel
* Return clean carbon to ground for possible future use as CO_2 environment permits

Fossil fuel	Heat of combustion HHV Kcal/mol	Cracking process	Atomic H fraction in fossil fuel	Heat of cracking Kcal/mol C	Net energy in H % of fossil fuel
Nat Gas	-212	CH_4 $C+2H_2$	80	+18	64-5 = 59%
Petroleum					
Alkanes	-165	CH_2 $C+H_2$	67	+6	41-4 = 37%
Aromatics	-142	CH $C+1/2H_2$	50	-3	24+1 = 25%
Coal	-116	$CH_{0.8}O_{0.08}$ 0.96 C+0.04 CO_2 + 0.4 H_2	43	0	24±0 =24%

Note: Heat of Combustion of C = - 94 Kcal/mol.
Heat of Combustion of H_2 = - 68Kcal/mol.

The HYDROCARB Process, conceived principally for coal utilization is a thermally efficient process for conversion of any carbonaceous material (carbon, hydrogen and oxygen) to carbon black and a hydrogen-rich co-product fuel (Straus et al., 1988). The carbonaceous material is first hydrogasified in excess hydrogen to form a methane-rich gas (CH_4) with small amounts of CO and H_2O. This is an exothermic process and does not require steam or oxygen as in conventional gasification. The methane-rich gas is then thermally decomposed to carbon and hydrogen in an endothermic process and the hydrogen is recycled to the first step. By catalytic reforming, methanol co-product can be produced from the CO formed during hydrogasification in an intermediate step. As previously stated, photosynthesis is the only energy reasonable process for fixing CO_2 from the atmosphere. By tying the thermo-chemical HYDROCARB process to the biochemical photosynthesis process the organic material from the photosynthesis process can be co-processed with fossil fuel and converted to carbon black; a practical storable material and a hydrogen-rich gas or methanol. It is physically easier to sequester solid carbon than gaseous CO_2. Figure 5 shows a schematic of the overall system. With this system, the trend toward increasing CO_2 in the atmosphere can not only be slowed down but can also be reversed. The methanol which is hydrogen-rich can be used as fuel and the carbon can be stored in the earth. Because of their physical properties it is much easier to store carbon than it is to store CO_2.

However, to achieve reversal, we do have to balance the rate of photosynthesis versus the rate of consuming fossil fuel. For reversal, the rate of the former must be greater than the rate of the latter. The key is turning the lignocellulose and other bio-organic material formed by photosynthesis back to carbon in an efficient manner. The following outlines several options in utilizing HYDROCARB for that purpose. The overall stoichiometry of the process is represented by the equations.

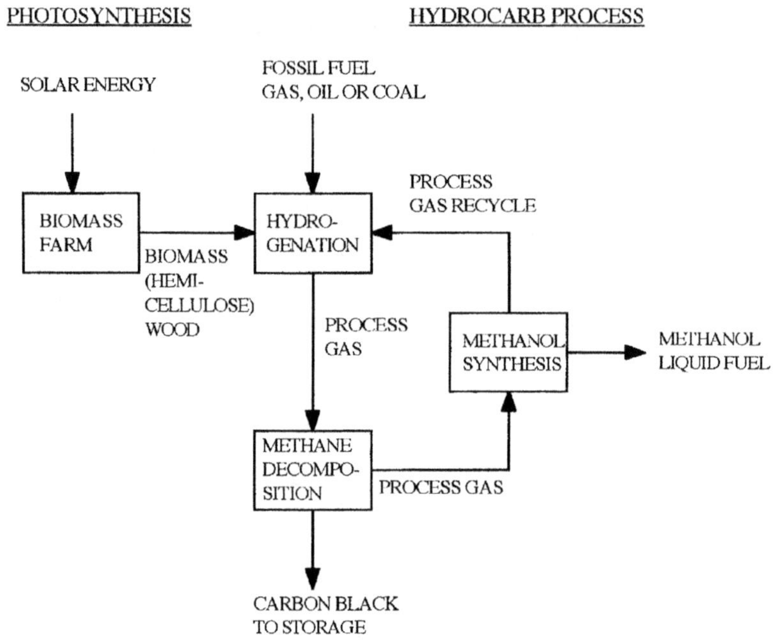

Figure 5. Coprocessing biomass with fossil fuel via the HYDROCARB Process for removal of atmospheric CO_2 and production of methanol fuel.

The stoichiometric reaction for fixation of CO_2 with water by photosynthesis is as follows:

$$CO_2 + 0.72\ H_2O = CH_{1.44}O_{0.66} + 1.03\ O_2 \qquad (1)$$

The $CH_{1.44}O_{0.66}$ referred to as ligno-cellulose is essentially biomass material or wood.

The following are several methods for combining the biomass with HYDROCARB for converting the biomass to carbon and co-producing hydrogen rich fuel.

a. Conversion of Biomass to Carbon and Burial of the Carbon:

The HYDROCARB Process for the formation of carbon from the biomass can be characterized by the following overall stoichiometric equation.

$$CH_{1.44}O_{0.66} = C + 0.66\ H_2O + 0.06\ H_2 \qquad (2)$$

In this system, all of the carbon formed from the atmospheric CO_2 is removed and stored as carbon. The amount of hydrogen available as a fuel source is very limited.

b. Co-processing of biomass with methane to form carbon and methanol using the HYDROCARB Process can be stated as follows:

$$CH_{1.44}O_{0.66} + 0.3\ CH_4 = 0.64\ C + 0.66\ CH_3OH \qquad (3)$$

The carbon is stored and the methanol is used as fuel.

c. Co-Processing of biomass with oil to produce carbon and methanol with HYDROCARB is as follows:

$$CH_{1.44}O_{0.66} + 0.7\ CH_{1.7} = C + 0.66\ CH_3OH \qquad (4)$$

The carbon is stored and the methanol is used as fuel.

d. Co-processing biomass with bituminous coal by HYDROCARB can be represented as follows :

$$0.32\ CH_{1.44}O_{0.66} + CH_{0.8}O_{0.1} = C + 0.32\ CH_3OH \qquad (5)$$

The carbon is stored and the methanol is used as fuel.

These systems fix CO_2 from the atmosphere and process the wood either alone or in combination with the fossil fuels, natural gas, coal and oil (or rubber, plastics and paper derivatives, as in municipal solid waste and even sewage sludge), to put the carbon black back into the ground. The hydrogen-rich gas or methanol co-products can be utilized as fuels. The only fuel which represents a significant amount of energy in the coproducts, as indicated in the above stoichiometry, is methanol. Therefore, we will concentrate on methanol as the principal fuel product. It should be noted that in system I, not very much energy can be derived from the small amount of hydrogen produced while cracking the wood to carbon. System I is efficient for removing CO_2 but not for generating energy.

An interesting situation now presents itself. If we consider the photosynthetic fixation of CO_2 as an energy-free (no cost) process because it is driven by the sun and that methanol, as a fuel, is derived solely from the fossil fuel reserves (natural gas, oil or coal), then we can calculate the efficiency of energy utilization and the quantity of CO_2 generated by the combustion of the methanol solely on the basis of fossil fuel feed to the process. the biomass acts as an intermediate making the C derived from atmospheric CO_2 accessible for reaction with the fossil fuels. The net CO_2 generated or removed is the difference between that fixed in the biomass from the atmosphere and that generated by the combustion of methanol returned to the atmosphere. This has been calculated for the above four processing options. the results are shown in Table 4 and are compared to the amount of CO_2 generated per unit of energy produced using conventional processes.

What we are doing in the above co-processing systems II, III and IV for methanol synthesis is utilizing the natural hydrogen contained in the fossil fuel resource, gas oil or coal to reduce the biomass (ligno-cellulose) to carbon and methanol. As Table 4 shows. The most efficient system for CO_2 reduction is coprocessing biomass with natural gas. This system can yield 66% more energy in the form of methanol when combined with wood than the energy in natural gas alone and more than twice the energy than when natural gas is steam reformed by the conventional process to form methanol. Furthermore, a net removal of up to 78 lb/MM Btu of methanol is actually achieved from the atmosphere, while conventional reforming of natural gas increases the emission of CO_2 gas into the atmosphere to the extent of 170 lb/MM Btu. A similar situation holds for oil in terms of CO_2 removal per unit energy produced (-78 lb CO_2/MM Btu methane). The energy yield in the form of methanol per unit of oil is 15% greater than energy in the oil alone and the efficiency is more than twice that obtained from oil by the partial oxidation process. With bituminous coal (System IV), since there is some oxygen in the coal, the methanol energy utilization efficiency drops to 50% which is about 80% of that obtained in conventional steam-oxygen coal reforming process. But of greater importance is the fact that there is essentially no net geneartion of CO_2, compared to conventional reforming process which produces up to 330 lb/MM Btu. The latter is of real concern when producing synthetic fuels from the vast resorces of coal by conventional means.

Because of the still higher oxygen contents in lignite, it is no longer possible to co-process with ligno-cellulose and, therefore, the lignite via the HYDROCARB Process can convert only 35% of the energy to methanol at a CO_2 emission of 130 lb/MM Btu. In the above calculation, a 90% thermal efficiency is assumed for the HYDROCARB Process based on previous process designs estimates. It is assumed that the energy of growing and harvesting biomass is small compared to the combustion of the natural fuel feedstocks. This has been shown based on estimates of multiple rotational biomass plantations.

There is still another caveat to this process approach. The energy in the carbon stored is not destroyed. Should the greenhouse effect not take place in the future because of other natural feedback phenomena, then the clean carbon can be removed from storage and used in an environmentally acceptable manner. The above indicates that the system is worthy of future exploration an development.

TABLE 4. CO_2 generated or removed from the atmosphere by various methanol synthesis and coprocessing systems using fossil fuel feedstock.

Feedstock	Methanol Process	Carbon Utilization methanol only based on fossil fuel feedstock only %	Energy Utilization efficiency methanol only based on fossil fuel feedstock only %	CO_2 generated (+) CO_2 removed (-) lb CO_2/MM Btu of methanol generated energy
Conventional- Produces CO_2				
Natural gas	Steam reforming	82	68	+170
Oil	Partial oxidation	50	64	+280
Coal Bitum.	Steam-oxygen reforming	42	64	+330
HYDROCARB- Store carbon				
Coal Bitum. (added H_2O)	HYDROCARB	27	40	+130
Lignite	HYDROCARB	18	30	+130
HYDROCARB Coprocessing with biomass store carbon				
II. Biomass + nat. gas	Photosynthesis + HYDROCARB	200	166	+78
III. Biomass + oil	Photosynthesis + HYDROCARB	85	115	+78
IV. Biomass + bit. coal	Photosynthesis + HYDROCARB	30	50	0

Note: 1. Combustion of natural gas generates 110 lb CO_2/MMBtu, oil 160 lb CO_2/MMBtu and bit. coal 215 lb CO_2/MMBtu.

2. Assume 90% conversion feedstock to methanol in HYDROCARB processes.

9. Conclusions

After studying CO_2 greenhouse mitigation technology to reduce CO_2 emissions from the combustion of fossil fuels over the past decade the following conclusions have been reached.

1. A positive approach to reduced CO_2 emissions is to improve the utilization of fossil fuels for energy purposes by a) improving energy conversion efficiency, b) Practising conservation. This approach not only reduces CO_2 emissions per unit energy produced, but is economically important since it results in a return on investment in improved technologies through savings in fuel.

2. Fuel substitution by using mainly oil and gas instead of coal will slow the increase in CO_2 emission to the atmosphere. However this is a temporary procedure since the reserves of oil and gas are limited and we still require the use of the vastly greater resource of coal.

3. Stack gas scrubbing removal, recovery and disposal of CO_2 from fossil fuel burning plants, is not a desirable method. Not only is it economically unattractive but there is no certain method of storing the CO_2, in the large quantities generated to keep it from reentering the atmosphere (Steinberg, 1983a). The physical properties of CO_2 make it most difficult to store alone. There is no known natural alkaline material in the earth to chemically bind CO_2 so it can be securely stored essentially forever. Even ocean disposal has its technical and ecological uncertainties.

4. There is no useful commodity that can be produced from CO_2 that would absorb the large quantities of CO_2 resulting from the combustion of fossil fuel.

5. Nuclear and direct solar energy can be substituted for fossil energy but this would be costly and in the case of nuclear, result in potentially hazardous conditions in the case of an accident.

6. The recycling and reduction of CO_2 for producing synthetic fuels would require the use of large quantities of non-fossil energy, i.e., direct solar or nuclear energy which would be expensive and would result in potentially hazardous conditions.

7. Nuclear and direct solar energy can be used to extract and reduce CO_2 from the atmosphere by physical and chemical processes for returning carbon to the earth for long-term storage. However, these processes would be much more costly and hazardous than photosynthesis.

8. Because of its physical properties, it is much simpler and more practical to store carbon than it is so store carbon dioxide. It is possible to treat fossil fuels by pyrolysis processes to form a carbon-rich material such as char or carbon black which can be readily stored in the earth while utilizing only the hydrogen-rich fraction CH_4 obtained as fuel. Although this can be applied to all carbonaceous fuels, the main resource to consider is coal because this energy resource is vastly greater than oil and gas. The unique technology for accomplishing reduced CO_2 emissions while generating energy from fossil fuels is to apply the HYDROCARB Process which efficiently cracks the carbonaceous material to carbon black and a hydrogen-rich fuel. The carbon black can be stored in the earth (land or sea), and only the hydrogen-rich portion can be used as fuel.

Thus, we can utilize coal as an energy source without generating any CO_2. This process approach will decrease the utilization of the available energy in the coal significantly; however, from an energy conversion efficiency point of view, it is about the same as using nuclear and direct solar power for production of hydrogen as a fuel. Furthermore, with improved energy conversion technologies, such as combined cycle, fuel cells and MHD, the coal derived hydrogen-rich energy fraction for power production can be significantly improved.

10. References

Dang, V.D. and Steinberg, M. (1980) 'Value of Forestation in Absorbing Carbon Dioxide Surrounding a Coal Fired Power Plant', BNL 51279, Brookhaven National Laboratory Upton, N.Y.

Steinberg, M. (1983a) 'An Analysis of Concepts for Controlling Atmospheric Carbon Dioxide', DOE/CH/00016-1 TR007 U.S. Dept. of Energy, Carbon Dioxide Research Division, Washington, D.C.

Steinberg, M. (1983b) 'Advanced Technologies for the Recovery, Disposal and Reuse of Carbon Dioxide for Atmospheric Control', BNL 33441, Brookhaven National Laboratory, Upton, N.Y.

Steinberg, M. (1987) 'Clean Carbon and Hydrogen Fuel from Coal and Other Carbonaceous Raw Materials', BNL 39630, Brookhaven National Laboratory, Upton N.Y.

Steinberg, M. and Cheng, H.C. (1984) 'A Systems Study for the Removal, Recovery and Disposal of Carbon Dioxide from fossil Fuel Power plants in the U.S.', DOE/CH/ 00016-2 TRO16, U.S. Dept. of Energy, Carbon dioxide Research Division, Washington, D.C.

Strauss, C.H. et al. (1988) 'An Input Analysis of Multiple Rotation SRIC Biomass Plantation', Contributions from Pennsylvania State University presented at the Annual Solar Technology Conference, Cambridge, MA .

THERMOCHEMICAL HYDROGEN PRODUCTION; OCEAN RAFT SYSTEM FOR HYDROGEN ENERGY

TOKIO OHTA
Yokohama National University
4-8-15 Inamuragasaki
Kamakura, JAPAN 248

ABSTRACT. Thermodynamics of water-splitting is considered and thermochemical water-splitting method is reviewed . The thermal efficiency and the concrete thermochemical cycles are introduced. Feasibility study on the ocean raft systems for large scale hydrogen production using tropical solar heat and solar photons is discussed.

1. Thermodynamics of Water-Splitting [1,2,3,4,5]

A free energy as well as the provision of heat energy are necessary in order to split water. The enthalpy change, ΔH, in splitting water can be expressed by

$$\Delta H = \Delta G + Q \qquad (1.1)$$

$$Q = T\Delta S, \text{ for reversible processes,} \qquad (1.2)$$

where ΔG, Q, ΔS and T are the change of Gibbs' free energy, amount of heat energy, the change of entropy, and the absolute temperature, respectively. Equations (1.1) and (1.2) are valid when water in liquid or gaseous state is decomposed into hydrogen gas and oxygen gas with the same temperature and pressure as in the initial state under reversible conditions.

Two important facts should be noticed about equations (1.1) and (1.2). The first is that the sources of free energy are

(1) electrical energy,
(2) chemical energy,
(3) photon energy,

all of which can activate the electrons binding the atoms of the water molecule and separate the hydrogen and the oxygen. Among these three sources of free energy, electrical energy can be considered as unstable in the natural world, chemical energy is mostly available in fossil fuel, while photon energy is plentifully available as solar beam. Therefore, it seems reasonable to utilize solar photon energy to split water, just as nature carries it out in a part of the photonsynthesis process of plants.

The second fact is that when higher temperature is available then

Y. Yürüm (ed.), Hydrogen Energy System, 69–82.

less free energy is needed. This is the fundamental basis of high temperature water electrolysis, thermochemical water decomposition, and the hybrid water decomposition systems. Solar photon energy is a kind of free energy but is readily converted to heat whenever it is absorbed into substances. This is the reason why hybrid utilization of photon and heat energy is indispensable in order to raise up the utilization efficiency of solar energy.

When water in a liquid state is split into gaseous hydrogen and oxygen under the condition that the temperature and the pressure of the initial and the final states are the same, then the numerical values for the thermodynamic parameters in equations (1.1) and (1.2) are,

$$\begin{aligned} \Delta H &= 285.57 \text{ kJ/mol} \\ \Delta G &= 244.82 \text{ kJ/mol} \\ \Delta S &= 149.2 \text{ J/K.mol.} \end{aligned} \qquad (1.3)$$

The initial and the final states are assumed to be the normal state, namely, T = 298 K and the pressure = 1,013 hPa. The values show more than 85 % of the energy needed to split water at the normal state is due to free energy change.

If **photon energy** is applied to the free energy, then the relationship to the wave length L should be

$$hc/(eL) > G/96.39. \qquad (1.4)$$

where h and e are, respectively, Planck's constant and the electronic charge. According to this formula, photons whose wave length is shorter than 507 nm are, in principle, able to split water.

However, it is impossible practically because the transfer of photon energy to the binding electrons of the water molecule is very in efficient unless an appropirate catalytic process exists. Good catalysers are required in order to attain reasonable efficiency, and to overcome irreversible reactions.

As the temperature rises, the entropy change ΔS decreases slightly. Figure 1 shows the enthalpy H vs temperature T diagram for both water ($H < 0$) and the gaseous states of the split hydrogen and oxygen ($H > 0$). Two points should be noted from the figure.

The first point is that at the temperature $To = \Delta H/\Delta S$, where $\Delta G = 0$, we get a gaseous mixture of hydrogen and oxygen. An exact calculation using the appropirate partition function gives the result that not only hydrogen and oxygen molecules, but also hydroxyl radicals and hydrogen and oxygen atoms can exist at such high temperatures. A major concern of the direct thermal water decomposition method is thus how to separate hydrogen from this mixed gas.

The second point is that more than 83 % of the energy needed to split water at room temperature is the free energy. This is the reason why large quantity of energy in the form of electrical energy, photon energy or chemical energy, are necessary in order to carry out this process. Considering the alternatives, this is also the reason why

water electrolysis is the traditional and practical technology for water splitting at room temperature.

At moderate temperature considerably lower than To (= 4,150 K under 1,013 hPa), we need the free energy corresponding to the distance between the broken line (horizontal) and the solid line of the enthalpy of water in Fig. 1.

If this free energy is provided by electrical energy, the method of water splitting is called water electrolysis. The electrical energy corresponding to the the value of the Gibbs free energy given by Eq. (1.3) is equivalent to electrical energy of E = 2.46 V. However, in the water decomposition process two electrons can participate, so that the needed voltage for water electrolysis is ideally 1.23 V. In the practical electrolyzer, there will be overvoltage whose amount is typically about 0.7 to 0.8 V at room temperature. This overvoltage depends upon the current density, temperature, the electrode material, and so on; reducing its magnitude is the goal of much R & D effort in this field.

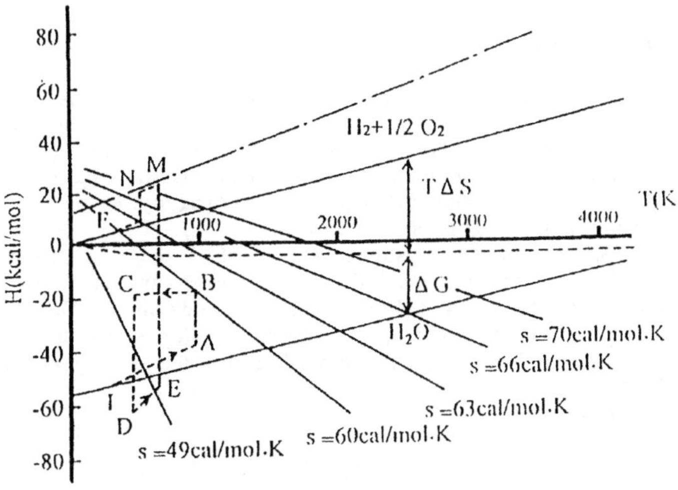

Fig. 1. Enthalpy vs temperature diagram for water-splitting

2. Thermodynamics of Thermochemical Water-Splitting Cycle

2.1. OUTLINE OF THERMOCHEMICAL WATER-SPLITTING CYCLE

Heat energy producing tempeartures around 1,000 K can be provided by solar radiation and by nuclear fission and fusion processes without great difficulty. Therefore, if chemical energy can be made to supply the required free energy at such a temperature, then this could be an effective technique for water decomposition. The method of thermochemical water decomposition was born from such an idea.

Consider, for example, one set of three chemical equations:

$$\begin{aligned} A + B + H_2O &\rightarrow AH_2 + BO, \\ AH_2 &\rightarrow A + H_2 , \\ BO &\rightarrow B + \tfrac{1}{2}O_2 , \end{aligned} \qquad (2.1)$$

where A is an oxidizing agent and B is a reducing agent. They are the catalyst. The key reaction is the first in Eq. (2.4), which splits H_2O into AH_2 and BO. This means that chemical energy provides most of the necessary free energy in this scheme. The remaining two reactions can be advanced by heat, at a lower temperature, or by using electrical energy with a lower voltage.
The process denoted by IABCDEMNF in Fig.1 expresses a thermochemical cycle for water decomposition. The compensation of the free energy such as AB, or DEM are due to the released chemical energy and the change in free energy with temperature.
Multistage water-splitting cycle as Eq.(2.1) has entropy-increasing as well as entropy-decreasing processes. However, the entropy change is ideally zero or very small as the overall system.

2.2. AN IDEAL THERMOCHEMICAL WATER-SPLITTING CYCLE

Thermochemical water-splitting cycle is defined by way that the free energy in Eq. (1.1) is provided by the chemical energy. The chemical free energy is usually given to the system by the catalytic function of the oxidizing and the reducing agents as shown in Eq. (2.1).
We show a simplified thermochemical water-splitting cycle by plotting entropy S vs temperature T diagram in Fig.2.
Let us start from the state A where water of 1 mol is assumed in a vapor state with the temperature T_2K and the pressure is 1.013 hPa. The entropy S_A at A-state is taken as the standard of the present water-splitting cycle. Heat input Q_i is given to the vapor then the entropy S_A increases to S_B and is accompanied by increasing temperatur from T_2 K to T_1 K.
The entropy change is

$$S_B - S_A = C \ln(T_1 / T_2), \qquad (2.2)$$

where C, the heat capacity of the vapor, is assumed to be temperature independent.
As the temperature of the vapor increases, the chemical energy is provided to release the binding between the hydrogen atom and the oxygen atom of the water vapor by the catalytic function of chemical agents, A and B. Then the state of B becomes

$$H_2O \rightarrow H_2 + \tfrac{1}{2}O_2 .$$

The entropy difference of (2.2) can also be expressed by

$$Q_i(1 - \varepsilon_c)/T_1 , \qquad (2.3)$$

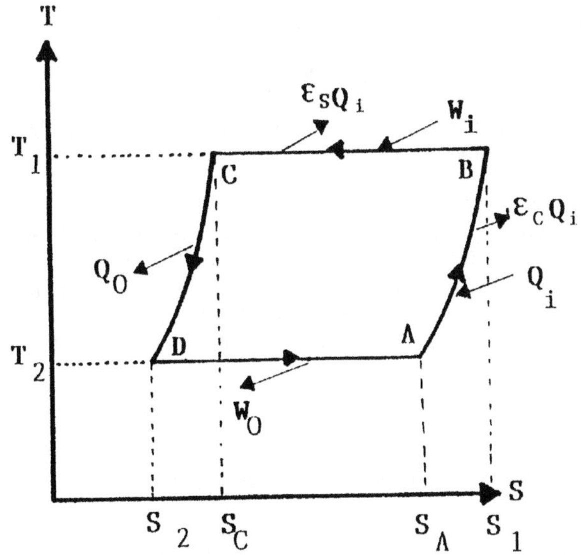

Fig.2. Thermochemical water-splitting cycle

where $Q_i \varepsilon_c)/T_1$ is a reduced entropy by decomposing the vapor into H_2 and O_2 gases. $Q_i \varepsilon_c$ is the free energy expressed in heat equivalence. Now, we apply an external work to the ($H_2 + \frac{1}{2}O_2$) gas and separate H_2-gas from the mixed gas. This is an isothermal process and we have

$$H_2 + \tfrac{1}{2}O_2 \rightarrow (H_2) + (\tfrac{1}{2}O_2).$$

The entropy S_C is given by the relationship:

$$S_C - S_B = -(1/T_1)\ \varepsilon_S Q_i \qquad (2.4)$$

where $W_i = \varepsilon_S Q_i$ is the applied free energy from the external system and expreesed in heat equivalence.
From the state C to the state D, the system exhausts heat Q_o and is cooled down to the temperature T_2. The entropy change is quite similar to the right hand side of Eq. (2.2), where C due to ($H_2 + \frac{1}{2}O_2$) gas must be replaced by C' due to [$(H_2) + (\frac{1}{2}O_2)$] gas.
The separated gases in two vessels are capable to do work W_o against the external system in an isothermal process. Fuel cell system is a typical example of obtaining the useful work W_o.

2.3 EFFICIENCY

We shall discuss briefly the three kinds of efficiency. The first is the system efficiency, the second is the heat efficiency, and the third is the actual efficiency.

2.3.1. System Efficiency. An ideal reversible cycle is shown in Fig. 2, where input heat , input free energy, output heat, and work are

denoted by Q_i, W_i, Q_o and W_o, respectively.
The quasi static reversible cycle requires the conditions that

$$Q_i + W_i = Q_o + W_o. \tag{2.5}$$

Referring Eqs (1.1) and (1.2), we have $\Delta G = W_o$ and $Q_o = T_1 \Delta S$ and the overall efficiency is obviously 100 %. No loss appears in such an ideal cycle.

2.3.2. Heat Efficiency. The input and the output heat are shown by the area $S_1 BCDS_2 S_1 = \oint TdS$ and the area $ABCD = \oint TdS$, respectively, in Fig.2. Then the heat efficiency can be expressed by the ratio of the area $\oint TdS$ of ABCD to the area $\oint TdS$ of S_1BCDS_2.
Then it can be readily shown that the ratio becomes the Carnot's efficiency;

$$e = (T_1 - T_2)/T_1 \tag{2.6}$$

if the area $S_A S_1 BAS_A = \oint TdS$ is equal to the area $S_C CDS_2 S_C = \oint TdS$.

2.3.3. Actual Efficiency. Actual system of thermochemical water-splitting is not so simple as described so far. The input and output heat losses due to heat conduction through vessel materials, convection by air, and heat radiation increase the entropy. Thus any actual cycle becomes irreversible process so that the conditions described in 2.3.2 are not satisfied.
It must be emphasized that the process from the state B to the state C is not always isothermal because the heat of gas will escape in the separating operation. The temperature at C will actually be lower than that at B and the length between B and C will be actually shorter than that as shown in Fig. 2.
The separation of AH_2 from BO and also A and B from AH_2 and BO utilize phase difference (solid, liquid, and gas), boiling point difference, and so on. The separated chemical materials are stored in separated vessels. Not only the separation process but also the conveyance need unexpected more energy, which reduces the efficiency.
Nevertheless the efficient utilization of the heat output Qo requires many contrivances. Usually Qo is applied to heat up liquid water to generate vapor but unavailable heat loss is unavoidable, i.e.,

$$Q_o < Q_i. \tag{2.7}$$

This relationship means that the process is irreversible.
The overall actual efficiency is roughly estimated to be below 50 %.

3. Concrete Examples of Thermochemical Cycle

About 800 billion cubic meter of hydrogen gas have been manufactured

in the world every year. The 76 % and 23 % of them are produced from natural gas and naphtha, respectively. Only 1 % is manufactured by water electrolysis, from by-product of caustic soda manufacturing, and from off-gas of iron making.
The thermochemical water-splitting has been proposed since the oil crisis age, because water electrolysis is not suited to production on a large scale that could be the alternative to these fossil fuels. More than one hundred kinds of the thermochemical cycles as shown by Eq. (2.1) have been proposed and studied by many laboratories in 1970s. However, the R and D decayed since the price of oil became less expensive. Today only few research laboratories have studied the thermochemical water-splitting cycle. Nevertheless it is worth while to review the essence of the system.

3.1. SULFUR-IODINE CYCLE [5,6]

The sulfur-iodine thermochemical cycle has been developed since 1974. The process can be described by the following three chemical equations

$$xI_2 + SO_2 + 2H_2O \rightarrow H_2SO_4 + 2HI_x \quad (370\ K), \qquad (3.1)$$

$$2HI_x \rightarrow H_2 + xI_2 \quad (393\ K), \qquad (3.2)$$

$$H_2SO_4 \rightarrow H_2O + SO_2 + \tfrac{1}{2}O_2 \quad (1144\ K), \qquad (3.3)$$

where HI_x represents the average of several HI_n compounds in the Eq. (3.1).
This process is conducted as an all-liquid and a gas - phase process and the model shown by Fig. 2 cannot be applied.
The characteristic feature of this cycle is the following:
(a) The reaction of Eq. (3.1) is done in liquid phase and the separation of H_2SO_4 from HI_x takes place under gravity, since the two acids are nearly immiscible. The upper phase contains most of the H_2SO_4 and the lower phase contains most of the HI_x. This means that the energy necessary to the separation process is far less than that in the case of gas separation from a mixed gas.
(b) The separation of O_2 and SO_2 gases is not carried out cryogenically but rather, in the main solution reaction where the SO_2 is reacted and O_2 leaves the system. This separation process can save energy.
(c) The efficiency of this cycle can be calculated from Eqs. (3.1) and (3.2). The value is theoretically 67.7 %. On the other hand, G. A. group has empirically estimated the heat efficiency to 48 %. However, the overall system- efficiency including all necessary procedures is supposed to be near 40 %.
The efficiency of water electrolysis is about 75 % and the efficiency of electric power generation is, on average, 34 % , therefore the system efficiency is only 25.5 %, where the higher heating value of hydrogen is applied in the calculation of efficiency.
The comparison of the two values is not always meaningful because the electrolysis hydrogen is purer than that produced by thermochemi-

cal cycle.

(d) Most difficult problem of the thermochemical hydrogen production facility is the materials. In a process plant of commercial size, the different process fluids will require large containment vessels, heat exchangers, and pumping hardware. Industrial experience in handling some of these corrosive fluids is limited. Most iron- and nickel-base alloys exhibits moderate to rapid general corrosion even at room temperatures, and thus they are not available.

Zirconium, tantalum, Durichlor 51, and Hastelloy B-2 are considered to be candidates for practical design.

(e) Process engineering is carefully needed, which includes the heat source (temperature and heat quantity), heat transport, heat storage, heat recover, hydrogen and oxygen storages, safety, and automation.

The problems described in (d) and (e) are common to all thermochemical water-splitting cycles.

3.2. SULFUR-BROMINE CYCLE [7]

Variations of the pure thermochemical cycle have been introduced,which is called the hybrid cycle. The examples are thermochemical process combined with electrolysis, or photochemical process [2].

Here we briefly introduce the sulfur-bromine cycle which has been developed by CEC joint center at Ispra since 1974, and is called Euratom Mark 13.

The present cycle consists of the following three reactions:

$$2HBr \rightarrow H_2 + Br_2 \quad \text{(electrolysis)}, \tag{3.4}$$

$$SO_2 + Br_2 + 2H_2O \rightarrow 2HBr + H_2SO_4 \quad (363\ K), \tag{3.5}$$

$$H_2SO_4 \rightarrow H_2O + SO_2 + \tfrac{1}{2}O_2 \quad (1144\ K), \tag{3.6}$$

The characteristic feature of this cycle is the followings:

(a) Electrolytic process is necessary to proceed Eq. (3.4). The necessary voltage at 200 mA/cm and 350 K is 0.85 V. This value is compared to the conventional water electrolyzers, e.g., by Life System Inc., which shows 1.5 V at 1,600 mA/cm at 355 K. The energy of HBr electrolysis is about half of that due to water electrolysis.

Another comparison is with the data: 1.03 V at 500 mA/cm , at 350 K, of HBr electrolysis to the data: 1.07 V at 300 mA/cm at 355 K of the Dornier/Lurgi conventional water electrolysis.

In all the experiments of electrochemical reaction of Eq.(3.4), platinum is used as a homogeneous electrocatalyst. The application of this noble metal considerably reduces the overvoltage at the cathode. On the other hand, the usage of such noble metal must be avoided from the economical point of view.

(b) Theseparation of HBr from H_2SO_4 in Eq. (3.5) is not possible only under gravity as in Eq. (3.1), because the both specific weight are not so much different. The mixed acid of HBr and H_2SO_4 can be concentrated, and without separation of acids, gaseous HBr leaves the solution upon distillation.

(c) The decomposition of sulfuric acid is an important process common to sulfur-iodine and sulfur-bromine cycles. The decomposition of H_2SO_4 vapor into SO_2 , O_2 and H_2O needs good catalysts such as Pt/SiO , Pt/ZrO , Pt/TiO , and Pt/$BaSO_4$. These catalysts are well fuctional at the temperature as high as above 1,000 K [8].

3.3. CALCIUM-BROMINE-IRON CYCLE [9]

Research group of the University of Tokyo has developed the calcium-bromine-iron cycle since 1978, which has been called UT-3 cycle. This thermochemical process can be described by the following four chemical equations:

$$CaBr_2(s) + H_2O(g) \rightarrow CaO(s) + 2HBr(g) \quad (998\ K), \tag{3.7}$$

$$CaO(s) + Br_2(g) \rightarrow CaBr_2(s) + \tfrac{1}{2}O_2(g) \quad (823\ K), \tag{3.8}$$

$$Fe_3O_4(s) + 8HBr(g) \rightarrow 3FeBr_2(s) + 4H_2O(g) + Br_2(g) \quad (523\ K) \tag{3.9}$$

$$3FeBr_2(s) + 4H_2O(g) \rightarrow Fe_3O_4(s) + 6HBr(g) + H_2(g) \quad (848\ K) \tag{3.10}$$

where the temperatures are the average value around the written value. The notations (s) and (g) mean solid and gaseous phases, respectively. There exists no liquid phase but flowing gas mixture containing a few components and fixed solid reactants.

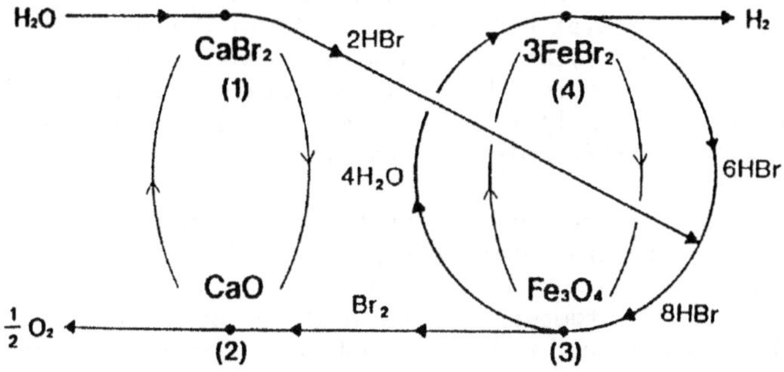

Fig. 3. Flow diagram of calcium-bromine-iron cycle [9]

The characteristic feature of this cycle is the followings:
(a) Four chemical processes take place, so that some demerit in efficiency cannot be avoided unless any contrivance is done.
(b) Schematic flow diagram is shown in Fig. 3. All the reactions can be performed contineously by transferring only the gas reactants from one reactor to another one while keeping solid reactants in the same reactors. This is so advantageous as to cover the demerit of four-step cycle.

4. Ocean Raft System for Hydrogen Energy [10,11,12]

4.1. INTRODUCTION

The option of energy source is one of the most important problems in the hydrogen energy system. Hydrogen itself is a clean and renewable energy medium, on the other hand the primary energy is not always environmental, for example, if hydrogen is manufactured by electrolysis due to the electric power by oil fire plant, then the total system cannot be clean and renewable.
The first hydrogen energy conference was called the THEME Conference which was held at Miami beach in the spring of 1974. The present author has advocated a solar-hydrogen energy system and proposed PORSHE plan at the conference, which means Project of Ocean Raft System for Hydrogen Energy. Since then, the R & D group in Japan has taken a feasibility study and the result has been taken into the Japanese national project "WENET" (World Energy Net Work) launched in 1994.
Here a general survey is introduced.

4.2. THE FIRST KIND RAFT SYSTEM

The solar energy at the ocean of tropical zone is as plenty as more than 2,200 kWh/m^2 per year.If a raft of surface area with 1km installs a solar heat collector and a heat reservoir, it can generate hydrogen gas of 6 X $10^7 m^3H_2$ per year, which is calorie-equivalent to 14,400 kl of oil. The calculation is done under the assumptions: (1) The overall heat efficiency is 10 % and (2) The necessary electric energy to evolve hydrogen by electrolysis is 3.5 kWh/m^3H_2.
The first kind raft system had been designed in the age when the cost of electric energy by solar cell is more expensive than that by solar heat-steam turbine system. Today this situation has been changed and the both became nearly same. However, the second kind raft system described in 4.3. is preferable from the view point of maintenance of the system.
A functional block diagram of the first kind system and an illustration of the raft structure are shown in Fig. 4 and Fig. 5, respectively.
The solar collectors installed on the surface of the raft generate heat energy with the temperature as high as about 773 K. which is given to alkali metal-heat medium stored in the vessels. The turbine generate electric energy even in the absence of solar radiation by the steam made by the stored heat.
A part of electric energy is used to purify the sea water using a membrane system and the pure water is supplied to the electrolysis system to evolve hydrogen and oxygen. The both gases are liquified and shipped to the industrial areas.
The construction design of a raft as well as the array of many rafts, their rotation, mooring, and the necessary elemental technologies are studied. The utilization of squall, wind and wave is also considered.
The raft system is constructed in the tropical area where the "eggs" of typhoon are born and typhoon itself seldom attacks the zone. The

system is also considered
The characteristic feature of this raft system is concretely shown in Fig.5, which was firstly invented by Escher Technology Inc. One big buoy whose total height is 450 m and the bottom part of the buoy with length of 200 m is equipped with the heat reservoir, generator of vapor, turbine, electrolyzer, gas liquifier, and other necessary facilities, which is under the sea surface and support the whole raft by its buoyant force.
The raft base is suspended by the ropes which are strained by fixing to the top of the buoy 250 m high above the sea surface level, and the raft base is 50 m higher than the sea surface level, so that ships can pass through under the raft.
The sea area under this raft is able to breed fish, while other areas are not because the solar energy is very strong enough to killing the nutritious living things, the fish's food.

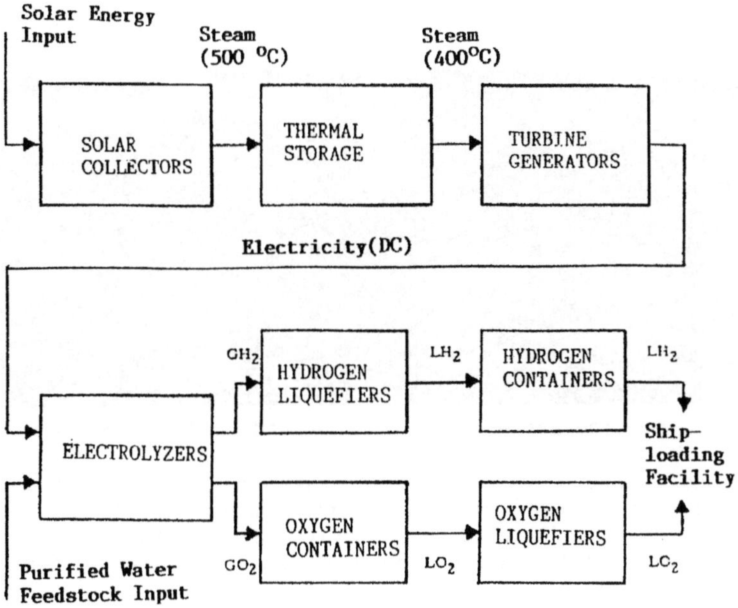

Fig 4. Functional block diagram of the first kind raft system

4.3. THE SECOND KIND RAFT SYSTEM

The development of the solar cell has been so progressive that the electric energy generated by it has become comparable or rather cheaper than that by the solar heat steam turbine generator. Thus the second kind raft system has been investigated since the autum of 1983.

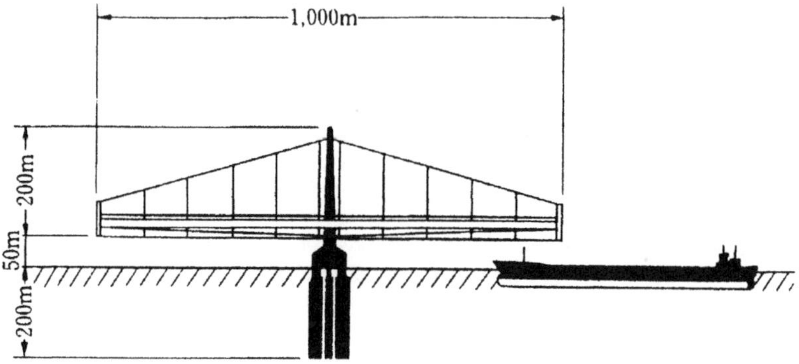

Fig. 5. Structure of the first kind raft system

Fig.6. Outside appearance of the second kind raft system

The solar cell panels are installed on the surface of raft instead of the solar collectors. Then no suspension system by the central huge buoy is needed and lighter and simpler raft is supported by 25 small buoys distributed among the raft base with equal distance, 250 m.

The merits of the second kind raft system is that the electric energy generation system is much simpler than that of the first kind. Nevertheless the weight becomes less than half of another. The huge suspension buoy system is no longer necessary, so that the stability of the raft is improved.

The cost of hydrogen manufactured by the second kind raft - system is estimated to be about 20 times more expensive compared to the hydrogen by naphtha cracking at the present time.

4.4. WENET and EURO-QUEBEC PROJECTS [13,14]

The high price of hydrogen produced by the raft system is originated from the construction of the raft. Therefore it is obvious that the cheaper hydrogen will be provided by the establishment on land.
The economical demerits of solar energy are due to (1) its rare energy density and (2) its intermittence. Hydropower is one of the accumulated forms of the solar energy so that hydropower electric generation - electrolysis - hydrogen energy system is undoubtedly more feasible in the near future. The world's theoretical hydropower resources are estimated to be 60 trillion kWh/Y and the recoverable reserve is 20 trillion kWh/Y (world's total power generation is 13 trillion kWh/Y).
Conceptof the WENET (World Energy Network) Project of Japan is to use the abundant and inexpensive hydraulic power generation to produce hydrogen for delivery to oversea consumption areas. This project has been launched on the April of 1993 and a similar project named EURO-QUEBEC Project has started formerly started in 1989.
The latter has been planned up in order to supply the clean fuel hydrogen from Quebec of Canada, hydrogen of which is manufactured by the electrolysis due to the plentiful hydraulic power in Quebec.
The most of elementary technologies of both the Ocean Raft System and the World Energy Network are common to each other and have been put spurs to the goal internationally.

References

1. T. Ohta et al, Int. Jr. Hydrogen Energy, **10**(1985)571,
2. T. Ohta, Int. Jr. Hydrogen Energy, **13**(1988)333.
3. T. Ohta, ENERGY TECHNOLOGY,Sources,Systems and Frontier Conversion, Chap. 5, p.202 (1994,Pergamon Press, Oxford .
4. T. Ohta, Thermodynamics of Water-Splitting, Chap.2in Solar-Hydrogen Energy Systems Ed. by T. Ohta (1979, Pergamon Press, Oxford).
5. S.Satoh, Thermochemical Hydrogen Production, Chap.5, in Solar-Hydro en Energy Systems Ed. by T. Ohta (1979, Pergamon Press, Oxford).
6. G.E. Besenbruch et al, Hydrogen Production by GA Sulfur-Iodine Process, a Progress Report, p.243 in Hydrogen Energy Progress III, Ed. by T.N.Veziroglu, K.Fueki and T.Ohta (1980), Pergamon Press, Oxford).
7. D. Van Velzen et al,Status Report on the Operation of the Bench-Scale Plant for Hydrogen Energy Production by the Mark-13 Process, p.423 in Hydrogen Energy Progress III, Ed. by T.N.veziroglu,K.Fueki and T.Ohta (1980, Pergamon Press, Oxford).
8. H. Ishikawa et al, Catalyzed Thermal Decomposition of H_2SO_4 and Production of HBr by the Reaction of SO_2 with Br_2 and H_2O, p.298 in Hydrogen Energy Progress III, Ed. by T.N.Veziroglu,K.Fueki,and T.Ohta (1980, Pergamon Press, Oxford).

9. M. Aihara et al, Reaction Improvement in the UT-3 Thermochemical Hydrogen Production Process, in p.493 in Hydrogen Energy Progress VIII, Ed. by T.N.Veziroglu and P.K.Takahashi (1990,Pergamon Press, Oxford).
10. T. Ohta, Hydrogen Energy Systems as to be Aplied in Japan and the Key Technology, THME Conference Proc. S15-69, Miami (1974).
11. W.J.D.Escher and T.Ohta, Direct Solar Energy Conversion at Sea, in p.225 of Solar-Hydrogen Energy Systems, Ed. by T.Ohta (1979,Pergamon Press,Oxford).
12. T.Ohta, Chap.5 in Hydrogen Energy (1989, Morikawa Pub. Co., Tokyo).
13. T.Homma, Sunshine Jr. **4**(1993)7.
14. J.Gretz, Int. Jr. of Hydrogen Energy **15**(1990)419.

SOLAR HYDROGEN - WHY, POTENTIAL, WHEN?

J. GRETZ
Head of Energy Environment Interaction Unit. CEC Coordinator of the EQHHPP.
European Commission, Joint Research Centre, Ispra (VA) Italy.

ABSTRACT. Hydrogen is a carbonfree fuel which oxidizes to water as combustion product. The generated water becomes, together with renewable primary energy for splitting it, a source of clean and abundant energy in a carbon-free, natural cycle.

Hydrogen is a fuel which can be transported over long distances and stored so that solar energy can be transported from energy rich countries over long distances in ships to Europe, stored and used in gaseous or liquid form in industry, households, power stations, motor cars and aviation propulsion.

Solar energy as primary energy is discussed, a special form of it, the relatively cheap hydropower is stressed.

Techniques of hydrogen production, vectorization and end use is discussed as well as safety aspects, costs and strategy for its implementation.

Why Hydrogen?

Hydrogen is a carbon-free fuel which oxidizes to water as combustion product. The generated water becomes, together with renewable primary energy for splitting it, a source of clean and abundant energy in a carbon-free, natural cycle.

The hydrogen technology is ecologically benign, entailing little or no local pollution.

Hydrogen can be transported and stored in gaseous and liquid form and used as fuel in industry, households, power stations, motor car and aviation.

The intermittence and the geographical distribution of solar energy require means of storing and transporting it to the user's place. An almost ideal means of doing this it to split water with this primary energy source in order to obtain hydrogen.

The characteristics of hydrogen assign to it a role which is complementary to that of electricity. But unlike electricity, hydrogen is a fuel which can, with by now established techniques, be transported over long distances and stored so that solar energy can be transported from energy rich countries over long distances in ships to Europe, stored underground or in containers and used in the different forms.

Hydrogen can thus open up clean and abundant renewable energy sources distributed all over the World and give countries of the Third World the opportunity to export energy to the developed countries. It leads by that virtue to more independence from the actual fossil fuel exporting countries and contributed thus to the economic cooperation of nations and the "North-South Compensation".

The transport of hydrogen entails no tanker oil spills (1989: Approx. 3 Mio To oil went

Y. Yürüm (ed.), Hydrogen Energy System, 83–94.

into the oceans from tanker wrecks (13), tanker cleaning, pipeline and drilling platform leaks).

Hydrogen from hydropower is relatively cheap and in some cases already competitive (steel fabrication with Brazilian hydro/hydrogen).

Hydrogen allows smooths transition from the existing energy systems to clean and renewable energy systems since hydrogen energy systems need virtually no basic engineering break-thorough; all components of such systems exist, they need continuous, gradual development; large pilot/demonstration projects can today be built and run.

Primary Energy

Hydrogen being a secondary energy - an energy vector - it has to be generated with primary energy i.e. direct or indirect solar energy such as hydropower and wind energy if the overall system is to be clean and renewable.

The most convenient process for the generation of hydrogen is electrolysis which requires electricity as primary energy.

DIRECT SOLAR ENERGY CONVERSION

The most developed technologies for helioelectricity generation are photovoltaic, thermal and wind conversion. All these technologies are today sufficiently developed and field-tested so that the erection and operation of pilot/demonstration plants in the 1-100 MW range can be undertaken.

A 100 MW photovoltaic power plant, situated in an equatorial country with global yearly insolation of say 2300 kWh/m^2, y and 7.5% global efficiency of conversion from solar radiation into hydrogen would have a collector surface of about 1 km^2, covering some 3 km^2 land surface and would produce yearly $0.5 \cdot 10^9\ m^3$ of hydrogen. New technologies like multiband solar cells, capable of utilizing photons in a number of adjacent frequency bands and bringing the theoretical efficiencies up to 25% or 45% would reduce costs, material and land requirements proportionally.

INDIRECT SOLAR ENERGY CONVERSION HYDROPOWER

Natural hydropower and hydropower of already existing dams are amongst the cleanest and most promising renewable energy sources. 23% of the solar energy falling on the earth operates the hydrological cycle, i.e. evaporation of water, precipitation and storage in water and ice.

Some characteristics of hydropower are given hereafter:

- 23% of the solar energy falling on the earth, i.e. $\sim 350 \cdot 10^6$ TWh/y, operates the hydrological cycle, i.e. evaporation of water, precipitation and storage in water and ice;
- the World's topological and technically exploitable estimated hydroenergy potential is $\sim 20 \cdot 10^3$ TWh/y, i.e. 0.0057% of the hydrological cycle energy;
- the yearly average utilization factor of hydroenergy is ~ 65% (in 1986, Worldwide installed 380 GW generated 2200 TWh electricity);
- today's hydropower electricity generation of about 2200 TWh/y represents ~ 21% of the World's electricity generation;
- hereafter is given the World's estimated hydropower potential and its approximate

distribution:

- Asia (without UDSSR) : $5.3 \cdot 10^3$ TWh/y;
- South America : $3.8 \cdot 10^3$ TWh/y;
- Africa : $3.1 \cdot 10^3$ TWh/y;
- North America : $3.1 \cdot 10^3$ TWh/y;
- Ex UDSSR : $2.2 \cdot 10^3$ TWh/y;
- Greenland : $1.5 \cdot 10^3$ TWh/y;
- Europe : $1.4 \cdot 10^3$ TWh/y;

(for comparison, Europe's (EC) electricity consumption (1984) was $1.3 \cdot 10^3$ TWh/y).

Oversea's Transportation of Hydrogen

Hydrogen can be transported by ships in liquid phase, in chemical bonds or stored in metal hydrides. Methanol and ammonia being dangerous to health and environment should be excluded as well as metal hydrides having small storage capacity.

Liquid hydrogen has emerged as the most suitable solution due to its high gravimetric energy density (lhv = 120 J/g).

Since the ambient temperature is considerably higher than the LH_2 temperature (20 K) heat flows into the tank leading to evaporation. Insulation is therefore essential. Using vacuum insulation the evaporation rate is in the order of 0.1%/d. The insulated tanks can be mounted on buoyant barges, as for instance within the Euro-Quebec Hydro-Hydrogen Pilot Project (EQHHPP), see Figure 1. Five barges are loaded on the ship. When the ship has to be loaded or discharged additional ballast water is pumped into the ship so as to lower it. The barges may then be loaded/discharged using a lifting platform and a barge transportation track. The tanks can, on land, act as storage containers.

A way of safe, simple and probably cheap oversea transportation of hydrogen could be the redox cycling of iron oxyde reduced with hydrogen. Iron would be shipped from the hydrogen production site (Quebec in the EQHHPP) - to the user's site (Hamburg in the EQHHPP) where it is oxydized by water-steam into hydrogen and oxydized iron. The latter is shipped back to the hydrogen production site for reduction. A rough calculation indicates an "energy transmission" of ~ 1 kWh/kg iron. The energy from the exothermic oxydation process can be used to produce 700-800°C water-steam so that the process may run without catalyst.

Hydrogen user's site (oxydation)	Hydrogen production site (reduction)
$3\,Fe + 4\,H_2O \,\varsigma\, Fe_3O_4 + 4\,H_2$ + energy	$Fe_3O_4 + 4\,H_2 \,\varsigma\, 3\,Fe_3 + 4\,H_2O$ - energy

The endothermal reduction process requires theoretically ~ 210 kJ/4 mol H_2 thermal energy at 700°C which is ~ 22% of the transported energy or 25-30%, including "industrial plant" losses.

The oxydation process, although exothermal, requires theoretically additional ~ 58 kJ/3 mol Fe thermal energy at ~ 150°C which represents ~ 6% of the transported energy, i.e. 8-10% including "industrial plant losses".

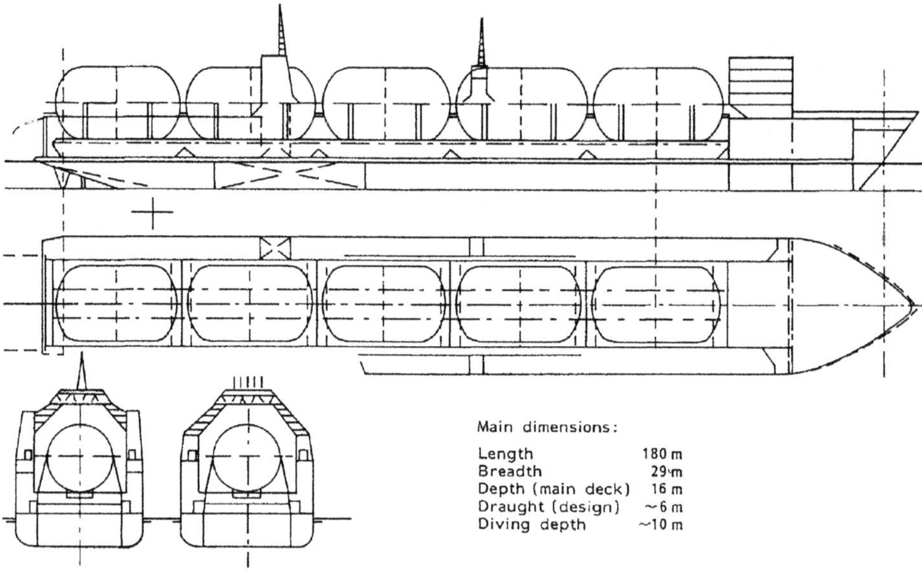

Figure 1. Barge carrier (EQHHPP).

Compared with LH_2 transport, about the same energy would be required in the redox system as for liquefaction, with the difference that the latter requires electricity (exergy) whereas the redox-iron system requires thermal energy. At the oxydation site some 10% low grade thermal energy is required.

With such a vectorization the rather costly and complicated cryotechnical ship transport would be replaced by iron bulk carrying with no safety and legislation problems.

A current iron bulk carrier of 200.000 DWT could carry about 100 GWh of energy @ "pore structure" yield of 50%. The back-shipping of the iron oxyde would require a 280.000 DWT ship. In the EQHHPP for instance, the 15 round trips per year with a LH_2 carrier would be substituted by about 6 trips by an iron bulk carrier.

Of course, the "price" of this kind of vectorization is that gaseous hydrogen will be arriving at the user's site instead of liquid hydrogen required by most applications. In order to have clean liquid hydrogen the gaseous hydrogen has to be liquefied with either the hydrogen itself or with nuclear energy, the latter leading to the solar - nuclear hydrogen mix.

End Use

Hydrogen is a carbonfree fuel with excellent combustion properties. This makes it a clean, universal fuel for use in industry, households, power stations, road vehicles and aviation.

The specific properties of hydrogen induce new end use techniques, like:

FUEL CELLS

The key element of an efficient electricity generation will be the fuel cell for which hydrogen is the best fuel. Being devices which transform chemical energy directly into electrical energy its generation is not subjected to Carnot cycle limitations, is done silently without moving parts and with high efficiencies. Efficiencies of future fuel cells like molten carbonate cells operating at 600-800°C, with total energy efficiencies of 60-80% with waste heat at 600°C, make the fuel cell concept a potential candidate for use in power stations, road and space vehicles. Progress in catalyst and electrode technology and cost reductions have to be made, however, before fuel cells can competively be considered for general use.

Hydrogen is an excellent fuel for fuel cells, its more or less cold combustion will reduce the NOx emissions even down to zero.

CATALYTIC BURNERS

Hydrogen, having a lower ignition energy than for instance natural gas, burns smokeless when in contact with a suitable catalytic surface at low temperatures between 20 and 400°C. This would be a preferred technology for generating low temperature heat in residential space heaters, cooking devices and industrial dryers and heaters.

Catalytic combustion has the advantage of high safety, very high efficiencies of up to 99% and negligible emissions of nitrogen oxides.

SUPER/HYPERSONIC AVIATION

Jet engines at super/hypersonic speeds are advantageously - if not necessarily - powered with liquid hydrogen (LH_2) for mainly two reasons, its almost 3 fold gravimetric heating value compared with kerosene and its capacity of cooling parts (turbine inlet rim, wing leading edges, passenger cabine) of the plane exposed to outside stagnation temperatures of ~ 1300°C at Mach 6, cryogenic storage temperature being at -252°C. Moreover, cooling of the wing skin induces laminar flow and therewith considerably reduces drag by up to, theoretically, 30% ("laminar flow control").

Water vapour emissions, however, are likely to form more or less long lasting ice clouds which enhance the greenhouse effect. The problem is of extreme importance and is seriously to be investigated not only in view of hydrogen - the combustion of hydrogen sets 2.5 times more water vapour free than that of kerosene - but also concerning kerosene combustion in general.

The elimination of CO, CO_2, SO_2 and VOC by using hydrogen not only is beneficial for the atmosphere, but reduces considerably airport ground pollution. For example, the NOx emissions resulting from the daily average 1259 landing/take-off cycles at Los Angeles International Airport are equivalent to the operation of about 1 Mio passenger cars.

ROAD VEHICLE PROPULSION

In Germany, for example, from the overall emissions, those resulting from road traffic are: 52% Nox, 70% CO and 49% VOC. The main problem of hydrogen utilization in vehicles is storage due to the low volumetric energy content of hydrogen of about one third of that of gasoline. From the three different storage technologies i.e. gaseous hydrogen in pressure vessels, cryogenic liquid hydrogen and hydrogen chemically bonded in metal and liquid hydrides (methylcycloexane), liquid hydrogen seems to be the most suitable solution.

The low ignition energy of hydrogen and its wide ignition range of hydrogen/air mixtures make gas turbines as well as piston engines well adaptable for hydrogen combustion. The ignition within a wide range of non-stoechiometric air/fuel mixture permits combustion with high amount of excess air and thus full and part load operation with low nitrogen oxides emission.

Since 1937, more than 30 vehicles with hydride storage have been built and operated in Europe by Daimler Benz, (more than 600.000 km), BMW and experimental trucks by Paul Scherrer Institute (Switzerland).

It should be emphasized that the building up the necessary infrastructure is at least as lengthy as the development of the technology itself.

STEEL FABRICATION

With ~ 2 kg CO_2 emissions per kg steel fabricated with carbon as reductant, the World's steel fabrication contributes 10-12% to the world's total anthropogenic CO_2 production and is therewith amongst the most intensive single CO_2 polluter.

Hydrogen is an excellent and clean reductant emitting water vapour instead of CO_2 and does not introduce extra impurities as coke does (sulphur in particular). If generated with relatively cheap and abundant hydropower, its application in steel fabrication is in some cases already competitive for instance Brazil which disposes of enormous resources of iron ore in the vicinity of large hydropower installations.

The penetration of hydrogen into the steel industry would be favorized by the fact that steel fabrication is rather centralized, the power per iron fabrication plant being in the order of GW compared with, for instance, 100 kW per motor car, which eases the logistics of the use of hydrogen in steel fabrication.

Safety

Each energy carrier has its specific dangers and risks. Compared with natural gas and liquid fuel hydrogen scores less well since it:

- is flammable within a wide range of hydrogen - air mixtures of between 4 and 75 Vol. % vs. methane between 5 and 14%;
- has a low ignition energy of up to 15 times lower than methane;
- is much more prone to leak;
- has high flame velocities of about 7 times that of methane and therewith reacts violently once it is inflamed.

Hydrogen is advantageous since:

- it has a high diffusivity of about 4 times larger than air so that spills disperse rapidly to below flammability; there is therefore a higher probability of surviving an airplane crash - if one is not killed by the impact if the airplane - if it is fuelled with hydrogen rather than with kerosene;
- air flames are nearly invisible i.e. they release little heat by radiation implying both hazards in fire fighting because one does not feel the flame as well as advantages since

fire fighters can work closer to the flame.

Experience from the use during decades of town gas containing 60 Vol. % hydrogen is available as well as from the operation of a 210 km long pipeline network operating with pure hydrogen at 25 bar in Germany since 1936, and by a 80 km hydrogen carrying pipeline in Houston, Texas. Considerable experience in liquid hydrogen storage and road/rail transportation has been gained in the course of space activities, in which no specific hydrogen fatality has been experienced yet.

Hydrogen is undoubtedly a hazardous material but there is general agreement that with proper use under adequate safety measures hydrogen will be no more hazardous than the fuels currently used.

On 18.12.1990 the BAM (German Federal Institute for Material Research) declared that liquid hydrogen would be not more dangerous than LNG and LPG and that it has no objection to the transport of liquid hydrogen in 2 G type ships (IMO rules).

Environmental Aspects

The combustion of hydrogen does not produce CO_2, CO, SO_2 VOC and particles, but entails emission of water vapour and NOx.

Water vapour emissions from airplanes may be harmful since they generate - depending on the cruising altitude and latitude - ice clouds with ensuing greenhouse effects and ozone depletion, the problem is of great importance and actually under investigation.

The formation of NOx is a function of flame temperature and - duration. Considering the wide flammability range of hydrogen its combustion can be influenced by the design of the engine so that the NOx emission can be reduced.

The worldwide water evaporation from the oceans and rivers is $\sim 5 \cdot 10^{14}$ m^3 per year. If mankind's today's total energy consumption of sustained 11 TW would be effectuated by hydrogen, the ensuing yearly water evaporation would be $\sim 2.5 \cdot 10^{10}$ m^3 i.e. about 1/20000 of the natural evaporation. Once hydrogen will be massively used, local considerations are obligatory like it is the case for today's wet cooling towers.

Economics

An adequate comparison of hydrogen costs with costs of the actual hydrocarbon energy has to take into account the external costs i.e. costs for pollution abatement/prevention and climate effects of fossil fuel burning. If the costs for pollution abatement originating from fossil energy use are available only very approximatively the cost of the negative drawbacks on climate changes are clearly imponderable i.e. impossible to evaluate numerically.

The hydrogen cost of 15 cents$_{ECU}$/kWh are pictured in Figure 2 together with the costs of taxed gasoline prices in Europe (average of the 12 EC countries, August 1990) of 8.5 cents$_{ECU}$/kWh which are made up of 3 cents$_{ECU}$/kWh for the crude oil itself, its transportation, refinement, manipulation and distribution and of 5.5 cents$_{ECU}$/kWh for taxes.

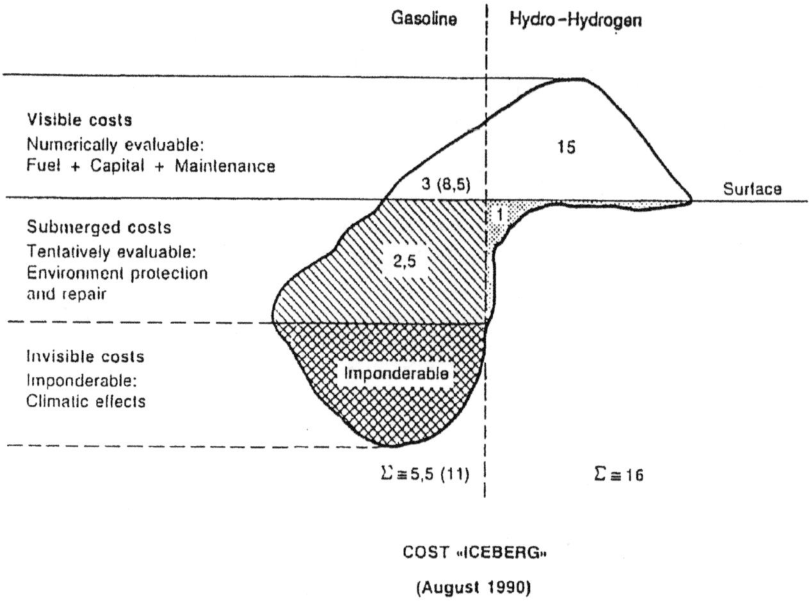

Figure 2. Fuel cost comparison ($¢_{ECU}$/kWh; figures in ( ) are for taxed fuel.

With the above values and including the submerged costs of the protection/repair of the damaged environment resulting from the use of fossil fuel (the value of 2.5 $cents_{ECU}$/kWh is the average from various literature sources), the "FUEL COST ICEBERG" would look as shown in Figure 2.

With the internalization of external costs, hydrogen energy would thus be about (1.45) 2.9 times higher than hydrocarbon energy costs, not taking into account the imponderable costs of climate effects. In the absence of carbon atoms, hydrogen fuel would not be subjected to mineral oil taxes.

When?

Given the long lead times and the dynamics of the introduction of energy technologies, time is likely to become a precious raw material in view of the menacing deterioration of the environment.

Analysis of the molecular composition shows an increasing ratio of hydrogen to carbon molecule of the fossil fuels used worldwide since 1860, from wood (H/C = 0.1) to coal (1), to oil (2) and finally - during the 90's - to methane (4). If that trend is to continue, extra hydrogen has to be provided from an external source i.e. water, since there is no natural hydrocarbon product with a higher H/C ratio, see Figure 3. This leads to the concept of water splitting using non-fossil fuels.

The analysis of the market penetration curve of new technologies, see Figure 4, indicates that the conquest of the first 10% of the market is not very different from the one between 10 and 90%. In the past it took about 100 years for a market penetration

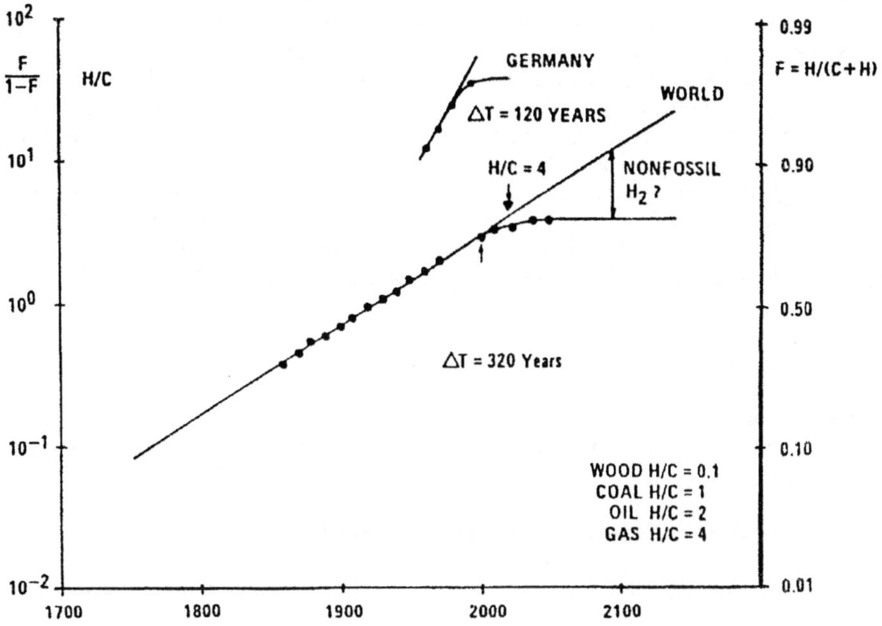

Figure 3. Evolution of the hydrogen/carbon molecule ratio of the fossil fuels.

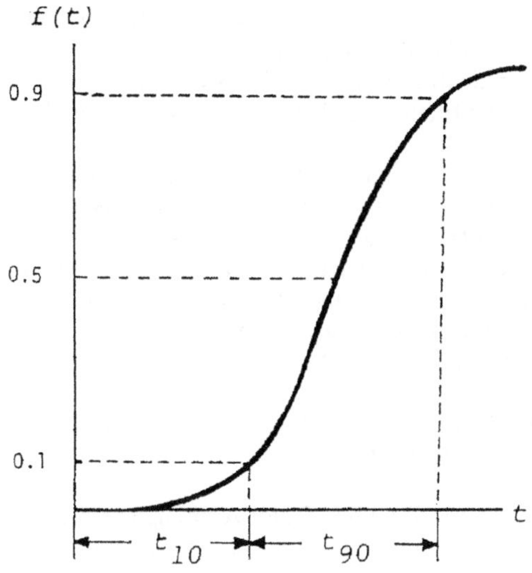

Figure 4. Penetration time of technologies.

from 10 to 50%, which means that solar hydrogen should cover 10% of the energy market by the year 2000 if by the year 2100 it is assumed to cover 50% and if things would go at the same pace as in the past. Even if things move faster now solar hydrogen will not cover 10% of the market by the year 2000. Works on solar hydrogen should - by judiciously analyzing the penetration curve - have started before the beginning of the century in order to have 50% of mankind's energy consumption covered by solar hydrogen by the year 2100. The least to say is that it is not too early to start working on it now.

Although it is not possible to predict neither the evolution of fossil fuel costs nor the moment the costs of fossil fuel pollution become internalized there are indications that the onset of a hydrogen technology within new clean and renewable energy supply systems will take place by 1995-2000, with pronounced deployment by 2020-2040.

Strategy for the Implementation of Solar Hydrogen

In the light of the existing state of the art of the technology, the time is ripe to undertake, in parallel with R&D for components and systems, pilot and demonstration projects.

Pilot/Demonstration Projects, Programmes and Projects

In Europe, essentially three pilot/demonstration projects/programmes are actually under way.

EQHHPP, EURO-QUEBEC HYDRO-HYDROGEN PILOT PROJECT

The concept of a hydrogen-based clean and renewable energy system, conceived by the Joint Research Centre Ispra of the European Commission has been investigated jointly by European and Quebec industry and the JRC-Ispra.

The scope of the investigation has to establish the main characteristics of the system and its components, to give first cost indications and then to build the system of 100 MW rating.

System layout and first cost evaluation studies started in 1986.

The 100 MW pilot project is to demonstrate the provision of clean and renewable primary energy in the form of Quebec hydropower converted via electrolysis into hydrogen and shipped to Europe, where it will be stored and used in different ways: electricity/heat cogeneration, vehicle and aviation propulsion and hydrogen enrichment of natural gas for use in industry and households.

At this moment demonstration projects on the use of hydrogen are under way.

HYSOLAR, A JOINT GERMAN-SAUDI ARABIAN RESEARCH, DEVELOPMENT AND DEMONSTRATION PROGRAMME

The HYSOLAR Programme is a joint Saudi-Arabian/German research, development and demonstration undertaking, carried out within the framework of the "Agreement between the Government of the Federal Republic of Germany and the Government of the Kingdom of Saudi Arabia on Cooperation in the Field of Scientific Research and Technological Development".

The investigations started in the early eighties, the agreement being signed in 1986.

The project management is undertaken by the German Aerospace Research Establishment DFVLR, the University Stuttgart and the King Abdul Aziz City for Science and Technology KACST.

The objective of this Programme is to accomplish the scientific and technological prerequisites for future production and utilization of solar hydrogen. In particular it is:

- to demonstrate the reliable operation of a 350 kW photovoltaic/electrolytic hydrogen production plant at Riyadh;
- to investigate and test advanced components and systems in a 10 kW test facility at Stuttgart for photovoltaic/electrolytic hydrogen production;
- to carry out fundamental research for new and improved hydrogen production methods and the reconversion of hydrogen;
- to demonstrate the utilization of the hydrogen produced by the 350 kW plant;
- to carry out system analysis of solar hydrogen production and utilization systems;
- to install and operate a laboratory photovoltaic/electrolytic hydrogen test facility at Riyadh for scientific and educational purposes;
- to carry out an exchange of personnel and an educational program.

SWB, BAYERNWERK SOLAR HYDROGEN PROJECT

The German utility Bayernwerk in Munich, Germany, together with BMW, Linde, Messerschmitt-Bölkow-Blohm and Siemens is investigating, with 500 kW solar photovoltaic power generation, the potential of solar hydrogen production under Central European insolation conditions. The rationale is that under Central European weather/clima conditions, i.e. with less solar radiation than in the distant desert regions, but considering:

- lower ambient temperature, yielding higher photovoltaic conversion efficiencies;
- more rain on more vertically orientated panels, reducing soiling;
- cheaper demineralized water for electrolysis;
- no transport to Europe;

solar hydrogen production in Europe might not be that disadvantageous.

The pilot project implies operation of the complete solar hydrogen energy system i.e. photovoltaic electricity generation, hydrogen production, storage and application: fuel cell operation, catalytic combustion, heat production and gas engine operation. The project has programme character in that emphasis is given to permanently incorporating the latest technological development of the components in order to have the system constantly à jour.

References

[1] Funk, J.E. and Reinström (1966), Energy requirements in the production of hydrogen of hydrogen from water. I&EC Process Design and Development, Vol. 5, No. 3, 1966.

[2] Giacomazzi, G., "Prospects for Intercontinental Seaborne Transportation of Hydrogen", 7th World Hydrogen Energy Conference, Moscow, September 1988, Published in "International Journal of Hydrogen Energy", Vol. 13, 1989.

[3] Gretz, J., Baselt, J., Ullmann, O. and Wendt, H., The 100 MW Euro-Quebec Hydro-Hydrogen Pilot Project, International Journal of Hydrogen Energy, Vol. 15, No. 6, pp. 419-424, 1990.
[4] Marchetti, C. (1985), When will hydrogen come? Int. J. Hydrogen Energy, Vol. 10, No. 4, 215-219, 1985.

BIOPHOTOLYSIS, HYDROGEN PRODUCTION AND ALGAL CULTURE TECHNOLOGY

O.M.SKULBERG
Norwegian Institute for Water Research
P.O.Box 173Kjelsas
N-0411 Oslo
Norway

ABSTRACT. In this essay the importance and place of biophotolysis in the hydrogen energy system is described. The biophotolysis of water is achieved by two biochemical processes carried out by the activity of chlorophyl containing reaction centres coupled to hydrogenase and nitrogenase. Microalgae belonging to the classes Chlorophyceae and Cyanophyceae can produce molecular hydrogen by the decomposition of water using solar energy. Among Anoxyphotobacteria organisms of the families *Chromatiaceae* and *Chlorobiaceae* are also used for the bioengineering development of biophotolysis. A review is presented of the organisms and the processes involved in the context of their applications for algal culture technology.

1. Introduction

Supply of energy from the sun offers fascinating perspectives for the future of mankind. The sun radiates to the earth in half an hour that quantity of energy, which is annually used in the form of primary energy throughout the world. The solar hydrogen cycle (Johansson et al. 1993) opens the biggest theoretical range of possibilities for the direct exploitation of insolation. Hydrogen is close to being the ideal energy carrier for the purpose and is potentially the least polluting fuel (Ogden and Nitsch, 1993). Biophotolysis of water provides a means for the capture and utilization of solar energy on a global scale. The practical and economical technology for converting the

Y. Yürüm (ed.), Hydrogen Energy System, 95–110.

absorbed and hydrogen-stored energy into electricity or fuel will in the proximate future have a profound impact on world energy availability and economics.

2. Biophotolysis - the Process

Photochemical reactions are initiated by the absorption of energy in the form of radiation (infrared, visible or ultraviolet). The process by which a photochemical reaction is carried on is designated photolysis. Photochemical processes are sometimes extremely efficient in the conversion of light energy into chemical energy.

In photosynthesis in green plants, light energy is converted into stored chemical energy by the production of cellular constituents from carbon dioxide and water. The process is initiated with the breaking of the stable bond between hydrogen and oxygen in the water molecule by radiant energy in the primary photochemical reaction (Brody & Brody 1962):

$$[1] \quad H_2O \rightarrow 2\,H^+ + 2\,e^- + \tfrac{1}{2}\,O_2$$

The water molecule can not - under usual conditions - be cleaved directly into H- and OH-radicals by the action of sunlight. A photodissociation of this type requires a high-energy radiation of at least 114 kcal per Einstein. (1 Einstein = 1 mole of quanta, i.e. 6×10^{23} quanta; Giesel & Schilling 1982). This corresponds to a wavelength of 260 nm. Photodissociation requires this or even shorter wavelengths if no suitable sensitizer is present. Radiation of these wavelengths are, because of their high energy content fatal for living cells (Hurlbut 1976).

Light energy is a flow of photons whose energy content is inversely related to the wavelength of the radiation. Photosynthetically active organisms (Ormerod 1992) have acquired appropriate pigments to enable them to use the biologically less dangerous part of the solar spectrum for the cleavage of water. The energy of effective red quanta amounts to about 40 kcal per Einstein. In this case, more than one quantum is required to split one molecule of water, in contrast to the photodissociation of water by ultraviolet radiation.

The various kinds of photosynthesis among higher plants and microorganisms show common fundamental features. The structural and functional basis of the physiological process has been intensively studied (Richter 1978). The series of events takes place in a lipid bilayer membrane - the thylakoid. Special pigment - protein complexes known as antennae, are situated on the surface of or in the membrane and capture light energy. By resonance the energy is transferred to reaction centres (Reed & Clayton 1968) where the energetically decisive photochemical reactions take place. By combining photochemically passive light-harvesting chlorophylls with more active reaction centre chlorophylls, the phototrophic organisms have achieved an efficiency of the initial reactions of photosynthesis of 50-70% over a large range of intensities of incident solar radiation.

Chlorophyll molecules inside the reaction centres become activated by the transferred energy from the antennae. They form powerful reducing agents. High-energy electrons are donated to a series of electron and hydrogen carriers in the thylakoid. A charge separation across the membrane is thus established. In accordance with the vectorial nature of oxygenic photosynthesis, the lumen side (inside) of the thylakoid becomes positive charged relatively to the stroma side (outside). This energy-rich state is the chemical energy source for the overall reactions of photosyntesis.

$$[2] \qquad n\,CO_2 + n\,H_2O \xrightarrow{hv=n\ 2821kJ} (CH_2O)\,n + n\,O_2$$

Photosynthesis is interpreted as a two stage process. The first stage produces a potential or voltage, while the second stage is a result of chemical reactions driven by this potential, and resulting in the formation of high energy molecules such as adenosintriphosphate (ATP) and nicotinamide adenine dinucleotide phosphate ($NADPH_2$).

In the reaction sequence of photosynthesis there are three cardinal sites from which energy might be obtained for technological purposes:

- The initial harvesting of the solar radiation with the converting of light energy into a voltage.
  This process is very efficient on a per photon basis. But the scale of this process is in the microdimension, and the pigment protein complexes involved are only stable *in vivo* (Richter 1978).

- The harvesting of plant material and the organic substances synthesized by phototrophic organisms.
  This source of energy is widely used (Hall et al. 1993). But when burned as fuel the material produces carbon dioxide and other polluting substances.

- The harvesting of by-products of photosynthesis.
  Some microorganisms can use light energy to form hydrogen (Mitsui et al. 1983). This process is the most promising for the future development of energy production on a

practical scale. However large surface areas will be required to produce useful quantities of energy.

When considering the question of possible photobiological systems for producing energy on a commercial scale , two options are therefore in forefront.

The first is to develop a system that mimicks the charge separation reaction of photosynthesis for use in a more efficient new generation solar cells (Porter 1989, Bergene & Skulberg 1994). Cyanophytes can be suitable mediators in the conversion of solar energy into electricity in fuel cells. For example intact living films of *Phormidium* sp. and *Mastigocladus laminosus* immobilized on SnO_2 semiconductor electrodes are capable of light-dependent electron transfer to the electrode (Ochiai et al. 1983). Other fuel cells using cyanophytes have been developed in which the electrical current is dependent on endogenous glycogen reserves and light (Tanaka et al. 1988).

The second option is to harness the ability of photosynthetic microorganisms to support light-driven hydrogen production (Greenbaum 1991, Markov et al. 1993). This solution is definitely the most promising as a technology to generate energy on a global, viable scale. This will be discussed in the following pages.

3. Photobiological Hydrogen Production

A variety of microorganisms can evolve molecular hydrogen according to the reaction:

$$[3] \qquad \mathbf{2\,H^+ + 2\,e^- \leftrightarrows H_2}$$

Two types of enzymes form the basis for the production of hydrogen at the expense of light energy. They are respectively hydrogenase and nitrogenase (Cammack et al. 1985, Schlegel 1985). Hydrogenase activity is present in most prokaryotic and eukaryotic microalgae and anoxyphotobacteria. Nitrogenase activity is present in e.g. prokaryotic microalgae (cyanophytes) and photosynthetic bacteria.

The hydrogen-producing systems accept electrons from electron donors which are thereby oxidized in the process. For any commercial application a cheap electron donor is essential. Water is ideal, but also various industrial waste products with negative cost are suitable (e.g. organic wastes, sodium sulphide).

3.1. HYDROGEN PRODUCED BY THE ACTIVITY OF HYDROGENASE.

Solar energy drives the light-harvesting, photochemical and electron transport systems of the photosynthetic reaction in green algae (e.g. species of *Chlamydomonas* and *Scenedesmus*). Under normal, aerobic conditions, $NADPH_2$ transports reducing equivalents into the Calvin cycle for the enzymatic reduction of atmospheric carbon dioxide. But in a carbon dioxide-free anaerobic atmosphere, green algae are capable of synthesizing hydrogenase and evolving molecular hydrogen (Gaffron & Rubin 1942). In this case the water molecule is split into H^+ and O_2, and the electrons are carried through the two photosystems involved. The reduced electron carriers are then reoxidised by hydrogenase, and hydrogen gas is formed.

There are different types of hydrogenase present in the relevant microorganisms (Cammack et al. 1985, Serebriakova et al. 1994). They can be categorized into three groups:

- uptake-hydrogenase. This enzyme catalyses the consumption of hydrogen in the cells

- reversible hydrogenases. These evolve or consume hydrogen depending on the presence of specific electron donors or acceptors

- nitrogenase-associated hydrogenase. The irreversible evolution of hydrogen is made possible by the acting of this enzyme group.

Most hydrogenases are intrinsic membrane proteins. They can be isolated and purified from many bacteria. They are iron-sulphur proteins (FeS), with molecular weights in the range of 65-67, 000 . Regulation of the synthesis of hydrogenase has been studied in a few cases (e.g. *Rhodobacter capsulatus*). The genetics of hydrogenases in photosynthetic bacteria are at present being only marginally studied (Tandeau de Marsac & Houmard 1987, Saunders 1992).

3.2. HYDROGEN PRODUCED BY THE ACTIVITY OF NITROGENASE.

Nitrogenase is the key enzyme for the ATP-dependent reduction of molecular nitrogen to ammonia (nitrogen-fixation):

$$[4] \qquad N_2 + 8H^+ + 8e^- + n\,Mg\,ATP \xrightarrow{\text{NITROGENASE}} 2NH_3 + n\,Mg\,ADP + n\,P\,i + H_2$$

Identification and isolation of nitrogenase have been achieved from several prokaryotic organisms, both among chemoheterotrophic bacteria (Scotobacteria, e.g. *Azotobacter*), phototrophic bacteria (Anoxyphotobacteriae, e.g. *Rhodospirillum*) and cyanophytes (Oxyphotobacteriae, e.g. *Anabaena*). The enzyme is structurally composed of a larger protein moiety and a smaller one (Schlegel 1985, Van Baalen 1987), both of which are essential for the activity. An electron donor, ATP and Mg^{2+} are required for the reduction of N_2 or other appropriate substrate under anaerobic conditions. Nitrogenase is particularly sensitive to oxygen. Preparations of nitrogenase are inactivated in the presence of oxygen even though they originate from aerobic organisms (e.g. *Azotobacter*).

The production of hydrogen occurs through the action of nitrogenase (Ormerod et al. 1961). In the absence of nitrogen can the enzyme reduce several compounds including protons. In the latter case hydrogen gas is produced:

$$[5] \qquad 2H^+ + 2e^- + n\,Mg\,ATP \xrightarrow{\text{NITROGENASE}} H_2 + n\,Mg\,ADP + n\,P\,i$$

The physiological mechanism of action is still under scientific investigation (Stal 1988).

Although the regulation of nitrogenase is associated with end-product inhibition, which causes repression of enzyme activity, it is important to note that the reaction is not affected by the accumulation of hydrogen gas. The availability of fixed nitrogen, reductants and light intensity all influence the regulation of nitrogenase, and the activity is therefore modulated by several environmental factors.

The genetics of nitrogenase have been intensively studied (Saunders 1992).

3.3. COMPARISON OF HYDROGENASE AND NITROGENASE FOR PHOTOBIOLOGICAL HYDROGEN PRODUCTION

The lysis of water via hydrogenase using cyanophytes for example, is an attractive process, provided that solar energy could be used. But hydrogenases and their electron donors are particularly unstable. The catalytic activity of hydrogenase is blocked by molecular oxygen which is an unavoidable end-product of the water photolysis reaction itself. In addition the enzyme is inhibited by hydrogen, such that as little as one atmosphere pressure of molecular hydrogen completely stops further hydrogen evolution (Lambert & Smith 1981). These problems are at present setting limitations for photobiological systems employing hydrogenase. There is a need to investigate more the structure and catalytic mechanisms of hydrogenase (Smith et al. 1992).

The use of nitrogenase for hydrogen production has been investigated for applied purposes. Relevant photosyntethic prokaryotes can utilize solar energy over a wide spectral range to produce the energy for nitrogenase activity. Fortunately the nitrogenase reaction is not subject to end-product repression by molecular hydrogen. Thus hydrogen gas evolution can continue even at one atmosphere hydrogen or more. However some sources of fixed nitrogen (e.g. NH_3) severely inhibit nitrogenase and so hydrogen production.

The cost effectiveness of using nitrogenase for commercial production of hydrogen will depend above all on the selection of suitable mutants of relevant organisms, and the construction of genetic modifications, based on new insights into prokaryotic gene organization and expression (Hunter & Mann 1992, Smith et al. 1992).

4. Organisms for Experimentation on Practical Hydrogen Production

Miroorganisms with the ability to evolve molecular hydrogen (Table 1) are physiological specialists with potential for industrial applications (Mann & Carr 1992). Their exploitation for biotechnology made possible through recent progress in molecular biology and genetics. Such organisms are in the forefront of microbiological and biochemical research because of their intrinsic interest.

Hydrogenase activity has been detected in a variety of eukaryotic algae (Kessler 1974, Harris 1989) and prokaryotes (Schlegel & Schneider 1978, Cammack et al. 1985). Among eukaryotic algae both unicellular and multicellular types posess hydrogenase. But only unicellular algae are capable of catalyzing the hydrogen-related reactions connected with evolution and consumption of molecular hydrogen. The photoevolution of hydrogen has not been demonstrated experimentally in macroscopic algae. Important examples of chlorophytes used for experimental purposes in research on photoproduction of hydrogen are listed in Table 2.

In cyanophytes, hydrogen metabolism is rather complex (Lambert & Smith 1981). Cultures of N_2-fixing cyanophytes can under appropriate conditions evolve molecular hydrogen catalyzed by nitrogenase. Enzymes with hydrogenase activity participate simultaneously (Cammack et al. 1985, Rao & Hall 1988):

- nitrogenase, which catalyzes H_2 evolution

- uptake-hydrogenase, which provides reductant for electron transport in the membrane

Table 1. Taxonomy and properties of photosynthetic microorganisms used in hydrogen production studies.

Common names	**Systematic designation**	**Photosystem type [1)]**	**Nature of metabolism**	**Substrate or eletron donor**
Green algae / chlorophytes	Eukaryotae: Chlorophyceae Mattox and Stewart 1984	PS I and PS II	Photolithotrophic Aerobic	Water
Blue green algae / cyanophytes / cyanobacteria	Prokaryotae: Oxyphotobacteria Gibbons and Murray 1978	PS I and PS II	Photolithotrophic Aerobic Facultatively anaerobic	Water
Purple and green bacteria / photosynthetic bacteria	Prokaryotae: Anoxyphotobacteria Gibbons and Murray 1978	PS I or PS II	Photoorganotrophic Anaerobic Facultatively aerobic	Organic substrate Hydrogen sulphide

[1)] PS I, i.e. photosystem I, reaction centres operating in a strongly electronegative redox range

PS II, i.e. photosystem II, reaction centres operating in a less strongly electronegative redox range.

Table 2. Major microorganisms being used for biotechnological research on hydrogen production.

Chlorophytes[1)]	**Cyanophytes**[1)]	**Photosynthetic bacteria**[2)]
Chlamydomonas reinhardtii Dang.	Anabaena cylindrica Lemm.	Rhodobacter capsulatus (Molisch) Imhoff, Trüper & Pfennig
	Anabaena flos-aquae Bréb. ex Born. et Flah.	
Chlorella vulgaris Beij.	Anabaena variabilis Kütz. ex Born. et Flah.	
	Chlorogloea fritschii de Mitra	Rhodobacter sphaeroides (van Niel) Imhoff, Trüper & Pfennig
Scenedesmus obliquus (Turp.) Kütz.	Fischerella muscicola (Thuret) Gom.	
	Leptolyngbya laminosa (Gom.) Anagn. & Kom.	
	Mastigocladus laminosus Cohn	Rhodospirillum rubrum (Esmarch) Molisch
	Nostoc muscorum Ag. ex Born. et Flah.	
	Pseudanabaena limnetica (Lemm.) Kom.	
	Synechococcus elongatus Näg.	Thiocapsa roseopersicina Winogradsky

1) Süsswasserflora von Mitteleuropa (Ettl et al. 1994).

2) Bergey's Manual of Systematic Bacteriology (Staley et al. 1989).

- a soluble hydrogenase, which catalyzes H_2 evolution or H_2 consumption when provided with relevant electron donors or acceptors.

The cyanophytes include many species that carry out the reduction of the dinitrogen molecule to ammonia (Stewart 1980, Van Baalen 1987). Selected physiological strains of laboratory cultured cyanophytes have been preferred for experiments with photoevolution of hydrogen (Skulberg & Skulberg 1990). Examples are listed in Table 2.

The N_2-fixing photosynthetic bacteria carry out H_2 evolution utilizing a wide spectrum of soluble organic substrates. In these organisms nitrogenase is not protected in specialized cells (e.g. heterocysts), and the photoproduction of molecular hydrogen must be performed under anaerobic conditions. They are unable to use water as an electron donor. Only a minority of species in this extremely heterogeneous eubacterial group (Pfennig 1977) have been experimented on as candidates for practical hydrogen production. However, the results so far indicate their many advantages for the purpose. Examples of relevant organisms are listed in Table 2.

5. Algal Culture Technology and Hydrogen Production

The basic knowledge about photosynthetic microorganisms make possible the functional operation of bioreactors for algal culture (Pirt et al. 1983).

Microalgae have so far scarcely been commercially exploited, in contrast to macroalgae (Lembi & Waaland 1988). Several advantages of culturing microalgae for economic purposes are readily recognizable, e.g.:

- cultivation of microalgae represents an efficient biological system for utilizing solar energy to produce organic matter (Zaborsky 1982)

- population density of the culture may be readily adjusted for optimal utilization of intense light irradiance (Grobbelaar 1982)

- the life cycle of relevant microalgae is completed within hours, making selection and improvements (genetic recombination) in the species relatively easy. This property is particularly at hand for the prokaryotes, a feature which will in time become of cardinal importance (Richmond 1988)

- the microalgae provide a range of primary and secondary metabolites of vital interest for mankind (Benemann 1989)

- the possibility of harvesting hydrogen as a by-product of photosynthesis is promising (Mitsui et al. 1983).

Microalgae are sources of many useful products, including hydrogen. Extensive research has taken place during the past two decades in the relevant fields of biotechnology. Many laboratories have initiated experimental work on the use of clones from algal culture collections for purposes of this kind (Kumazawa & Shimamura 1993). Properties and products of selected strains have been investigated. The algae may be used for food or fodder. Some strains have very high contens of proteins, vitamins and other essential growth factors. Vital pigments of current interest are produced, in particular carotenoids with provitamin-A activity, and astaxanthin (Borowitzka & Borowitzka 1988). Some cyanophyte strains produce substances of potential pharmaceutical interest (e.g. antibiotics). These examples are only a few of the possible applications of algal cultures for economic development (Gudin & Thepenier 1986, Kerby & Rowell 1992).

The basic concept is the development of an integrated system using algal culture technology for the combined production of biomass by nitrogen fixing microorganisms, utilizing their photobiological hydrogen production and properties for the biosynthesis of various useful metabolites (Skulberg 1994).

6. Prospects and Perspectives

Light-induced hydrogen evolution in species of green algae and cyanophytes has been investigated in several laboratories (Skulberg 1992). Relatively high rates of hydrogen production for sustained periods of time have been obtained using both indoor and outdoor photobioreactors (Mitsui et al. 1983, Richmond 1992). The quantity of hydrogen obtained is in the order of 25 ml H_2 m^{-2} day^{-1}. This is still too low for commercial exploitation.

The results of continuous photoproduction of hydrogen by free-living cells of the anoxyphotobacterium *Rhodospirillum rubrum* indicate better rates. In experiments lasting up to 80 days, with lactic acid-containing waste, the rate of hydrogen production was 65 ml H_2 $hour^{-1}$ litre $culture^{-1}$ (Solar Energy Research Institute, Colarado, U.S.A.; pers. comm.). By optimizing growth conditions, rates up to twice that much could be obtained. For comparison, the minimal rate necessary for large scale production of hydrogen based on a reactor process is stated to be 130 ml H_2 $hour^{-1}$ litre $culture^{-1}$ (Zürrer & Bachofen 1979).

A combined use - in separate photobioreactors - of anoxyphotobacteria and oxyphotobacteria should be tried out as a convenient practical solution to produce commercially significant quantities of hydrogen gas. The anaerobically operated photobioreactor produces hydrogen at the expense of a reduced organic carbon source. The carbon dioxide released simultaneously during the production of hydrogen gas, can be removed by growing cyanophytes. The second light-driven reactor with cyanophytes is used for the purpose of effectively cleaning the hydrogen of the accompanying carbon dioxide by converting it into harvestable biomass. The research required to overcome the engineering problems involved may be the most difficult to achieve.

The commercial use of biophotolysis for hydrogen production is today only at the conceptual stage of development, but promising results of fundamental and applied research strongly indicate that the basic requirements of a practical process can be achieved. Considerable costs may be involved in the construction of reactor systems adapted to maintain the producer organisms under optimum conditions. By fruitful combination of scientific research (e.g. genetic and metabolic manipulations of the relevant microbes) and engineering skill (e.g. light concentration and process coupling), resolution of the problems and constraints is getting closer.

The photobiological system implies obvious advantages compared to the conventional chemical or physical technologies (except electrolysis of water based on hydropower) used for hydrogen production. The biological system can be operated at low temperatures (e.g. 10-40°C), the only major input is solar energy and a hydrogen donor (e.g. water), and the process will not produce offending pollutants.

In further development of algal culture technology (Skulberg 1994) a potentially beneficial and economic program of multiple utilization may be realised. Uniformity of reactor equipment and methods for culturing and harvesting can be adopted. Access to a wide spectrum of products of vital human and commercial interest offers economic advantages for the total industrial operation of the photobiological hydrogen production system.

One strategy emerges for future research: Well established experimental plants should be built and operated as forerunners for commercial plants that will be economically attractive and environmentally sound.

7. References

Benemann, J.R. (1989) 'The future of microalgal biotechnology', in R.C. Cresswell, T.A.V. Rees & N. Shah (eds.), Algal and Cyanobacterial Biotechnology, Longman Scientific & Technical, Harlow, pp. 317-337.

Bergene, T. & Skulberg, O.M. (1994) 'Naturen som veiviser mot morgendagens energisamfunn', Naturen 118, 14-19.

Borowitzka, M.A. & Borowitzka, L.J. (1988) 'Micro-algal Biotechnology', Cambridge University Press, Cambridge, 477 pp.

Brody, M. & Brody, S.S. (1962) 'Light reactions in photosynthesis', in R.A. Lewin (ed.), Physiology and Biochemistry of Algae, Academic Press, New York, pp. 3-23.

Cammack, R., Hall, D.O. & Rao, K.K. (1985) 'Hydrogenases. Structure and applications in hydrogen production', in R.K. Poole & C.S. Dow (eds.), Microbial Gas Metabolism, Academic Press, London, pp. 75-102.

Ettl, H., Gerloff, J., Heynig, H. & Mollenhauer, D. (1994) 'Süsswasserflora von Mitteleuropa. Cyanophyceae', Band 19, Gustav Fischer Verlag, Stuttgart, (in press).

Gaffron, H. & Rubin, J. (1942) 'Fermentative and photochemical production of hydrogen in algae', J. gen. Physiol. 26, 219-240.

Giesel, H.B. & Schilling, H.-D. (1982) 'Das kleine Energilexikon', Verlag Glück auf GMBH, Essen, 222 pp.

Greenbaum, E. (1991) 'Hydrogen production by photosynthetic water splitting', in T.N. Veziroglu & P.K. Takahashi (eds.), Hydrogen Energy Progress VIII, Proceedings of the 8th World Hydrogen Energy Conference, Honolulu and Waikoloa, Hawaii, USA, Pergamon Press, New York, pp. 743-754.

Grobbelaar, J.U. (1982) 'Potentials of algal production', Water SA 8(2), 79-85.

Gudin, C. & Thepenier, C. (1986) 'Bioconversion of solar energy into organic chemicals by microalgae', Advances in Biotechnological Processes 6, 73-110.

Hall, D.O., Rosillo-Calle, F., Williams, R.H. & Woods, J. (1993) 'Biomass for energy: Supply prospects', in T.B. Johansson, H. Kelly, A.K.N. Reddy & R.H. Williams (eds.), Renewable Energy. Sources for Fuels and Electricity, Earthscan Publications Ltd, London, pp. 593-651.

Harris, E.H. (1989) 'The *Chlamydomonas* Sourcebook. A Comprehensive Guide to Biology and Laboratory Use', Academic Press, London, 780 pp.

Hunter, C.N. & Mann, N.H. (1992). 'Genetic manipulation of photosynthetic prokaryotes', in N.H. Mann & N.G. Carr (eds.), Photosynthetic Prokaryotes, Plenum Press, New York, pp. 153-179.

Hurlbut, J. (1976) 'The Planet we Live on', Harry N. Abrams, Inc., New York, 527 pp.

Johansson, T.B., Kelly, H., Reddy, A.K.N & Williams, R.H. (1993) 'Renewable Energy. Sources for Fuels and Electricity', Earthscan Publications Ltd, London, 1160 pp.

Kerby, N.W. & Rowell, PO. (1992) 'Potential and commercial applications for photosynthetic prokaryotes', in N.H. Mann & N.G. Carr (eds.), Photosynthetic Prokaryotes, Plenum Press, London, pp. 233-265.

Kessler, E. (1974) 'Hydrogenase, photoreduction and anaerobic growth', in W.D.P. Stewart (ed.), Algal Physiology and Biochemistry, University of California Press, Berkely, pp. 456-473.

Kumazawa, S. & Shimamura, K. (1993). Photosynthesis-dependent production of H_2 by marine N_2-fixing cyanobacterium, *Anabaena* sp. TU37-1', J. Mar. Biotechnol. 1, 159-162.

Lambert, G.R. & Smith, G.D. (1981) 'The hydrogen metabolism of cyanobacteria (blue-green algae)', Biol. Rev. 56, 589-660.

Lembi, C.A. & Waaland, J.R. (1988) 'Algae and Human Affairs', Cambridge University Press, Cambridge, 590 pp.

Mann, N.H. & Carr, N.G. (1992) 'Photosynthetic Prokaryotes', Plenum Press, London, 275 pp.

Markov, S.A., Lichtl, R., Rao, K.K. & Hall D.O. (1993) 'A hollow fibre photobioreactor for continuous production of hydrogen by immobilized cyanobacteria under partial vacuum', Int. J. Hydrogen Energy 18(11), 901-906.

Mitsui, A., Philips, E.J., Kumazawa, S., Reddy, K.J., Ramachandran, S., Matsunaga, T., Haynes, L. & Ikemoto, H. (1983) 'Progress in research toward outdoor biological hydrogen production using solar energy, sea water, and marine photosynthetic microorganisms', Annals N.Y. Acad. Sci. 413, 514-530.

Ochiai, H., Shibata, H., Sawa, Y., Shoga, M. & Ohta, S. (1983) 'Properties of semiconductor electrodes coated with living films of cyanobacteria', Appl. Biochem. Biotechnol. 8, 289-303.

Ogden, J.M. & Nitsch, J. (1993). 'Solar hydrogen', in T.B. Johansson, H. Kelly, A.K.N. Reddy & R.H. Williams (eds.), Renewable Energy. Sources for Fuels and Electricity, Earthscan Publications Ltd, London, 1160 pp.

Ormerod, J.G. (1992) 'Physiology of the photosynthetic prokaryotes', in N.H. Mann & N.G. Carr (eds.), Photosynthetic Prokaryotes, Plenum Press, New York, pp. 93-120.

Ormerod, J.G., Ormerod, K.S. & Gest, H. (1961). 'Light dependent utilization of organic compounds and photoproduction of molecular hydrogen by photosynthetic bacteria, relationship with nitrogen metabolism', Arch. Biochem. Biophys. 94, 449-463.

Pfennig, N. (1977) 'Phototrophic green and purple bacteria: A comparative systematic survey', Ann., Rev. Microbiol., 31, 275-290.

Pirt, S.J., Lee, Y.K., Walach, M.R., Pirt, M.W., Balyuzi, H.H.M., & Bazin, M.J. (1983) 'A tubular bioreactor for photosynthetic production of biomass from carbon dioxide. Design and performance', J.Chem. Technol. Biotechnol. 33B, 33-50.

Porter, G. (1989) 'Solar energy from photochemistry', in J. Barber & R. Malkin (eds.), Techniques and New Developments in Photosynthesis Research, Plenum Press, New York, pp. 3-7.

Rao, K.K. & Hall, D.O. (19888) 'Hydrogenases: Isolation and assay, in L. Packer & A.N. Glazer (eds.), Cyanobacteria, Methods in Enzymology, Volume 167, Academic Press, London, pp. 501-509.

Reed, D.W. & Clayton, R.K. (1968) 'Isolation of a reaction center fraction from *Rhodopseudomonas sphaeroides*, Biochem. Biophys. Res. Commun. 30, 471-475.

Richmond, A. (1988) '*Spirulina*', in M.A. Borowitzka & L.J. Borowitzka (eds.), Micro-algal Biotechnology, Cambridge University Press, Cambridge, pp. 85-21.

Richmond, A. (1992) 'Mass culture of cyanobacteria', in N.H. Mann & N.G. Carr (eds.), Photosynthetic Prokaryotes, Plenum Press, New York, pp. 181-210.

Richter, G. (1978) 'Plant metabolism. Physiology and Biochemistry of Primary Metabolism', Georg Thieme Publishers, Stuttgart, 475 pp.

Saunders, V.A. (1992): 'Genetics of the photosynthetic prokaryotes', in N.H.Mann & N.G. Carr (eds.), Photosynthetic Prokaryotes, Plenum Press, New York, pp. 121-152.

Schlegel, H.G. (1985) 'Allgemeine Microbiologie', Georg Thieme Verlag, Stuttgart, 571 pp.

Schlegel, H.G. & Schneider, K. (1978). 'Hydrogenases: Their Catalytic Activity Structure, and Function', Erich Goltze KG, Göttingen.

Serebriakova, L., Zorin, N.A. & Lindblad, P. (1994) 'Reversible hydrogenase in *Anabaena variabilis* ATCC 29413', Arch. Microbiol. 161, 140-144.

Skulberg, O.M. (1992) 'Hydrogen som energibeærer. Biofotolytisk produksjon av hydrogen med bruk av blågrønnalger/alger', Norsk institutt for vannforskning, Oslo, 27 pp.

Skulberg, O.M. (1994) 'Oscillatorialean cyanoprokaryotes and their application for algal culture technology', Arch. Hydrobiol./Suppl., Algological Studies 75, (in press).

Skulberg, R. & Skulberg, O.M. (1990) 'Forskning med algekulturer - NIVAs kultursamling av alger. Research with algal cultures. NIVA's Culture Collection of Algae', Norsk institutt for vannforskning, Oslo, 32 pp.

Smith, G.D., Ewart, G.D. & Tucker, W. (1992) 'Hydrogen production by cyanobacteria', J. Hydrogen Energy 17(9), 695-698.

Stal, L.J. (1988) 'Nitrogen fixation in cyanobacterial mats', in L. Packer & A.N. Glazer (eds.), Cyanobacteria, Methods in Enzymology, Volume 167, Academic Press, London, pp. 474-484.

Staley, J.T., Bryant, M.P., Pfennig, N. & Holt, J.G. (1989) 'Bergey's Manual of Systematic Bacteriology', Vol. 3, Williams & Wilkins, Baltimore.

Stewart, W.P. (1980) 'Some aspects of structure and function in N_2-fixing cyanobacteria', Ann. Rev. Microbiol 34, 497-536.

Tanaka, K., Kashiwagi, N. & Ogawa, T. (1988) 'Effects of light on the electrical output of bioelectrochemical fuel cells containing *Anabaena variabilis* M-2: mechanisms of the postillumination burst, J. Chem. Technol. Biotechnol. 42, 235-240.

Tandeau de Marsac, N. & Houmand, J. (1987) 'Advances in cyanobacterial molecular genetics', in P. Fay & C. Van Baalen (eds.), The Cyanobacteria, Elsevier, Amsterdam, pp. 251-303.

Van Baalen, C. (1987) 'Nitrogen fixation', in P. Fay & C. Van Baalen (eds.), The Cyanobacteria, Elsevier, Amsterdam, pp. 187-198.

Zaborsky, O.R. (1982)' CRC Handbook of Biosolar Resources', Volume 1, Part 1, Basic principles, CRC Press Inc., Boca Rator, 480 pp.

Zürrer, H. & Bachofen, R. (1979) 'Hydrogen production by the photosynthetic bacterium *Rhodospirillum rubrum*', Appl. Environ. Microbiol. 37, 789-793.

ATOMIC HYDROGEN PRODUCTION AND MODELLING REVISITED

E. EKINCI
Istanbul Technical University
Department of Chemical Engineering
Ayazağa-Istanbul
Turkey

ABSTRACT. In this study the microwave discharge technique for atomic hydrogen production is reviewed starting with the earlier work of Wood in 1921 uptill the extensive electrical discharge era of 1970's. The evaluation is completed with the recent studies which concentrated on atomic hydrogen utilization in semiconductor manufacturing, thin film formation, surface treatment and modification, nano particle formation, sintering and hydropyrolysis of hydrocarbons.

A model reported in literature is modified and used to predict the production of atomic hydrogen in a 2.45 GHz MW system. The effect of MW power, hydrogen flow rate, moisture, ageing of reactor surface, surface coating and diameter of discharge tube are investigated experimentally and theoretically.

1.0 Introduction

The low pressure electrical discharge is universally recognized as an effective and convenient means for the production of radicals by dissociation of molecular gases. Highly active species formed through a suitable discharge medium have been considered by chemical engineers as a potential synthetic reagent. With the expectation of cheap electricity a big research programme was built in USA and Europe and voluminous literature has been accumulated on gas discharges. One branch of these attempts involved breaking of molecules into atoms for their chemical activity. Some examples of the dissociation studies were on oxygen, Mearns and Morris[1], and Bell and Kwong[2], on hydrogen, Mearns and Ekinci[3], on oxidation of hydrogen chloride, Cooper, Mickey and Baddour[4], on formation of hydrogen cyanide from methane and ammonia Juul-Dam and Brockmeier[5], and on oxidation of carbon monoxide Brown and Bell[6].

The reasons why this study on production of atomic hydrogen was carried out using a microwave (MW) discharge are:

Y. Yürüm (ed.), Hydrogen Energy System, 111–133.

a) Hydrogen being the simplest of all molecules, will serve as a good model for the understanding of discharge phenomena. Also data on many model parameters are available for hydrogen.
b) Hydrogen as a molecule plays a key role in chemical industries. Hydrogen atoms being more reactive is expected to open new horizons such as in reaction between salts of heavy metals.
c) Hydrogen discharge systems are finding upto date applications in semiconductor processing, especially in film formation and etching[7].
d) Chlorine industries are finding thighter controls due to environmental reasons.
Hydrogen discharge system may be an alternative compact chlorine production method from HCl or from iron chloride[8].

In this report experimental results, on effect of variation of different parameters on production of atomic hydrogen in a 2.45 GHz discharge is given. The experimental results are compared with a hydrogen dissociation model developed by Bell[9].

2.0 Historical Development of Atomic Hydrogen Production

2.1 EARLIER WORKS

In 1921 Wood[10] produced atomic hydrogen by means of a glow discharge tube is known as a Wood's tube.Using a Wood's tube, Poole[11,12] made a systematic study on hydrogen to determine the effect of power and flow rate on the yield, and investigated surface effects. Poole used both silica and pyrex discharge tubes. He found that the optimum pressure for dissociation of hydrogen was approximately 0.6 mm. The maximum value of the energy efficiency of hydrogen atom production obtained was $4.5x10^{-2}$ molecules per electron volt. The Wood's tube was improved by Weinrich and Hughes[13] in 1954, by the introduction of water cooled electrodes which made it more versatile for studies of hydrogen and oxygen dissociation.

Direct current or low frequency alternating current discharges was later replaced by radio-frequency or microwave techniques. Adoption of the high frequency discharge is attributed to a combination of factors such as ease of control, simplicity of construction of the discharge tube, improved purity and greater stability.

The development of microwave sources during the second world war led researchers to use microwave power for dissociation of gases. The early users of microwave power sources in discharge were Nagle, Julian, and Zacharias[14] and Broida and Moyer[15]. For a hydrogen discharge operated at 2.45 GHz, the highest yield of atomic hydrogen obtained by McCarthy[16] was 1.0 g.atoms/kw-hr or $3x10^{-2}$ molecules per electron volt. Rose and Brown[17] have measured the high frequency electric field required to maintain a hydrogen plasma electron density. Shaw[18] made a more extensive study of atomic hydrogen production than any previously done, using a 2.45 GHz microwave power source. He found a maximum conversion of 90% at 0.5 Torr pressure, 10^{-5} moles/sec flow rate and when 100 watts of 3.0 GHz power was dissipated. His experiments, using discharge tubes of 1.75 cm. and 0.9 cm. inside diameter, resulted in a higher yield of atoms for the smaller tube at the same power input and flow rate. A marked effect of temperature on yield was found by Shaw; for a flow rate of 140 μgmoles/sec of hydrogen, the yield of hydrogen atoms decreased about six-fold when the temperature was increased from 60°C to 350°C. In 1965 Rony[19] reported a study of hydrogen atom reactions in the gas phase and at solid surfaces where the atoms were produced by microwave techniques at low pressure. Mezey and Oxley[20] in 1967 reported on the production of hydrogen atoms in high and low pressure microwave discharge. Information was given on the extent of dissociation, pressure, frequency and chemical aspects of production by microwave techniques. However, the reported conversions of 92% and 2% at respective pressures of 2.5 and 26 Torr are considerably higher than reported else where in the literature. Skidan and Schreider[21] investigated the degree of hydrogen in the positive column of a high frequency glow discharge in H_2-He mixtures.

During his pioneering work on atomic hydrogen production, Wood[10] noted that a trace water vapour or oxygen served to reduce the ability of the walls to catalyze recombination. Later Poole[11] reached the conclusion that the use of water vapour as an anti-catalyst in glass or silica discharge tubes is very unreliable and likely to yield erratic results. Tubes coated with metaphosphoric acid when used with wet hydrogen gave reproducible and high yields of atomic hydrogen. In this connection Dingle and LeRoy[22] in their work with atomic hydrogen, obtained no hydrogen atoms in clean apparatus. A satisfactory yield was obtained, however, when the walls of the apparatus were coated with a mixture of orthophosphoric acid and phosphorus pentoxide. However, the efficiency of this substance as an anticatalyst was reduced if water vapour or oxygen were present. Wittke and Dicke[23] in 1965 employed dry-film (mixture of dichlorosilane and methyl chlorosilane) which served to reduce the extent of surface recombination. This material was found to provide a film that was heat stable, had a very low vapour pressure at room temperature, and was more effective than water vapour or acid in preventing recombination of hydrogen atoms at glass surfaces. Shaw[13], and Zletz[24],1958 dry-film was used effectively to prevent recombination of hydrogen atoms in glass, quartz and Vycor tubes used in a microwave discharge.

2.2 RECENT APPLICATION OF DISCHARGE PRODUCED ATOMIC HYDROGEN

There is increasing interest in the usage of electrical discharge in processing. One of the upto date usage is in semiconductor processing in halogen plasmas on details of surface etching mechanism,Chou and Phillips[25]. In this context, hydrogen atoms have been found to etch silicon gallium arsenide and gallium antimony semiconductors.On the other hand the oxides of these semiconductors also etched in discharge medium but the rate was significantly lower compared to hydrogen. The mechanism of hydrogen involved is believed to be hydrogen atoms reaching to from volatile hydrides. In a high frequency (hf) discharge system Nirmatov et. al[7], investigated interaction of hydrogen atoms with tin chloride. It was found that in this discharge flow system tin films are obtained. The mechanism of film formation is also explained as the dissociation of the hydride of tin on the substrate.

Kato et.al[26] used hydrogen or argon plasma with a double-tubed, coaxial-line-type, MW plasma (chemical vapour deposition) CVD system. Hydrogenated amorphous silicon (a-Si:H) films that are good quality and thermally stable are obtained even at room temperature. Hydrogen atoms are explained to have good covering effect which helped the silane radicals to have large surface mobility.

Hydrogen discharge systems are used to modify resistors prepared from polysilicon on integrated circuits. Treatment of various polysilicon circuit elements can be decreased more than three order of magnitude in RF generated hydrogen discharge.

Mathsushita, et.al[27] used hydrogen plasma to produce epitaxial growth of semiconductors. The layers were GaSb, GaAs, InP, Si and sapphire substrates at relatively low temperatures by plasma-assisted epitaxy (PAE).

Surface structural changes of high density polyethylene (HDPE) induced by treatment by hydrogen and argon plasma. Electron conduction of HDPE conduction affected due to the transvinylenes and free radicals remaining on the surface after plasma treatment of boron-implanted CVD and high-pressure synthesized (HPS) diamonds improved the electrical properties substantially[28]. Hydrogen plasma used successfully as an insitu remote cleaner of Si (100) surfaces[29]. Brecelj and Mozetic[30] used 27.12 MHz 700W RF generator and hydrogen pressure between 0.05 and 50 Pa to discharge clean metal oxide thin layers. The efficiency of glow discharge cleaning was reported to be maximum at 1 Pa pressure .

Ferrocene was fed into the afterglow region of a low pressure, low power MW discharge system for the purpose of obtaining nanoparticles. In hydrogen discharge metallic iron nanoparticles were obtained. Low yields of small particles (~10nm) were obtained when the injection of ferrocene was done into the afterglow region but yield impaired considerable with large particles (~ 50nm) when the injection was done through the coupler[31].

In smooting of (Al,Ga) As/GaAs quantum well growth interruptions are obtained by many researcher. These growth interruptions may result in reduction of integrated emissions by an order of magnitude. The quality (life time) and luminescence efficiency can be improved by annealing the quantum wells in a hydrogen discharge for 2 hours. Hydrogen discharge treatment improves the luminescence efficiency and does not degrade the surface smoothness[31].

One new application of hydrogen plasma is in the field of pyrolysis of gaseous and liquid hydrocarbons[32]. Experiments carried out on the pyrolysis of coal in hydrogen plasma jet at atmospheric pressure and temperatures between 1000 and 2000°C showed that the pyrolysis mechanism is different than conventional pyrolysis system. Even though there is no economic claim at this stage the fact that hydrocarbons undergoing very fast reactions with highly reactive plasma species is an effective medium to work against the retrogressive char forming reactions generally operative under conventional pyrolysis processing systems. The volatiles released from coal are used in the production of acetylene in hydrogen plasma jet.

Another application of hydrogen-discharge to hydrocarbon processing is the hydrocracking of heavy paraffinic hydrocarbons[33]. Due to the special chemical activities associated with discharge induced species the cokification reactions were hoped to be avoided and decomposition rates increased. In this study a fluidised bed reactor was used which allowed good mixing of hydrocarbon and atomic hydrogen and to regulate the residence time of the gases. By thermal processing 95% efficiency was achieved together with high product selectivity. Adjustment of power input is a good control mechanism. Collection of differential splits of hydrocarbon by selective cooling is attempted for production of valuable products[33].

H_2 was utilized in a plasma (discharge) atmosphere in sintering of explosive shock activated Al_3O_3-ZrO_2 powder,Al_2O_3 powder-SiC whisker and Al_3O_3/ZrO_2 powder-SiC whisker mixtures. Alternative plasma gases used were air, O_2, and Ar[34]. In their study authors claim that for shock processing H_2 and air plasmas seem to be more effective in sintering compared to O_2 and Ar but in unshocked Al_2O_3-10ZrO_2 composites the effect is not significant. Same group of researches also used hydrogen in low energy plasmas for surface modifications in ion-nitriding and ion carburizing practices[35]. They report that ion-nitriding with the N_2/H_2 plasma has some advantages through the catalytic action of hydrogen which is required for the formation of nitrogen rich Fe_2N phase. For the case of hydrogen /nitrogen plasma thicker layers than pure nitrogen is obtained compared to pure nitrogen also for H_2/N_2 case larger case depths are obtained[35].

3.0 Theoretical Model

3.1 SURVEY OF REACTIONS INVOLVING ATOMIC HYDROGEN PRODUCTION

The most dominant elementary gas phase reactions responsible for the formation and loss of atomic hydrogen can be deduced from the above considerations to be:

$$e + H_2 \rightarrow 2H + e \qquad (1)$$

$$2H + H_2 \rightarrow 2H_2 \qquad (2)$$

$$3H \rightarrow H + H_2 \qquad (3)$$

Three additional reactions might be considered as contributors to kinetics:

$$H_2^+ + H_2 \rightarrow H^+ + H + H_2 \qquad (4)$$

$$H_2^+ + H_2 \rightarrow H_3^+ + H \qquad (5)$$

$$2H + H_2^+ \rightarrow H_3^+ + H \qquad (6)$$

The concentration of H_2^+ ions present in the system determines the rates of reactions (4)-(6) which are formed by ionisation of molecular hydrogen and consumed by reactions (4)-(6) plus the less important reactions not listed. The reported reaction rate constants for reactions (4) and (5) are comparable to that of reaction (1). Since $n_e = n_H + n_{H_2} + n_{H_3}$ and concentration of H_2^+ is one or two orders lower than that of free electrons, the formation of atomic hydrogen by reactions (4) and (5) is expected to be one or two orders of magnitude smaller than reaction (1). Reaction (6) can be neglected by comparison to reaction (2) since concentration of H_2 is much greater than H_2^+.

3.2 DEVELOPMENT OF MODEL EQUATION

Bell[9] found that the concentration profile in the direction perpendicular to the flow is essentially flat and becomes flatter for lower values of recombination coefficient.This means the heterogeneous recombination process is not severely limited by mass transfer. Thus, a one dimensional method will be adequately describe the reactor behaviour. Assuming the discharge tube to be a plug flow reactor, the rate of change of atomic hydrogen in terms of fractional conversion x, for a differential volume is:

$$r = F\frac{dx}{dV} \tag{1}$$

expressing the fractional conversion in terms of atomic and total gas densities,

$$r = Fd(\frac{n_1}{2N - n_1})/dV \tag{2}$$

Taking the differential of the term in parenthesis, a material balance on atomic hydrogen gives:

$$\frac{4N}{(2N - n_1)^2}F\frac{dn_1}{dV} = 2k_d\langle n_e\rangle(N - n_1) - \frac{1}{d}\gamma_1\bar{c}n_1 - 2k_2n_1^2(N - n_1) - 2k_3n_1^3 \tag{3}$$

This equation represents the net production and loss rates in the active discharge zone.Since no dissociation is assumed to take place between the discharge exit and titration point for the situation outside the discharge afterglow the model equation will reduce to:

$$\frac{4N}{(2N - n_1)^2}F\frac{dn_1}{dV} = -\frac{1}{d}\gamma_2\bar{c}n_1 - 2k_2n_1^2(N - n_1) - 2k_3n_1^3 \tag{4}$$

where F is the inlet flow rate of molecular hydrogen, N is the total gas concentration, V is the volume of the reactor, n_e is the volume averaged electron density , $\bar{c}$ is the mean hydrogen atom velocity, γ_1 and γ_2 are the wall recombination coefficients inside and outside the active discharge zone, k_d, k_2 and k_3 are the rate constants for electron-impact dissociation, and homogeneous recombination reactions. The initial value of the second equation n_1 is the discharge exit value of hydrogen calculated by first equation.

In order to be able to make comparisons with experimental results, the output of the model is expressed in terms of conversion and energy yield, which are defined as :

$$conversion = {n_1}/{2N - n_1} \tag{5}$$

$$\textit{energy yield} = \frac{7.2 \times 10^6 \times conversion \times F/V}{\overline{P}} \tag{6}$$

The two model equations were solved numerically using a fourth order Runge-Kutta integration method. Three different programs were developed for the prediction of conversion and energy yield of atomic hydrogen:

a) for variation with absorbed power,
b) for variation in inlet flow rate,
c) for variation with absorbed power when t_d is constant.

Using these programs, it is possible to investigate the effects changes in reaction constants, temperature of the gas, inlet flow rate absorbed power, electron number density, discharge volume and pressure on conversion and energy yield of atomic hydrogen.

In the model k_W recombination constant on the wall outside the discharge zone is determined from the relation $k_W = \gamma_2 \bar{c}/d$ .The value of γ_2 for orthophosphoric acid is taken as 7×10^{-5} obtained after temperature adjustment of Dingle and LeRoy[22] reported value. The value of k_2 employed is 7.9×10^3 m^6 mol^{-2} s-1 Schofield[36], and $k_3 = 1.6 \times 10^3$ m^6 $mol^{-2} s^{-1}$ Bell[9], was used in the model.

Values of flow rate F were measured, discharge volume V, is taken from length of glow times cross-sectional area and the number density of molecules was calculated at the measured pressure and temperature. No data was available for recombination coefficient γ_1 for active region of glow region but the ratio of $\gamma_1 = 100\gamma_2$ for oxygen reported in literature may be indicative Mearns and Morris[3].

Dissociation constant k_d is calculated for a Maxwellian distribution of electron energies :

$$k_d = \left(\frac{8}{\pi m}\right)^{0.5} (kT_e)^{-3/2} \int_o^\infty \in \sigma_1(\in) e^{-\in/kT_e} d\in \qquad (7)$$

Cross section data was derived by Khare and Moisewitch[37], who assumed excitation o ${}^1\sum_g^+ \rightarrow {}^3\sum_u^+$ as the only process leading to atom formation.Their solution is plotted in Figure 1.

The power density at a point x in discharge is given by Bell[9]

$$\text{and} \quad P(x) = \frac{n_e e^2 E_e^2}{2mv} \qquad (8)$$

$$E_e^2 = E_o^2 \frac{v^2}{v^2 + \varpi^2} \qquad (9) \qquad\qquad n_e = n_{e_0} \cos\frac{\pi x}{L} \qquad (10)$$

For hydrogen $\nu = 3.67x10^7 p_o$ here p_0 is reduced pressure $p_0 = 298p/T_g$ with units N/m2.

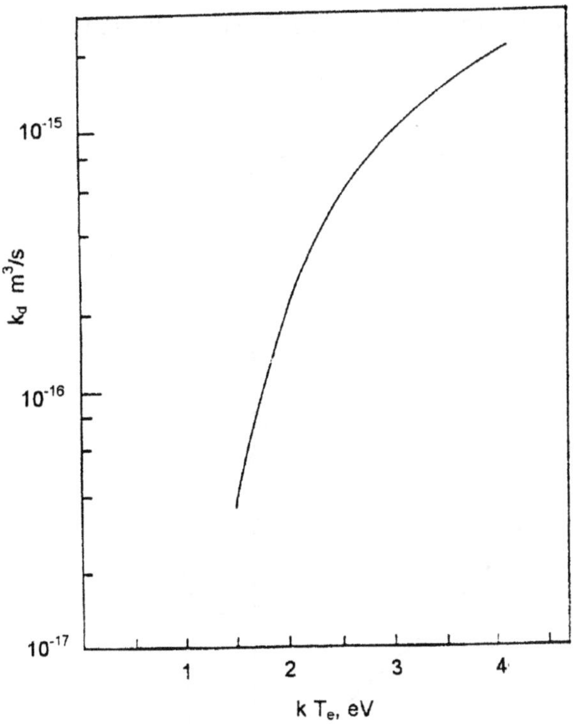

Figure 1. Dissociation rate constant as a function of k.T_e.

The average electron density in the discharge is given by

$$n_{e_{av}} = \int_{-L/2}^{L/2} \frac{n_{e_0} \cos \pi x}{L} dx = \frac{2}{\pi} n_{e_0} \tag{11}$$

Now the average power density may be expressed as

$$\overline{P} = \frac{\frac{2}{\pi} n_{e_0} e^2 E_e^2}{2m(3.64x10^7)p_0} \tag{12}$$

rearranging

$$n_{e_o} p_o = \frac{\overline{p}\pi m(3.64x10^7)}{e^2(E_e/p_o)^2} \tag{13}$$

for a cylindrical discharge with L>>R characteristic diffusion length,

$$\Lambda = \frac{R}{2.405} \tag{14}$$

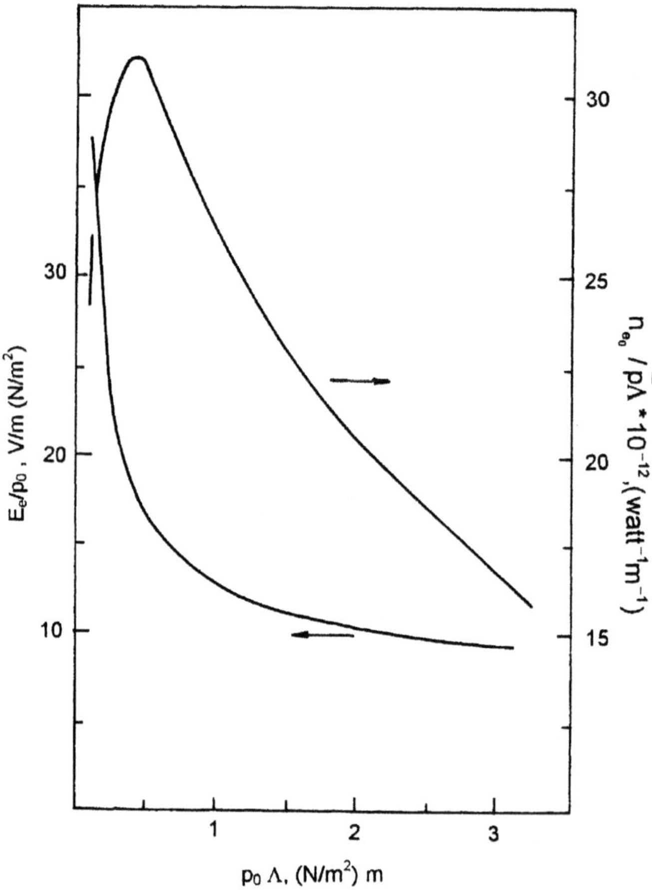

Figure 2. Plot of E_e/p_o and n_{e_0}/p against p_o/Λ.

The relationships between $n_{e_o}/\overline{p}\Lambda$, $p_0\Lambda$, and E_e/p_0 are plotted in Figure 2. Through relationship between E_e/p_0 and $p_0\Lambda$,k_d can either be read directly from Figure 1. Average electron density $n_{e_{av}}$ is calculated using $n_e = n_{e_0} \cos \pi x / L$ and the averaging equation.

Bell's[9] model used in this study was solved in two ways: The first was by values of k_d and n_e calculated from known values of P_0 , Λ, $\overline{P}$ $^-$and the second by measured values.

First set of data show the effect of variation is surface recombination coefficient, γ, for active discharge zone for which the temperature of the gas was assumed to remain constant at 300 ^{o}K. The effect of γ on conversion is relatively more pronounced at lower pressure than higher. The best fit conditions was determined to the γ_1 9.5×10^{-3}, Figure 3.

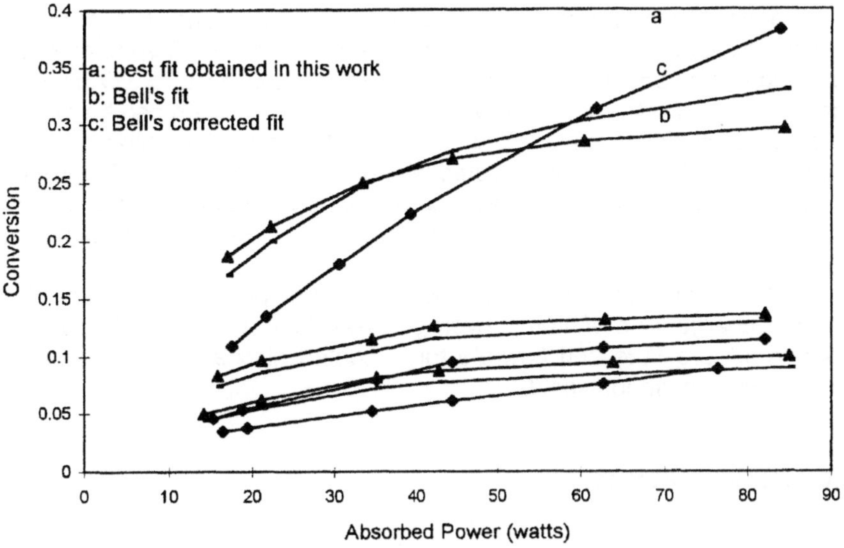

Figure 3. Comparison of the theretical model results for Bell and present work.

Comparison between the best fit results obtained by Mearns and Ekinci and those calculated using the theoretical electron number density values theoretically predicted by the method of Bell[9] are made in Figure 3. Conversion calculated using Bell's expression for electron density changes much less with absorbed power than do the experimental or as calculated by the model used in Mearns and Ekinci[3] although the difference become less at higher pressures. Correcting the theoretical electron number densities for dissociation improved the trend of the model results towards the experimental data . For this and reasons explained in their study of reaction of methane and ammonia to produce hydrogen cyanide in a MW discharge Juul-Dam and Brockmeier[5] replaced the residence time with specific energy W_a/V.

4.0 Experimental

Atomic hydrogen production is realized using a quarter-wave coaxial cavity which operated at 2450 MHz. The output of the microtron generator was of range 3-200 watts. It contained incident and reflected power and in the assessment of power losses at the coaxial line was incooperated. Two cavities accommodating 13 and 25 mm quartz tubes were used.

Hydrogen (99.99% pure) was provided by BOC and the reported impurities were oxides 10 vpm, CO_2 10vpm, carbon monoxide 10 vpm, nitrogen 0.01 vpm and water vapour 5 vpm. H atoms were measured using NOCl which exits light due to H and Cl reaction Clyne and Stedman[38], which can be picked up by a photomultiplier tube. The distance between the glowing and photomultiplier sensing zones was coated with orthophosphoric acid, Figure 4.

Electron temperature measurements are done by double cylindrical and planer Langmuir probes. The details of which were reported in Ekinci[39], Means and Ekinci[3].

In the production atomic hydrogen experiments in the 2.45 GHz microwave discharge yield of atomic hydrogen was determined as a function of power absorbed, pressure and flowrate for uncoated discharge tubes and for tubes coated with phosphoric acid.

5.0 Results and Discussion

The zone containing high concentration of charged particles in hydrogen discharge is the glowing region Chau and Phillips[25] . Indeed the length of glow of hydrogen discharge responds directly to increasing power, Figure 5.

Under similar conditions the largest discharge volume was obtained when clean tube and cylinder hydrogen were used. The smallest discharge volume was obtained for an orthophosphoric acid coated discharge tube and water vapour saturated hydrogen. Remembering that diffusion of electrons to the walls is major electron loss process for a low pressure hydrogen discharge and water having a tendency to capture electrons, Shaw[18], the experimental results are understandable.

For all pressures, an increase in absorbed power resulted in an increase in discharge volume, the reason being that increasing the power input to the discharge is the same as axially extending the applied electric field to increase breakdown volume.

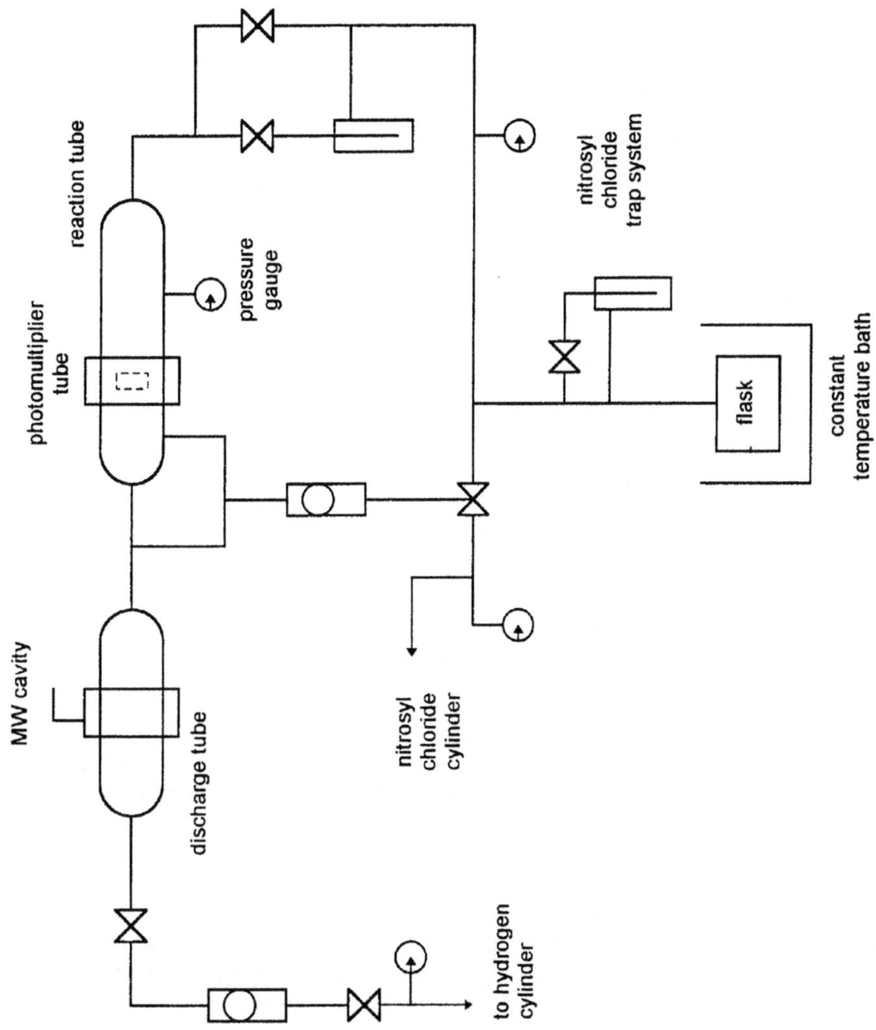

Figure 4. Schematic diagram of flow discharge reactor

An increase in discharge pressure at a fixed power resulted in a decrease in the glow volume. At higher pressures the mean free path of electrons becomes smaller and electrons travel shorter distance before they collide with a hydrogen molecule, which consequently results in smaller discharge volume.

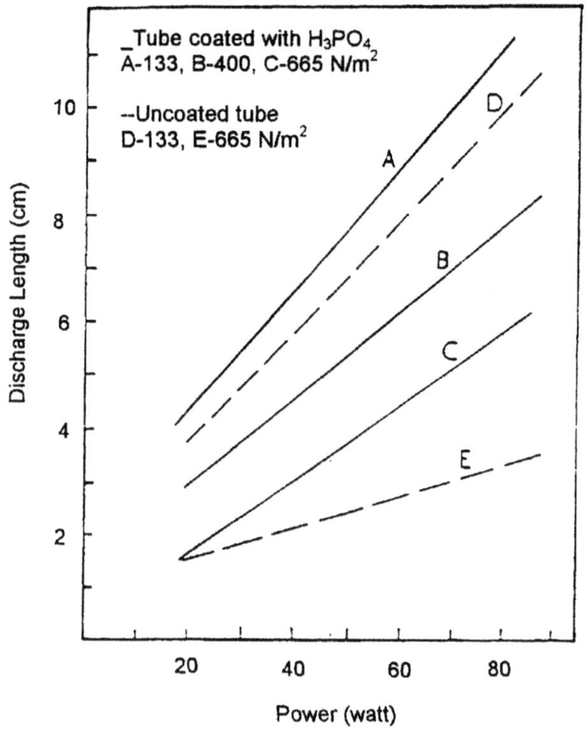

Figure 5. Discharge length as a function of power absorbed and pressure.

In this study two different electron temperature and density measurement method had to be used to be sure for the double probe results based on collisionless theory of Johnson and Malter[40] and double planar results based on theory developed by Cozens and von Engel[41]. The results of the measurements are shown in Figure 6. From the measurements it is obvious that volume averaged electron number density at constant pressure changes linearly with increasing absorbed power. The rate of increase of electron density becomes large as the pressure increases. The mean electron temperature of discharge increase but only slightly with absorbed power. The mean electron temperatures are highest for lowest pressures and the rate of change decreases as pressure raised.

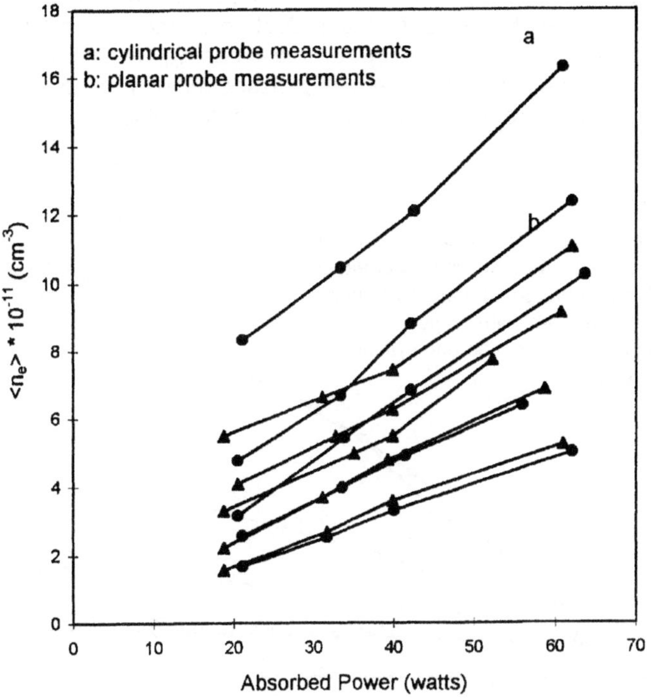

Figure 6. Volume averaged electron number density versus absorbed power for hydrogen.

5.1 EFFECT OF PARAMETER VARIATION ON CONVERSION TO HYDROGEN ATOMS

Figure 7 shows the effect of power variation on conversion to hydrogen atoms at 133, 400 and 665 N/m^2 for reaction in an uncoated discharge tube and in a discharge tube coated with ortho-phosphoric acid. Conversions are highest at the low pressure and for the coated tube, although there was little difference in conversion obtained in a coated tube from that achieved in an uncoated tube but with hydrogen saturated with water vapour. Conversions at 665 N/m^2 obtained in an uncoated tube are not shown because they were very small.

The increase in conversion with power at a given pressure is expected since electron density increases with power. The leveling off in conversion occurs because of the enhanced atom recombination at higher degrees of dissociation , and also because of the reduction in molecular concentration. Increasing the pressure reduces k_d but increases n_e. Their product declines with pressure increase resulting in reduced dissociation. The accelerated atom recombination which occurs at higher pressures also contributes to the reduction in conversion.

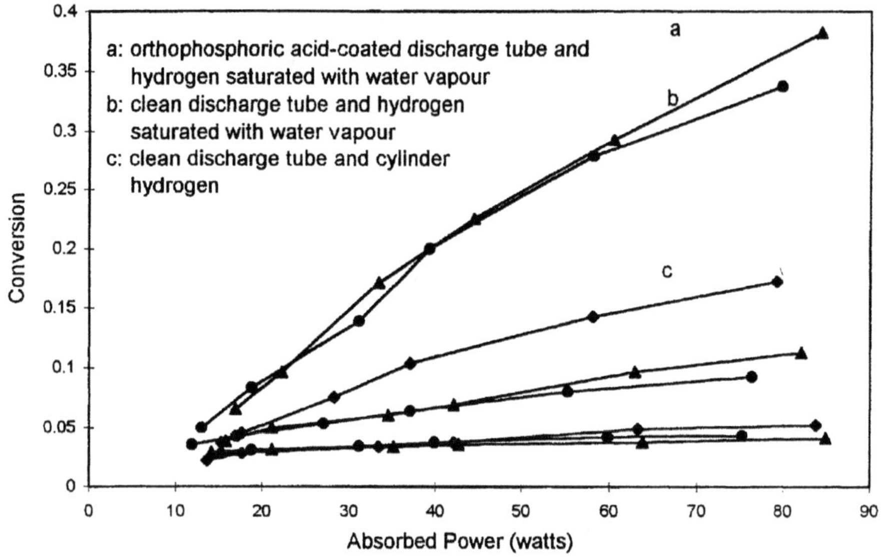

Figure 7. Effect of absorbed power on conversion at constant flow rate.

The sensitivity of degree of conversion to surface condition is illustrated in Figure 8. At the pressures under consideration surface recombination is an important process. The " aged " tube had been in operation for about 45 h prior to obtaining the results shown in this figure. Continued operation of this sort would appear to decrease the catalytic role of the surface for recombination considerably. Also shown in Figure 8 are results obtained with an 18 mm I.D. discharge tube in an " aged" condition. Conversion in the larger tube were always lower than those obtained in the smaller diameter tube at the same pressures and power absorption. Various factors dictate the effect of tube diameter on conversion: (1) The rate of surface recombination is slower in the larger diameter tube since γ is inversely proportional to diameter. (2) For the same gas flow rate, the space time in the discharge would be expected to be about 2.5 times greater in the large tube than in the small tube, the length of discharge in the two cases being similar. (3) The product k_d and ne_{av} is 2-3 times greater in the case of the 11mm diameter tube than for the 18 mm tube.

For example at 50 W absorbed power at a pressure of 133 N/m^2 the products are 42.5 and $20 s^{-1}$ respectively while at 665 N/m^2 they are 19.8 and 5.8 s^{-1}. Thus considering all three factors, it is deduced that the reduction in rate of dissociation in the large tube more than offsets any gain obtained through a longer exposure to electrons or through the effect of reduced surface recombination. Conversions decreasing with increase in tube diameter was also reported by Shaw[18], however for oxygen opposite effect is observed Stoker[42] .The contradiction may be explained in terms of much smaller k_d value for oxygen compared to hydrogen.

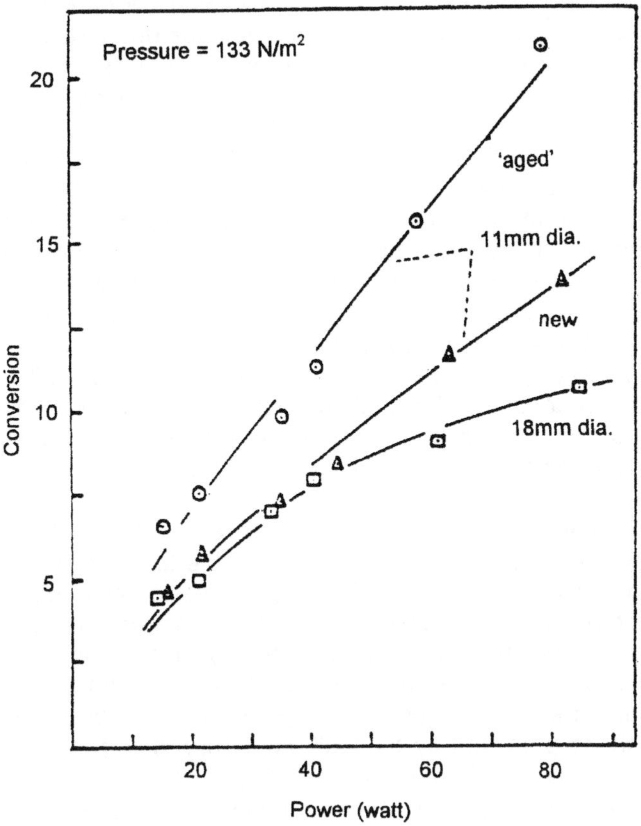

Figure 8. Effect of condition of surface and discharge tube diameter on conversion.

The sensitivity of conversion to parameter change was more pronounced for variation of γ_1 than of T_g especially at the lowest pressure. Results were also calculated assuming a gas temperature variation with power density according to (a) $T_g = 298 + 4\overline{P}$, (b) $T_g = 298 + 30\overline{P}$. In neither case was the " fit" better than that obtained when calculation was based on an isothermal discharge temperature of 420 ^{O}K.

The most noticeable difference between conversions predicted by the two methods is the much smaller spread indicated by the Bell's model. It is also apparent that the shapes of the calculated curves follow closely those of curves obtained for variation of calculated and experimental values of n_e with power. It would thus appear that values of conversions are greatly dependent on the values of n_e appearing in equation (1). Calculated values of n_e are a function of measured values of $\overline{P}$, the accuracy of which depends upon the accuracy of measurement of discharge volume. The discharge volume was assumed to be the volume of visible discharge , but the location of boundaries for this zone was somewhat inprecise by the visual method used, and in addition, uncertainty also exists in the assumption that the active volume of the discharge corresponds exactly to that of the visible volume. Although equation (11) used for calculation of n_e may be oversimplified in neglecting the effect of hydrogen dissociation on elastic collision frequency, experimental error in measurement of $\overline{P}$ may be the predominant feature which leads to the difference in values between experimental and calculated values of n_e.

Agreement between calculated and experimental values of T_e at 400 and 665 N/m^2 is very satisfactory , but, although values for T_e at 133 N/m^2, as calculated by the two double probe methods are similar, the large difference between them and the calculated value is difficult to explain. Values of k_d calculated at the experimental and derived values of T_e differ by less than a factor of two.

Since reliable values for k_3, γ_2, and to a lesser extent k_2, are not available, and because in any case it was not possible to make a meaningful measurement of gas temperature in the discharge , it would serve little purpose to analyze further the variation of, and to assess the errors in, the values of k_d and n_e . Bearing in mind the limitations in accuracy of data, the agreement between experimental conversions and those calculated from measured T_e and n_e may be considered extremely good. The agreement between experimental conversions and those predicted by the Bell model are less good, but this is more likely due to inaccuracy in measurement of power densities rather than from any inherent error in the model. The assumed value of γ_1, which effects a " good fit" to the experimental results is very similar to that used by Stoker[(42)] to describe oxygen dissociation in a microwave discharge , and the gas temperature of 420°K adopted here is not unreasonable.

6.0 Conclusions

Hydrogen in plasma leading to atoms and as a radical is receiving increasing interest due to its applications in pioneering technologies. Hydrogen discharge application received considerable success in semiconductor processing as surface etching agent. Production of this good quality and thermally stable films are obtained from H_2-discharge systems. Sintering,hydropyrolysis,ion-nitriding and chemical production applications are also new successful hydrogen-discharge applications.

For atomic hydrogen production in a microwave discharge has confirmed previous qualitative observations that conversion to atoms increases with increasing power absorption .This is most favourable at low gas pressures. Conversion to atomic hydrogen depends greatly on the surface condition of the discharge tube, being greatest in tubes coated to reduce atom recombination. At the same power absorbed, conversion has been found to be smaller in a large diameter tube than in a small tube.

A model, developed to relate experimental variables to observed conversion values, satisfactorily predicts conversion when measured electron energies and densities are employed in the model equations. When values of the electron densities, calculated from power densities , are employed the prediction is less satisfactory, the reason for divergence being uncertainty in assessment of discharge volume and temperature.

REFERENCES

1. Mearns, A.M.and Morris, A.J.,Oxygen Dissociation in a MW Discharge, Chem. Eng. Prog. Symp. Ser. 67 No. 112, 37, 1971

2. Bell, A.T.,and Kwong, K.,A, Model for the Kinetics of Oxygen Dissociation in a Microwave Discharge, Ind. Eng. Chem. Fundam, 12, 90, 1973

3. Mearns, A.W., and Ekinci,E., Hydrogen Dissociation in a MW Discharge, Jor. MW Pow. , 12, 2, 153, 1977

4. Cooper, M.W., Mickey, H.S., and Baddour,F.F,Oxidation of Hydrogen Chloride , Ind. Eng. Chem. Fundam, 7, 400, 1968

5. Juul-Dam, T. and Brockmeier, N.F., Electrical Probe Measurements in a Reacting Methane Ammonia Microwave Plasma , Ind.Eng. Chem. Prod. Res. Develop, 9, 388, 1970

6. Brown, L.C.and Bell, A.T.,Mass Spectrometric Study of Carbon Monoxide and the Decomposition of CO_2 in a Ratio Frequency Discharge, J. Chem. Phys. 61 (2), 666-671, 1974

7. Nirmatov, I. Sh., Shermatov, N., Mirsaidov, U., Spectroscopic and X-ray hase Examination of Products of Interaction between Hydrogen Atoms and Tin Chloride.Physics and Chemistry of Materials Treatment 26:1, 59-60, 1992

8. Nirmatov, I. Sh., Nateikina, E.V., Mirsaidov, U., Shermatov, N., Plasmo-chemical Reduction of Iron Chloride by Hydrogen Atoms, Physics and Chemistry of Materials Treatment, 25:1, 60-62, 1991

9. Bell, A.T., A Model for the Dissociation of Hydrogen in an Electric Discharge, Ind. Eng. Chem. Fundam, 11:2, 209, 215, 1972

10. Wood, R.W., Dissociation of H_2 into H by Low Frequency Discharge, Phil. Mag. 44,538,1922

11. Poole, H.G., Atomic Hydrogen 1. Calorimetry of Hydrogen Atoms, Proc. Roy. Soc., A163, 404, 1937

12. Poole, H.G., Surface Effects in the Discharge Tube, Proc. Roy. Soc., A163, 424, 1937

13. Weinrich, G., and Hughes, V.W., Hyperfine Structure of He^3 in the Metastable Triplet State, Phys. Rev., 95, 1451, 1954

14. Nagle, D.E., Julian, R.S.,and Zacharias, J.R., Hyperfine Structure of Atomic Hydrogen and Deuterium, Phys. Rev., 72, 971, 1947

15. Broida, H.P., and Moyer, J.W., Spectroscopic Analysis of Deuterium in Hydrogen-Deuterium Mixtures , J. Opt. Soc. Am., 42, 37, 1952

16. McCarthy, R.L., Chemical Synthesis From Free Radicals Produced in Microwave Fields , J. Chem. Phys., 22, 1360, 1954

17. Rose, D.J., and Brown, S.C., High Frequency Gas-Discharge Plasma in Hydrogen, Physical Review, Vol. 98, 310, No 2, 1955

18. Shaw, T.M., Techniques of Electrical Discharge for Radical Production, Gen. Elect. Microwave Lab. Report, No TIS R 58 ELM, 115, Unpublished ,1958

19. Rony, P.R., Atomic Hydrogen at Low Pressures, U.S. At. Energy Comm. UCRL, 16073, 215, 1965

20. Mezey, E.J., and Oxley, J.H., Microwave Hydrogen Plasma in Gas Solid Systems, The Journal of Microwave Power, 2-3, 79-86, 1967

21. Skidan, V.V., and Shreider, E., Determination of the Degree of Hydrogen Dissociation in a Discharge, Opt. Spektrosk, 27(3), 534-4, 1969

22. Dingle, J.R., and LeRoy, D.J.J., Kinetics of the Reaction of Atomic Hydrogen in Acetylene, Chem. Phys., 18, 1632, 1950

23. Wittke, J.P., and Dicke, R.H., Redetermination of the Hyperfine Splitting in the Ground State of Atomic Hydrogen, Phys. Rev., 103, 620, 1956

24. Zletz, A., Unpublished Report (1958) , (referred to it by Shaw, T.M., Formation and Trapping of Free Radicals, Acad. Press, 1960

25. Chou, C.H., Philips, J., The Foil Reconstruction in a Hydrogen Plasma, J. Var. Sci. Technol. A 8(6), 3941-3947, 1990

26. Kato, I., Hatanaka, K., Tatsumi, T., Deposition Mechanism of a-Si:H Films Fabricated by Coaxial-line-type MW Plasma Chemical Vapour

Deposition, Bulletin of Science and Engineering Research Laboratory, Waseda University 123, 1-12, 1989
27. Mathsushita, K., Hariu, T., Fang, S.F., and Shida,K., GaO, Q.2.,Plasma-Assisted Epitaxial Growth of Compound Semiconductors for Infrared Applications, Mat. Res. Soc. Symp. Proc. 90, 479-486, Mat. Res. Soc.,Pittsburg,1987

28. Mori, Y., Degushi, M., Okada, T., Eimori, N.,Yagi, H., Hatta, A., Nishimura, K., Kıtabatake, M., Ito, T., Hirao, T., Sasaki, T., and Hiraki,A., Electical Properties of Boron-Implanted Homoepitaxial Diamond Films, Jap. Journ. of Appl. Phys. Part 2: Letters 32:43, L601-L603, 1993

29. Hsu,T., Anthony,B., Qian,R., Irby,J., Kinsky,D., Mahajan,A., Banerjee,S., Magee,C., Tasch,A., Advances in Remote Plasma-Enhanced Chemical Vapour Deposition for Low Temperature In-situ Hydrogen Plasma Clean, Jour. Electronic Mat. Conf. 21:1,65-74,1992

30. Brecelj, F., Mozetic, M., Reduction of Metal Oxide Thin Layers by Hydrogen Plasma, Vacuum, 40:1-2, 177-178, 1990

31. Murray, R., Mookherjee, P.B., Yoshinaga, A., Dawson, P., Enhanced Luminescence Efficiency in Growth Interrupted Single Quantom Wells by Atomic Hydrogen, Jour De Physique 3:5 , 441-444, 1993

32. Beiers, H., Baumann, H., Bittner, D., Klein, J., Juentgen,H., Pyrolysis of Some Gaseous and Liquid Hydrocarbons in Hydrogen Plasma, Fuel 67:7, 1012-1016, 1988

33. Nikravech, M., Commarieu, A., Amouroux,J., Role des Especes Radicalaires Produits dans un Plasma Thermique, Reveu Int. des Hautes Temp. et des Refractaires 26:1, 23-38, 1990

34. Inal, O.T., and Metin, E.S., Surface Modification with Low Energy Plasmas, Plasma and Lasser Processing Materials Gd. K. Upadya, Minerals

35. Metin, E.S., and Inal. O.T., Formation and Growth of Iron Nitrides During Ion-Nitriding, Jour. Mat. Sci., 22, 2783, 1987

36. Schofield, K., Planet Space Science ,15, 643, 1967

37. Khare, S.P, and Moisewitch, B.L., Dissociation of Hydrogen Molecules by Electron Impact, Proc.Pyhs. Soc. , 88, 605, 1966

38. Clyne, M.A.A, and Stedman, D.H., Reactions of Atomic Hydrogen with Hydrogen Chloride and Nitrosyl Chloride,Trans. Faraday Soc., 62, 2164, 1966

39. Ekinci, E., Production of Atomic Hydrogen Using a Microwave Discharge, PhD. Thesis, University of Newcastle upon Tyne, 1976

40. Johnson, E.O., and Malter, L., Langmuir Probe Description and Analysis, Physical Review, Vol. 80,No1, 58-68, 1950.

41. Cozens, J.R., and Engel, Von A., Origin of Excessive Ionization in Flames, Int. Journal of Electronics, 19, 61, 1965

42. Stoker, A., Production of Atomic Hydrogen Using a 2.45 GHz MW Discharge, PhD. Thesis , University of Newcastle upon Tyne, 1972

HYDROGEN-METAL SYSTEMS

GARY SANDROCK
SunaTech, Inc.
113 Kraft Place
Ringwood, NJ 07456
USA

ABSTRACT. This chapter provides an introduction to and overview of the important scientific and practical aspects of hydrogen interactions with metals. In particular, it concentrates on the controlled absorption and desorption of hydrogen into and out of metals, especially the area of rechargeable metal hydrides. The subject of the hydrogen degradation of structural materials is also briefly discussed.

1. Introduction

The term "Hydrogen-Metal Systems" usually suggests three, somewhat overlapping technical areas:

A. Hydrogen in metals in random solid solutions
B. Metal hydrides
C. Hydrogen degradation effects in structural metals

Because of my own experience and interests, I will concentrate on areas A and B, especially B. Area C, although important, will be discussed briefly at the end of this chapter.

A basic understanding of the possible interactions between hydrogen and metals is important to those involved in many aspects of Hydrogen Energy Systems. Rechargeable metal hydrides offer a "solid" alternative to gaseous and liquid storage and handling of elemental hydrogen. A basic understanding of hydrogen degradation effects on structural metals is important for everyone involved in the design and fabrication of devices that are exposed to hydrogen. It is the purpose of this chapter to give an introduction and broad review of the science and phenomenology of hydrogen in metals. It will concentrate mainly on rechargeable metal hydrides and will serve as background for my other chapter in this collection, Applications of Hydrides [1].

2. Overview of Hydrogen-Metal Interactions

If there is a universal solute-solvent combination, it is the family of hydrogen-metal systems. Virtually all metals will dissolve hydrogen, many rapidly and in large amounts to form true chemical compounds called hydrides.

Y. Yürüm (ed.), Hydrogen Energy System, 135–166.

The science of hydrogen-metal systems is an old one. In 1866 Graham discovered that Pd metal "occluded" large amounts of H_2 gas at pressures near one atmosphere and temperatures near ambient [2]. In 1868 the same Graham also reported the ability of Pd to absorb H directly from an H_2O-based electrolyte in an electrolytic cell [3]. Since these early studies of the H-Pd system, the field of H-Metal systems has substantially grown and developed. In his classic book, "Hydrogen in Metals", Smith cited 1467 references on the subject between 1866 and the end of 1946 [4]. It is anyone's guess how many more papers and reports have been generated since 1946!

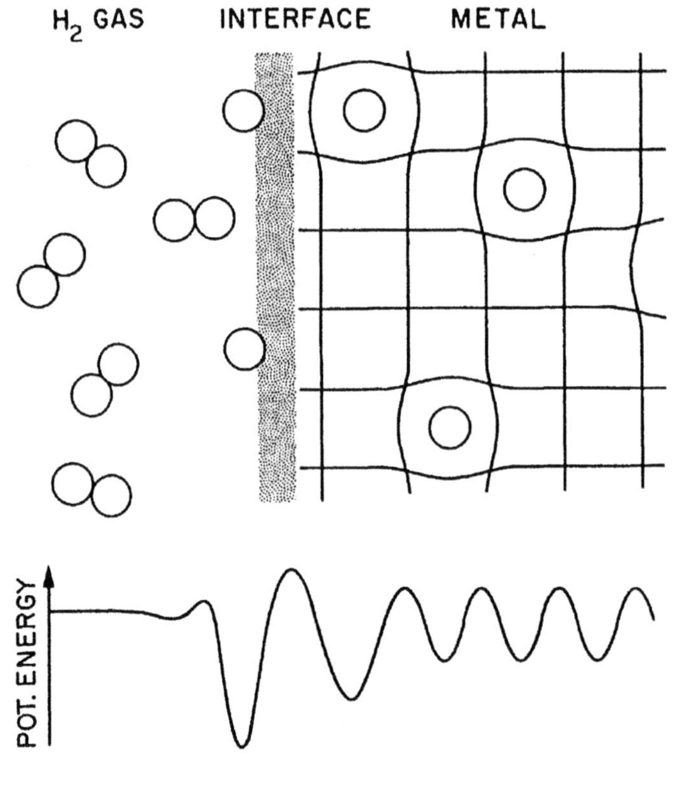

Figure 1 - Mechanism of H_2 Absorption in Metals [5]

The general mechanism of the absorption of H_2 gas (as well as its isotopes D_2 and T_2) by a metal is shown in Figure 1. The H_2 molecule is first weakly physisorbed on the surface and then dissociatively chemisorbed as strongly bound, individual H atoms. The H-atoms are light and small and can quickly diffuse away from the surface into periodic sites (often interstitial) in the metal crystal lattice. Interstitial diffusion rates of 10^{12} jumps per second are possible through the periodic lattice potential field, even near room temperature. Once in the crystal lattice, H-atoms can take the form of a random solid solution or an ordered hydride structure with distinct bonding to metal atoms and high volumetric packing density. As is implied by Figure 1, the surface plays an important role in catalyzing the H_2 --> 2H dissociation (and reassociation) reaction. The phenomenology of these and other important aspects of hydrogen- metal interactions will be discussed in some detail later. But first a brief description of the basic experimental techniques involved in loading metals with hydrogen might be useful to the reader.

2.1 EXPERIMENTAL TECHNIQUES

Although there are several ways to "implant" H in metals, the two techniques most commonly used today are basically the same as those a century ago - H_2 gas phase and electrolytic charging. Both are simple and require relatively inexpensive equipment.

2.1.1. *Gas Phase Charging*. This technique consists of simply applying H_2 gas at appropriate temperatures and pressures. A composite apparatus, designed for three different modes of charging (or discharging) is shown schematically in Figure 2. The

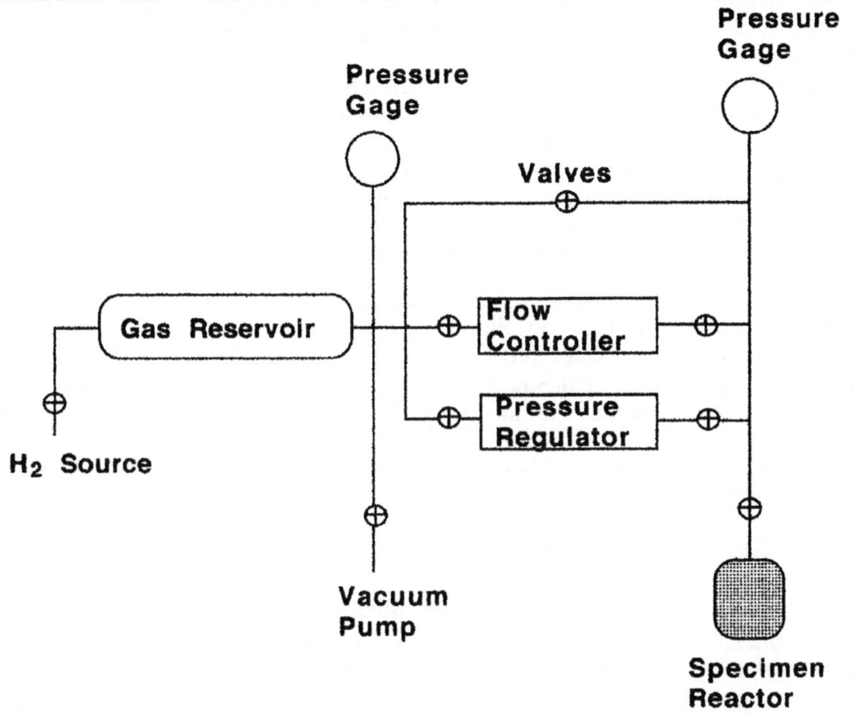

Figure 2 - **Multipurpose Gas Hydriding/Dehydriding Apparatus**

apparatus consists of a carefully calibrated gas reservoir and pressure gage connected to the specimen reactor (with its own pressure gage) by gas-tight tubing. All volumes must be carefully pre-calibrated using water or He filling. Often a few optional sizes of gas reservoirs and pressure gages are included for versatility. The specimen reactor should be designed so as to be easily opened for specimen replacement. The reactor is usually immersed in a water bath or furnace for temperature maintenance. The pressure gages and control instruments are usually electronic and the valves are often electrically operated so that the experiments can be controlled and analyzed by computer. If properly designed, such an apparatus can be used both for absorption and desorption experiments. In honor of his extensive and pioneering work on metal-hydrogen systems early in this century [4], such an apparatus is traditionally named after the German chemist Sieverts.

As shown in Figure 2, there are three possible means of transferring the H_2 gas from the reservoir to the specimen chamber. Using the top (open/close) valve, we can introduce (or remove) gas to (or from) the specimen on an aliquot by aliquot basis. For a manual absorption example, (1) the top valve is closed, (2) the gas reservoir is filled to an H_2 pressure P, (3) the top valve is opened and the H_2 allowed to react with the specimen until a static (quasi-equilibrium) pressure P' is reached. From the temperature, pressure change (P-P') and volumes, the amount of H that entered the specimen (usually quantified as atomic H/M ratio) can be precisely calculated. If this is done repeatedly at constant temperature we obtain a plot of P' vs. H/M called an absorption isotherm. Because we do this by a step-by-step procedure that waits for static quasi-equilibrium at each step we call the result a static isotherm. A similar procedure is used to obtain a static desorption isotherm.

Instead of the static aliquot-by-aliquot method, requiring the repeated opening and closing of the top valve, we can slowly and continuously introduce H_2 to the specimen chamber using an electronic flow controller, while recording pressure. The result is what we call a dynamic isotherm. Static and dynamic isotherms can differ a little.

Finally, we can use a pressure regulator to run an isobaric absorption experiment. The result is a plot of H/M vs. time. This technique is used to estimate absorption kinetics (or as is often more correct, reactor heat transfer properties).

2.2.2. *Electrochemical Charging*. This technique, shown in one of several possible experimental configurations in Figure 3, is often called cathodic charging or electrosorption. The metal specimen is placed in an acid or alkaline (e.g., KOH) H_2O-based electrolyte in series with a power supply and counter electrode (e.g., Pt). Often a reference electrode, such as saturated Hg/HgO, is placed near the sample electrode in order to monitor the applied potential relative to a well-established electrochemical standard. If electrons are added to the M electrode, water is partially split according the following charge transfer surface reaction:

$$M + H_2O + e^- \text{----->} M + H_{ads} + OH^-, \qquad \text{(Eq. 1)}$$

where H_{ads} is adsorbed atomic hydrogen resulting from the electrolytic water splitting. Applied cathodic potential is highly effective in driving (Eq. 1) to the right, thus increasing the thermodynamic activity of H_{ads}. This highly active atomic H is then readily absorbed into the sample, often far more effectively and rapidly than by gas phase charging. As mentioned earlier, Graham used this technique to charge Pd with H more than a century ago. Graham's basic electrolytic technique has been reused recently to charge Pd with the heavy-hydrogen isotope deuterium for "cold fusion" experiments [6]. It is also the basis for a new generation of nickel / metal hydride batteries I discuss elsewhere [1].

Because the subject of the electrochemistry and electrosorption is discussed in detail in other publications, e.g., [7], the rest of this chapter will be concerned only with gas phase phenomena.

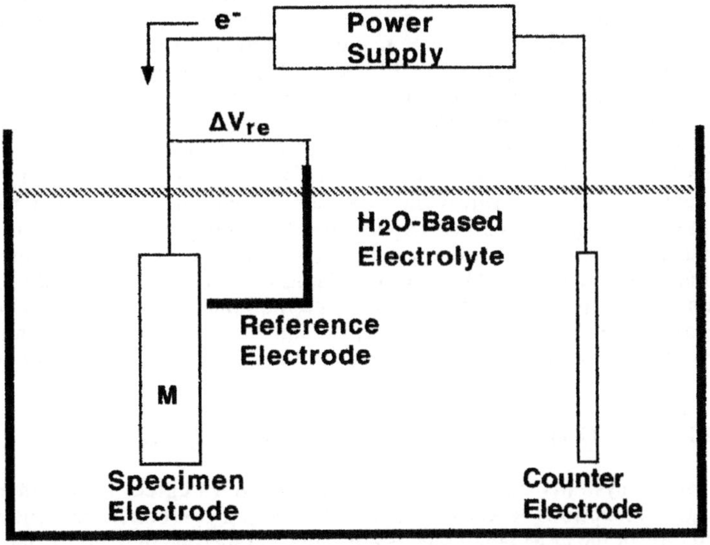

Figure 3 - Typical Electrochemical Hydriding Apparatus

2.2. SOURCES OF LITERATURE

The literature on the three traditional areas of hydrogen-metal systems is immense. This brief review cannot be very comprehensive. Therefore I offer those readers who want to go deeper into the subject some directions and suggestions for further, more detailed reading.

There have been two, somewhat interconnected, international series of symposia, run rather informally and often independently of organized technical societies. In all cases, summaries or detailed proceedings have been published. The first series of symposia was called "Hydrogen in Metals" and concentrated on solid solutions, although not universally so. Eight Hydrogen in Metals symposia were held, all in Europe, from 1968-1985 [8-15]. Another series was called "Properties and Applications of Metal Hydrides" and concentrated mostly on hydride phenomena. Five international PAMH symposia were held from 1977-1986 in Europe, USA, Japan and Israel [16-20]. In 1988 these two series merged and were renamed "Hydrogen-Metal Systems". Three combined international symposia were held in 1988 (Germany), 1990 (Canada) and 1992 (Sweden) [21-23]. The November, 1994 symposium of this series will be held in Japan.

In addition to those long-running series, other symposia can be found, usually associated with various technical societies, for example [24, 25] among others. The ten World Hydrogen Energy Conferences (International Association of Hydrogen Energy) have all had hydrogen-metal or metal hydride sessions that have been included in the published proceedings. Individual papers can be found in many physics, chemistry, metallurgy and engineering journals. Two especially rich sources are the Journal of Less-Common Metals (renamed about 1991 as the Journal of Alloys and Compounds) and the International Journal of Hydrogen Energy. Two excellent 2-volume collections on the subject were published in Topics in Applied Physics [26,27]. In addition to Smith's early monograph [4], the more recent ones on hydrides by Libowitz [28] and Buchner [29] are useful. The 1968 book dedicated mostly to elemental hydrides [30] is still very pertinent, as are some earlier monographs on hydrides [31-33]. A recent monograph covers the physics of M-H systems in detail [34]. There have also been three prior NATO conferences on the subject of H in Metals [35-37].

The third area of high activity, rather independent of the above two, is "Hydrogen Degradation Effects in Structural Metals". This is also a very large subject, to which I cannot do justice. It is metallurgical and fracture-mechanics in orientation, concentrating on things such as hydrogen embrittlement, weld cracking, stress-corrosion cracking, phase transformations among others. Papers can best be found in engineering or metallurgical journals, such as Metallurgical Transactions and Acta Metallurgica. This area has also seen a long-standing series of international symposia, published by the metallurgical societies ASM International [38] and TMS [39]. A large compilation of degradation effects concentrating on Fe-base alloys (e.g., steels) is specially noteworthy [40].

2.3 PHENOMENOLOGY, THERMODYNAMICS AND AND BONDING

Hydrogen-metal systems can be generalized to some degree, although numerous deviations and exceptions to those generalizations can be found. Depending on the strength and type of the H-M bond, metals are called exothermic or endothermic occluders of hydrogen [4]. If a given metal has a high affinity for hydrogen, then the H is absorbed exothermically with the attendant generation of heat (i.e. the enthalpy change of absorption, ΔH is negative). Conversely, if a metal has a poor affinity for hydrogen, then it will absorb H only when heat is added (i.e., ΔH is positive or endothermic).

Almost all practical M-H systems are exothermic. The phenomenology of hydrogen absorption (and desorption) is best seen by pressure-temperature-composition isotherms which in turn conform to well-established thermodynamic laws and principles. In most (but not all) cases a discrete hydride phase can be formed. Depending on the nature of the predominant chemical bond between the H and M atoms, several kinds of hydrides can form.

2.3.1. *Pressure-Composition Isotherms.* The best way to understand the absorption and desorption of hydrogen by metals is via pressure-composition isotherms which are determined experimentally by solid-gas reactions performed on a Sieverts apparatus such as shown in Fig. 2. A typical, somewhat idealized set of schematic P-C isotherms is shown in Figure 4. For discussion purposes, let us consider them to be absorption isotherms.

Starting with the H-free metal, generally called the α-phase, we apply an increasing pressure of gaseous H_2. If the surface of the metal is clean (i.e., free of oxide barriers) the H_2 will dissociate on the surface according to the dissociative chemisorption reaction

$$H_2 \dashrightarrow 2H \qquad \text{(Eq.2)}$$

and the individual H atoms will diffuse into the metal lattice to form a random M-H solid solution. That is, the H-atoms in the α solid solution do not take fixed and ordered

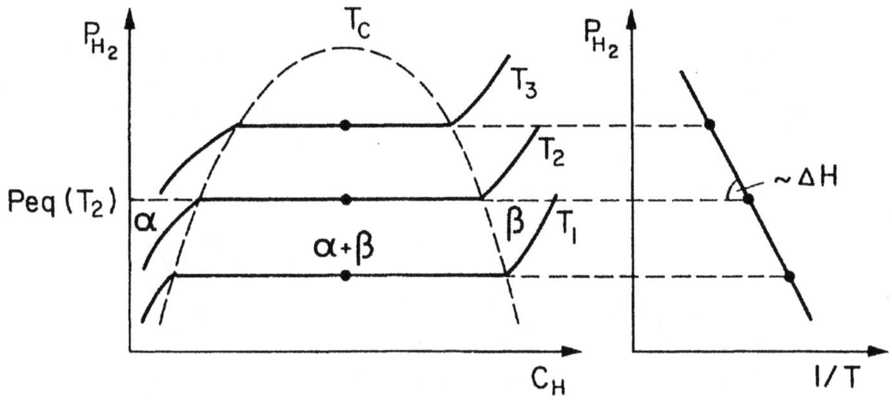

Figure 4 - Idealized Pressure-Composition Isotherms and van't Hoff Plot [5]

positions and in fact are usually quite mobile, even near room temperature. The H-concentration in the α-phase (C_H) is a strong function of pressure, often obeying a near square root pressure dependence called Sieverts' Law:

$$C_H = kP^{0.5} \quad \text{(Eq.3)}$$

where k is a constant dependent on temperature. Eq.3 is taken as strong experimental evidence that hydrogen exists in the metal lattice as individual H atoms, not as H_2 molecules (i.e., the surface reaction of Eq.2 occurs as a precursor for absorption). Within the α-phase, the P-C isotherms are largely reversible without hysteresis, as long as time is allowed for equilibrium to be established. That is, hydrogen can be in principal added or removed from the metal.

At some pressure or composition, the α-phase becomes saturated with H. If the metal in question forms a hydride, then we will observe a phase boundary as shown by the dotted line in Fig. 4. A discrete, ordered hydride phase will begin to form at nearly constant pressure, called the plateau pressure. The hydride phase is usually called the β-phase so that the plateau region of C_H represents a two-phase α + β field. As the plateau is traversed by adding more and more hydrogen the β-phase grows at the expense of the α-phase. The right end of the plateau represents 100% β-phase.

The β-phase hydride usually represents a discontinuous change in crystal structure (or at least lattice parameters) from the α-phase, although interstitial hydrides usually are closely related in structure to the parent metal phase (see Section 2.2.3). Many hydrides (especially interstitial hydrides) have a variable stoichiometry (i.e., x is somewhat variable in MH_x), so that further increases in pressure will result in slightly higher values of x, as shown by the sloping upper right legs of the isotherms shown in Fig. 4. The upper legs of the isotherms usually do not obey Sieverts' law very well.

The dotted envelope in Fig. 4 defines the limit of the two-phase field on a P-T-C phase diagram. In general (but not always), as temperature increases plateau width decreases.

In effect, the solubility of H in the α-phase increases and the H-content of the terminal β-phase decreases. Above a critical temperature, T_c, there is no two-phase plateau and the α- and β-phases continuously change with H-content into each other. Above T_c, the P-C isotherm tends to be continually sloping. Typical examples of all the features discussed above for Fig.4 are shown in the classic Pd-H system, Figure 5 [41-43]. In other systems, multiple plateaus can be observed (more than one hydride phase) and the absence of a critical temperature all the way to the melting point [30].

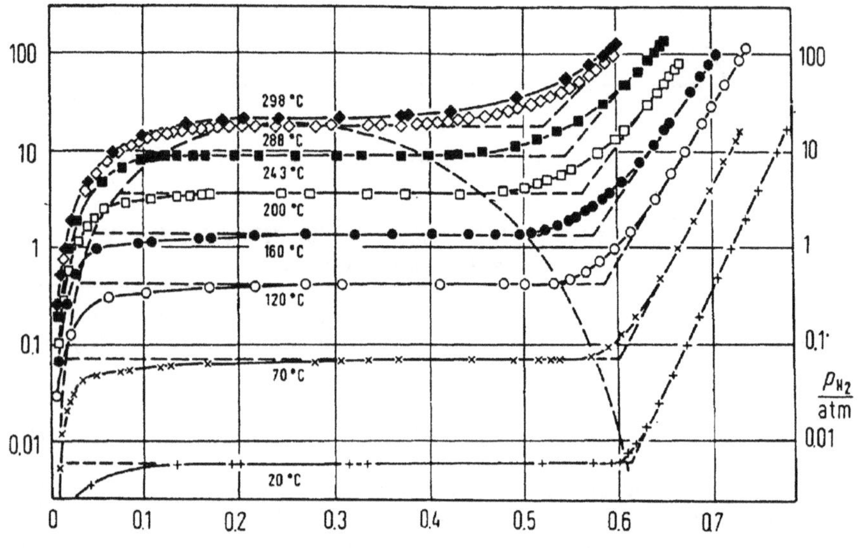

Figure 5 - **Pressure-Composition Isotherms for the Pd-H System** [43]

2.3.2. *The van't Hoff Equation.* Although the thermodynamics of M-H systems are thoroughly discussed elsewhere (e.g.,[44-45]), it is very important to note here the fundamental interconnection between pressure and temperature. As shown schematically in Fig.4 and experimentally in Fig.5, the pressure (especially plateau pressure) at a given H-content increases with temperature. This is the case with all exothermic H-occluders, and is a direct consequence of the thermodynamics associated with the hydriding reaction

$$M + \frac{x}{2} H_2 <\text{---->} MH_x \quad , \qquad \text{(Eq.4)}$$

which has enthalpy change, ΔH, and entropy change, ΔS associated with it. The equilibria of such gas-solid reactions are well understood. For pressures low enough (say less than 100 atm, 10 MPa), where the H_2 fugacity is nearly the same as pressure, the equilibrium H_2 pressure P is closely related to the absolute temperature T by the well-know van't Hoff equation

$$\ln P = \frac{\Delta H}{RT} - \frac{\Delta S}{R} \quad , \qquad \text{(Eq.5)}$$

where R is the gas constant. Thus, as shown on the right side of Fig.4, a plot of ln P vs. 1/T (van't Hoff plot) tends to be a straight line with slope a measure of ΔH and intercept (1/T=0) a measure of ΔS. ΔH can vary widely from metal to metal and is a measure of the strength of the M-H chemical bond. ΔS can also vary, but not nearly as much as ΔH. For most hydrides (but not all), ΔS tends to center around the configurational entropy change from gaseous to solid H [46]. The van't Hoff plot is a convenient way to graphically represent hydride P-T stability, especially to compare different metal-hydrogen systems. It is also very useful in the thermodynamic analysis of hydride devices [1].

2.3.3. M-H Bonding and Hydride Types. There are at least three basic kinds of classical metal hydrides that are based on the principal bonding between the M and H atoms [28, 31-33, 47]. Reference to the periodic table of metallic elements, shown as Figure 6, may be useful in reading this section.

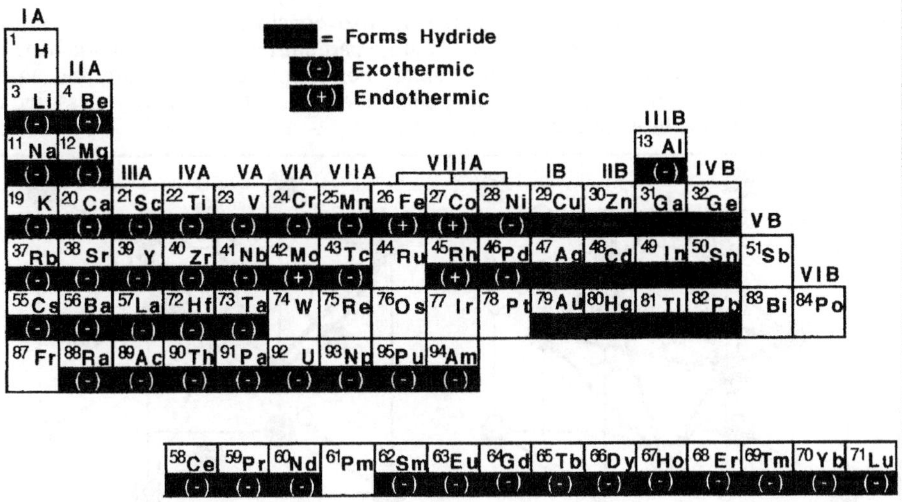

Figure 6 - Elements Known to Form Hydrides [28,30,48]

Ionic (Salt-like) Hydrides - In ionic hydrides the metal atoms exist as positive M^+ cations and the hydrogen exists as H^- anions. Examples include hydrides of the Groups IA (alkali) and IIA (alkaline earth) metals Ca through Ba. Examples include LiH and CaH_2 for Groups IA and IIA, respectively. They are highly exothermic on formation (very stable) and have salt-like properties: white, translucent appearance, brittleness and similar crystal structures to the corresponding halide salts (e.g., the Group IA alkali metals all form monohydrides with the NaCl-type structure). The H^- ions take substitutional positions in the hydride crystal lattice and the hydride is denser than the parent metal (i.e., hydriding results in lattice contraction).

Covalent Hydrides - These hydrides involve relatively weak, shared-electron bonds. Examples include hydrides of IB to VB metals. Most are quite unstable and some exist only in the gaseous form at room temperature (e.g., the VB hydrides of Ge, Sn an Pb).

They are generally difficult to make by reaction with gaseous H_2, but require various wet chemical techniques. In Group IIA, Be and Mg form form the relatively stable covalent hydrides BeH_2 and MgH_2, although MgH_2 probably also has a significant component of ionic bonding. The covalent hydrides that can be made in solid form are typically white. They are generally polymeric in nature, that is they consist of $(MH)_x$, $(MH_2)_x$ and $(MH_3)_x$ polymers held by multicenter hydrogen-bridge bonds [28]. A good example of a polymeric, covalent hydride is AlH_3.

Metallic Hydrides - The middle groups of the periodic table (transition metals) and most intermetallic compounds form hydrides with metallic properties: metallic luster and high electrical conductivity. At one time, the H-atom was thought give up its electron to the unfilled d-band and exist essentially as a proton (H^+) However, most of the metallic hydrides are brittle (i.e., lack the characteristic ductility of the parent metals), so some degree of ionic bonding (i.e., H^- anions) was also considered possible. It is now generally believed that the 1s electron of the H-atom participates in the conduction band of the metal to produce new M-H bonding states containing both protonic and anion components or characteristics [50]. The rare earth hydrides, although mostly classified as metallic, lack metallic luster and are dark.

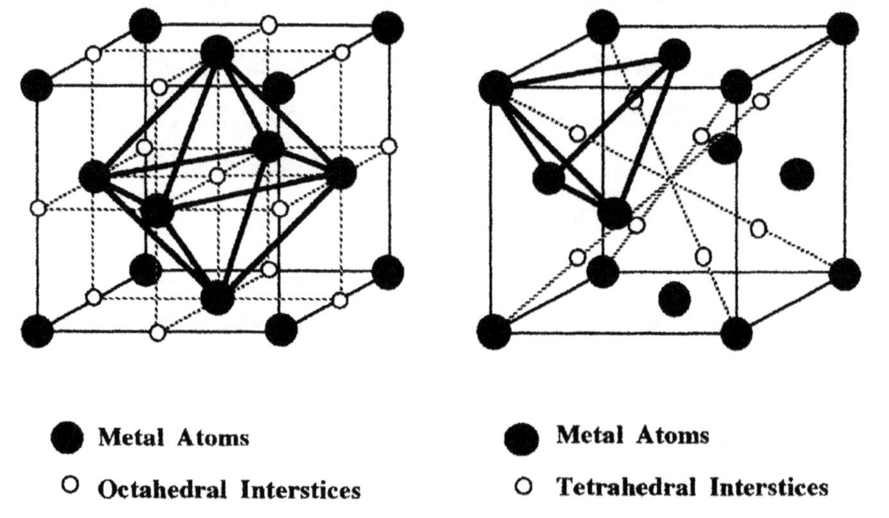

Figure 7 - Types of Interstitial Positions in a Face-Centered-Cubic Metal Lattice (after [49])

Many metallic hydrides are often called interstitial hydrides because the H-atoms occupy the interstitial positions (holes) of the host metal lattice. There are a variety of interstitial positions, depending on crystal structure, pure metal vs. intermetallic compound, etc. For example, the two types of interstitial positions in a simple face-centered-cubic (fcc) lattice (e.g., Pd) are shown in Figure 7. The larger interstices are called octahedral (Fig.7a) because they are surrounded by six atoms that form the corners of an octahedron. There are also smaller interstices called tetrahedral (Fig. 7b) that are surrounded by 4 atoms in a tetrahedron. An fcc unit cell, such as shown in Fig.7, contains 4 metal atoms, 4 octahedral interstices and 8 tetrahedral interstices. Thus in this particular crystal structure there are three interstices per metal atom for a potential limiting interstitial hydride composition of MH_3. In fact, all interstices cannot be filled because H atoms do not like to occupy positions that will be less than 0.21 nm from each other [50]. In the case of Pd, only the larger octahedral sites are preferred [42,43]; even then, the octahedral sites are not completely filled in $PdH_{0.6}$, the moderate pressure, room temperature hydride formed (see Fig.5).

Unlike Pd, most metallic hydrides can easily achieve H/M ratios if 1 to 2 (and in the case of the rare earth elements up to 3). In the formation of many elemental hydrides there is a distinct change of crystal structure from the parent metal [28], although there is usually some relation. Semantically speaking, perhaps such metallic hydrides should not always be called interstitial hydrides. Unlike ionic hydrides, the formation of a metallic hydride almost invariably results in a expansion of the parent metal lattice as H-sites are filled (typically 10-30%, by volume). This is usually accompanied by severe embrittlement and results in the cracking of solid particles into granules or powder. Exceptions include $PdH_{0.6}$, ZrH_2 and YH_2, which can be made in crack-free monolithic form [30]. Because the H-atom exists in a metallic hydride as a small, light atom or ion, it usually has excellent mobility in moving from one site to another. This is important for practical applications which require rapid hydriding and dehydriding [1]. As shown in Table I, metallic and other hydrides have high volumetric densities of H-atoms and offer

Table I
Gravimetric and Volumetric H-Densities [28, 51]

Hydrogen Form	H Weight Density, wt.%	H Volume Density, 10^{22} atoms/cm^3
Gas at 100 atm	100	0.5
Liquid (20K)	100	4.2
Solid (4.2K)	100	5.3
LiH	12.7	5.9
MgH_2	7.6	6.7
$PdH_{0.6}$	0.6	4.3
TiH_2	4.0	9.2
ZrH_2	2.2	7.3
VH_2	2.1	11.4
GdH_2	1.3	5.4
GdH_3	1.9	6.4
UH_3	1.3	4.0

a compact hydrogen storage option, compared to conventional gas and liquid H_2 storage. The metallic hydrides have seen the greatest applications-oriented R&D in recent years and will be discussed extensively below.

Complex Hydrides - In addition to the above classical metal hydrides, there are a number of more complex metal hydrides possible. These include the borohydrides, transition metal complexes, polyhydride complexes and metastable or "disordered" hydrides. These materials will be discussed later.

3. Families of Metal Hydrides

We now trace the evolution of metal hydrides into the materials that have been developed for application in hydrogen energy systems. This evolution has centered mainly on the need and search for appropriate P-T-C properties, as well as other appropriate engineering properties. The history starts with the elemental hydrides of the past, follows into the intermetallic compound hydrides of today, and finally projects into future possibilities in new, complex systems. The applications themselves are covered elsewhere [1].

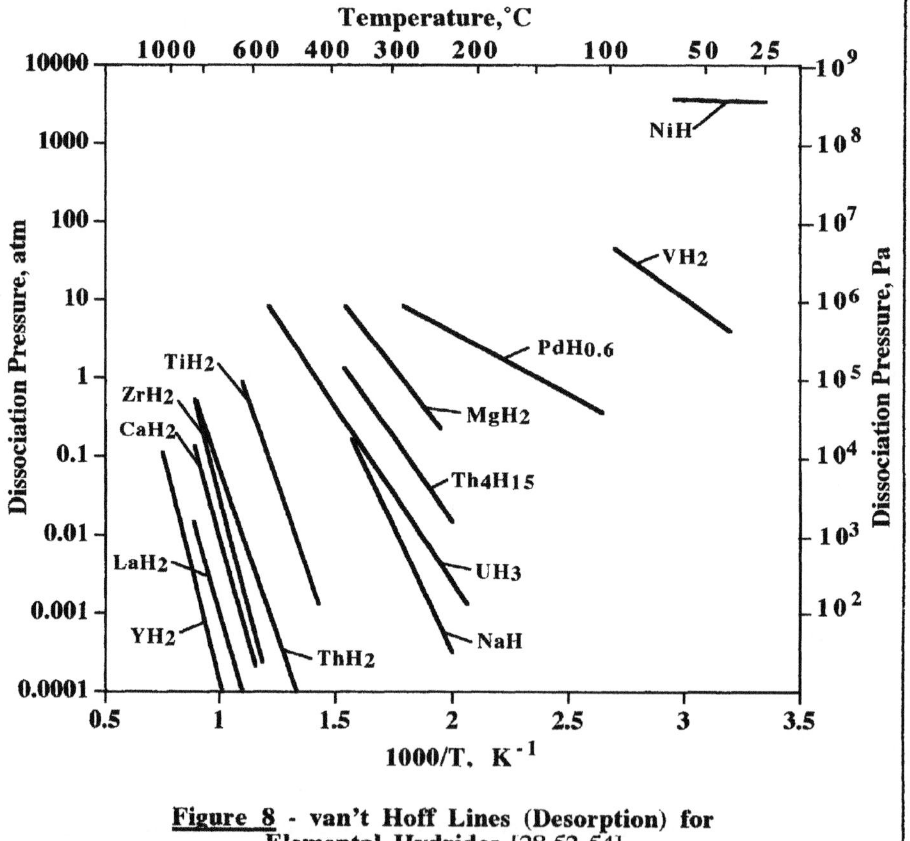

Figure 8 - van't Hoff Lines (Desorption) for Elemental Hydrides [28,52-54]

3.1 ELEMENTAL HYDRIDES

As shown in Fig.6, most of the metallic elements are known to form hydrides. However, from the points of view of ease of reaction, pressure, temperature or reversibility, few are really useful for practical (reversible) applications. The covalent hydrides (Groups IIB to IVB) cannot be easily formed from gaseous H_2 and are often liquids or gases at room temperature, so they can be generally discounted. The remaining ionic and metallic hydrides offer a wide range of stabilities, as is shown by the selected elemental van't Hoff lines of Figure 8. At the high stability extreme extreme of Fig.8, YH_2 has a ΔH_f of -214 kJ/mol H_2 [48], requiring more than 1000°C for a dissociation pressure of one atmosphere. At the low stability extreme, NiH has a ΔH_f of only -10 kJ/mol H_2 [54] and exhibits a dissociation pressure of 3000 atmospheres at room temperature.

The problem with elemental hydrides is the fact there are only a limited number of elements available and therefore a limited choice of van't Hoff lines. From the elemental hydrides shown in Fig.8, only VH_2 offers reasonable reasonable dissociation pressures near room temperature (NbH_2, not shown, is similar). From a practical point of view, the elements allow for little versatility. In some cases, solid solution alloys can be made for useful thermodynamic interpolation purposes, for example body-centered-cubic (bcc) alloys of Ti and V with other minor elements such as Fe [55,56]. But such situations are rather rare and still offer only a limited number of metallurgical possibilities. Historically speaking, hydride technology was limited almost entirely to elemental hydrides until the 1960s. Because of this, there was no effective, reversible, room temperature hydride technology developed until the discovery of intermetallic compound hydrides.

3.2 INTERMETALLIC COMPOUND HYDRIDES

Intermetallic compounds are strictly ordered (usually brittle) alloys of two or more metallic elements, usually having narrow, integer stoichiometry (A_aB_b) and crystal structures that are different from either of the parent elements (A and B). There are many hundreds of known intermetallic compounds of several classes [57] and they form the basis of a major branch of metallurgy that has occupied the careers of many, including this author. Intermetallic compounds (IMCs) form the basis of modern hard magnet technology and the practical strengthening of structural alloys [49].

The great breakthrough in hydride technology came in 1958 when Libowitz discovered the the intermetallic compound ZrNi could be easily and reversibly hydrided to form a true intermetallic hydride ($ZrNiH_3$) that had a thermodynamic stability that was intermediate between the very stable ZrH_2 and very unstable NiH elemental hydrides [58]. Although the major significance of this discovery was not fully appreciated for another decade, the stage had been set for the liberation of hydride technology from the very limited and restrictive domain of the elemental hydrides.

The key to all IMC hydride technology is the combination of strong hydride forming element A with a weak hydride forming element B in a natural integer ratio to form and intermetallic compound A_aB_b that can react reversibly with H_2 to form an intermediate strength hydride $A_aB_bH_x$. There are certain classes of IMCs that seem to work better than others for reversible H-storage, in particular AB_5, AB, AB_2 and A_2B phases. All of these IMCs will be briefly discussed, but with somewhat of an an emphasis on the interesting and versatile AB_5 hydrides. Far more complete reviews of the now very extensive field of IMC hydrides can be found elsewhere [27, 59].

3.2.1. *AB_5 Hydrides.* The AB_5 hydrides are based on AB_5 IMCs where A usually are rare earth (lanthanide) elements (at. no. 57-71) or Ca and B is Ni and other transition metals. The classic AB_5 hydride is based on $LaNi_5$, the remarkable hydriding properties of which were discovered about 1969 at the Philips Laboratories in The Netherlands [60]. It might be useful to use $LaNi_5$ as an example of the power and general characteristics of IMCs for H-absorption and desorption.

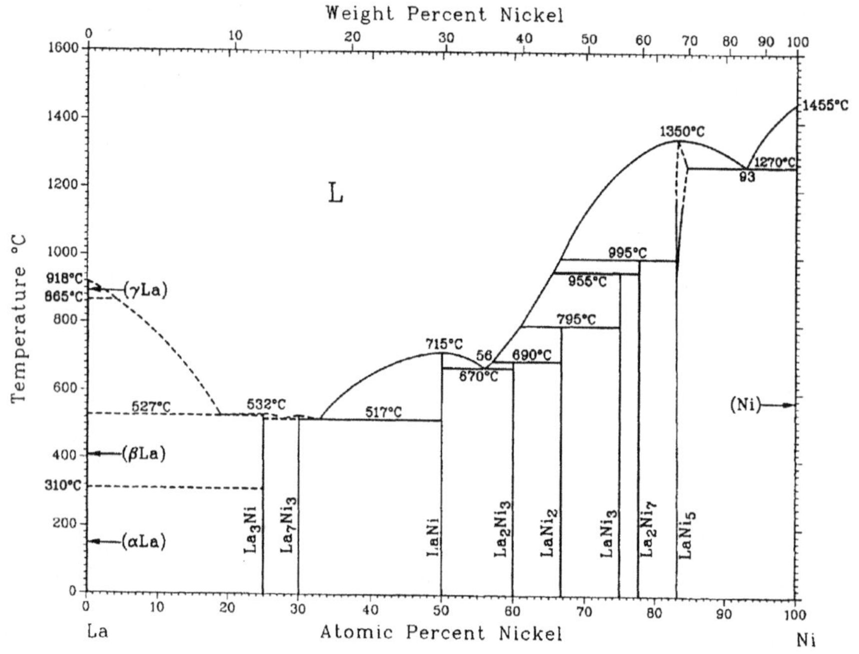

Figure 9 - **La-Ni Phase Diagram** [61]

Let us start with the La-Ni phase diagram, reproduced as Figure 9. In addition to the essentially pure La and Ni terminal phases, eight intermetallic compounds can be formed, depending on the alloy composition. From left (La) to right (Ni) these phases are La_3Ni, La_7Ni_3, LaNi, La_2Ni_3, $LaNi_2$, $LaNi_3$, La_2Ni_7 and $LaNi_5$. These IMCs are essentially line compounds (i.e., have no composition ranges), although $LaNi_5$ has some range of composition at high temperature.

Although all of these phases will absorb H_2, the most reversible, convenient and practical of the IMCs shown in Fig.9 is $LaNi_5$. $LaNi_5$ has the hexagonal $CaCu_5$ ($D2_d$) crystal structure, shown in Figure 10. There are a number of possible interstitial sites with differing La and Ni coordination [62,63] that can be easily occupied with H atoms to a level of about H/M = 1.08 ($LaNi_5H_{6.5}$). At very high pressures, additional interstices can be filled (up to H/M=1.5) [63]. As shown by the lower set of isotherms in Figure 11 [64], H_2 can be absorbed into and desorbed from $LaNi_5$ at near atmospheric pressures and

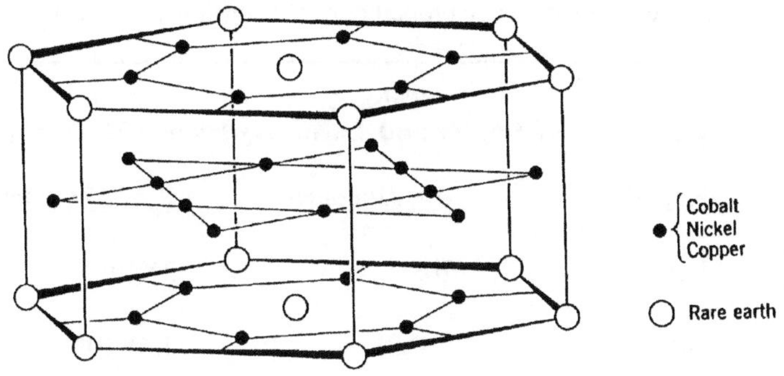

Figure 10 - $CaCu_5$ ($LaNi_5$) Crystal Structure [49]

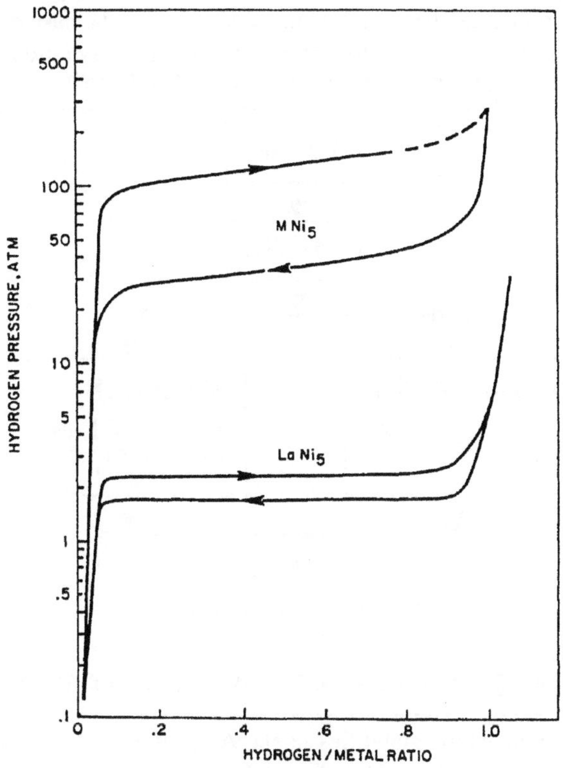

Figure 11 - 25°C Absorption/Desorption Isotherms for $LaNi_5$ and $MmNi_5$ [64]

25°C. Absorption/desorption pressure hysteresis is small and the plateaus are rather flat. Table 2 shows the effect of the using this intermetallic compound to "interpolate" or moderate the extreme thermodynamic properties of the constituent elements. In Table 2,

Table 2
Comparisons of La, Ni and $LaNi_5$ Hydrides [54,59, 65]

Hydride	ΔH_f, kJ/mol H_2	P_p at 25°C, atm
LaH_2	-208	$3X10^{-29}$
NiH	-8.8	$3.4X10^{3}$
$LaNi_5H_6$	-30.9	1.6

ΔH_f is the enthalpy of formation of the hydride and Pp is the desorption plateau pressure at 25°C (heavily extrapolated from high temperature data in the case of LaH_2[65]). The effect of moderating some 32 orders of magnitude of dissociation pressure is quite dramatic and forms the powerful basis of all practical IMC hydrides.

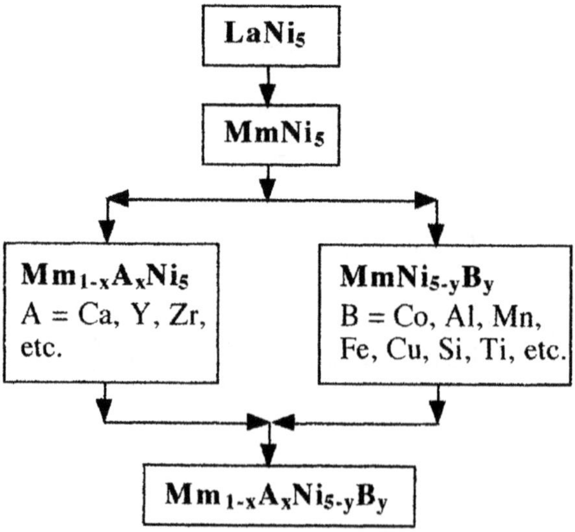

Figure 12 - Substitutional Modifications of $MmNi_5$

The exact thermodynamic stabilities of the AB_5 intermetallics can be adjusted by numerous possible substitutions for A and B atoms (summaries in [59,66,67], for example). Although the AB_5 $LaNi_5$ has very useful properties, it is rather expensive so early efforts centered around reducing AB_5 cost by substituting the low-cost commercial rare earth alloy mischmetal (Mm) for expensive elemental La. The composition of mischmetal is somewhat variable and depends on its source, but the most common mischmetal is derived from the fluorocarbonate mineral Bastnasite and ends up with the following approximate composition (in wt.%): 48-50 Ce, 32-34 La, 13-14 Nd, 4-5 Pr and 1.5 other rare earths. Unfortunately, the direct substitution of Mm for La in $LaNi_5$ results in uncomfortably high pressures and hysteresis, as shown in Fig.11. This led to a number of successful programs to partially substitute other elements for Mm and Ni in $MmNi_5$ [67-69]. The main substitutional elements are shown by the strategy tree of Figure 12. A practical example of partial Al substitution for Ni (i.e., $MmNi_{5-y}Al_y$) is

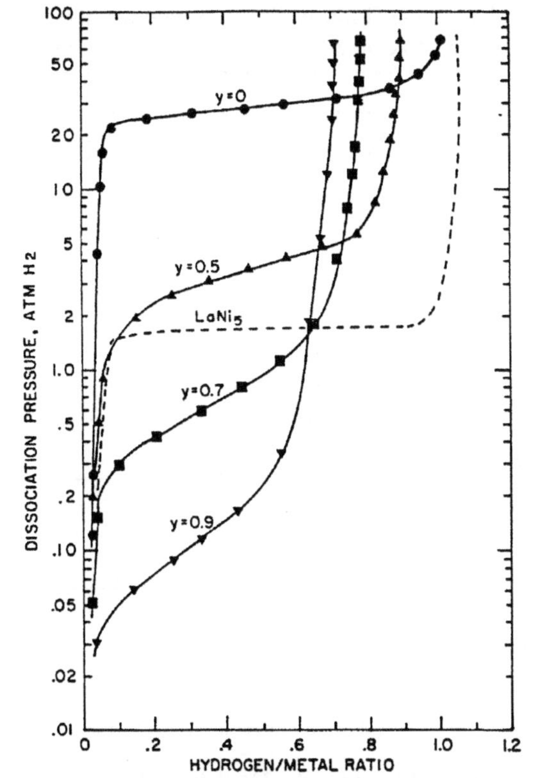

Figure 13 - Effect of Al-content y on the 25°C Desorption Isotherm of $MmNi_{5-y}Al_y$. As-cast. [68]

shown in Figure 13. The effect of Al (among other elements) is powerful in reducing hysteresis and allowing fine tuning of absorption/desorption pressures, very important for practical applications of hydrides [1]. Ternary substitutions to the $MmNi_5$ often produces a segregation-related slope to the plateau but this can usually be reduced by annealing [68]. Almost all AB_5 hydrides now used commercially (e.g., in Ni-MH batteries) are substituted $MmNi_5$ compositions represented in the branches of Fig. 12.

3.2.2. *AB_2 Hydrides.* The AB_2 hydrides are based on the AB_2 Laves phases which form under certain conditions of A/B metal atom size ratios and d-electron concentrations [57]. For the Laves phases that form useful hydrides, A is usually a Group IVA element (mostly Ti and/or Zr) and B is one or more transition elements in the range of V to Cu (with special emphasis on V, Mn and Cr). First studied in the 1960's [70], the AB_2 hydrides were the subject of renewed efforts that started in the mid 1970's [71,72] and have continued up to now (see 1986 review in [73]).

Like the AB_5 hydrides, numerous substitutions can be made for both A and B elements in AB_2 hydrides to control absorption/desorption pressures [71-73]. Unlike the AB_5 intermetallics, which are almost always line compounds, the AB_2 intermetallics often show significant ranges of stoichiometry. A good example is shown in the Mn-Zr phase

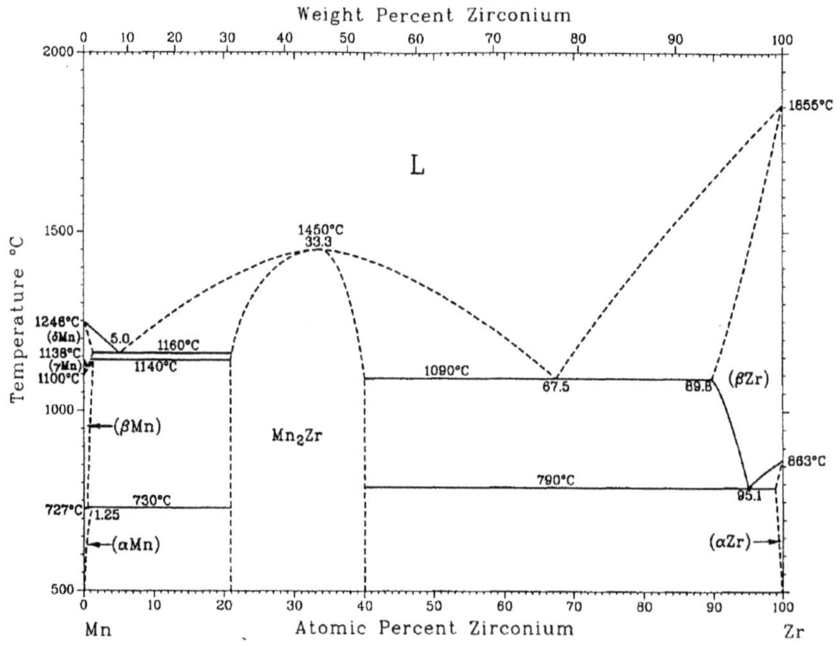

Figure 14 - **The Zr-Mn Phase Diagram** [61]

diagram, reproduced as Figure 14. The $ZrMn_2$ phase can be really expressed as $ZrMn_x$ where x varies from 1.5 ~ 3.8 within the single Laves phase field. As shown in Figure 15, this variable stoichiometry allows the control of plateau pressure without ternary substitutions to the composition [74,75]. In fact, most AB_2 IMC's used in applications are multicomponent and probably off-stoichiometric. Multiphase, disordered alloys based on at least one AB_2 phase have been promoted for Ni-MH batteries [76].

3.2.3. *AB Hydrides.* As was mentioned earlier, the first intermetallic hydride reported in detail was of the AB type, $ZrNiH_3$ [58]. However, $ZrNiH_3$ had high hysteresis and required about 300°C for a desorption pressure of 1 atm, so it was never really very practical for room temperature storage applications. The first practical room temperature AB hydride was based on the ordered body-centered-cubic (B2) IMC TiFe and discovered at Brookhaven National Lab in the U.S. [77]. As shown by the isotherms in Figure 16, two distinct hydrides are formed by the filling of two non-equivalent sets of interstitial sites and resulting in two separate plateaus. Like the AB_5 and AB_2 hydrides discussed earlier, the thermodynamics of TiFe can be modified by ternary and higher order partial substitutions, such as Mn, Ni, Cr, V, etc. [78].

TiFe and its modifications (especially Mn-modified) represent the lowest cost IMCs for hydrogen storage, and for that reason were often used for early demonstrations of

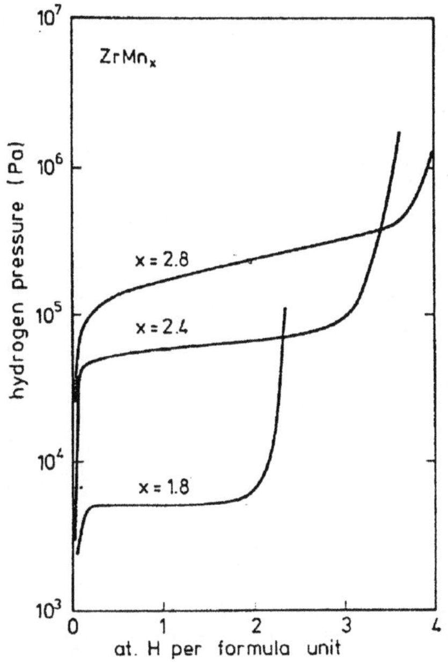

Figure 15 - Effect of Mn-content x on the 50°C Absorption Isotherm of $ZrMn_x$ [74]

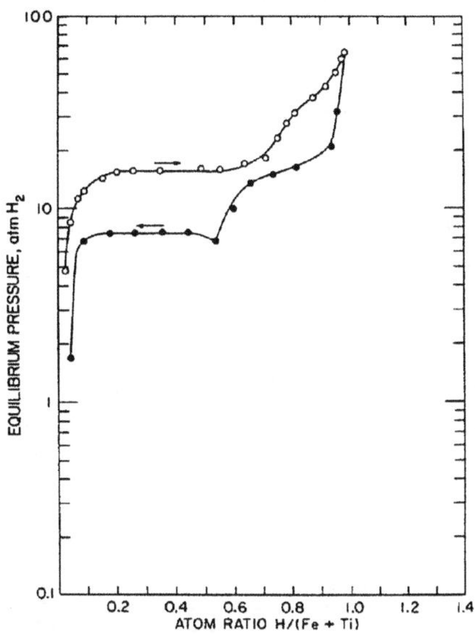

Figure 16 - 40°C A/D Hysteresis Loop for TiFe [77]

hydride applications [1]. Unfortunately, however, the AB family is rather sensitive to surface damage from trace impurities present in commercial purity H_2 [79] and is apparently little used any more.

3.2.4. *A_2B Hydrides.* Examples A_2B compounds that are historically important for hydrogen storage purposes include Ti_2Ni, important in early Ni-MH battery work [80], and Mg_2Ni , the earliest of the "lightweight" hydrides [81]. Mg_2Ni forms the hydride Mg_2NiH_4 (3.6 wt.% H_2) by direct and reversible reaction with H_2 gas around 300°C. Unlike all the intermetallic hydrides discussed above, Mg_2NiH_4 is not really an interstitial metallic hydride but rather a transition metal complex (to be discussed later in Section 3.3.2). Perhaps this, along with limited metallurgical opportunities for ternary substitutions, has resulted in the failure of several studies aimed at reducing the dissociation temperature of Mg_2NiH_4 much below 300°C.

3.2.5. *Summary of IMCs and Practical Details.* The world of intermetallic hydrides is far larger than the brief outline above suggests (see [27] for more detail). The van't Hoff diagrams of Figure 17 show a few of the representative AB_5, AB_2, AB and A_2B hydrides I have discussed above. Fig.17 shows how the limited world of elemental hydrides was profoundly expanded with the development of IMC hydrides. Note that only binary intermetallics are shown. Ternary substitutions can result in almost any degree of interpolation between the lines of Fig.17. Engineering properties of some of the more popular intermetallic hydrides are given in [82].

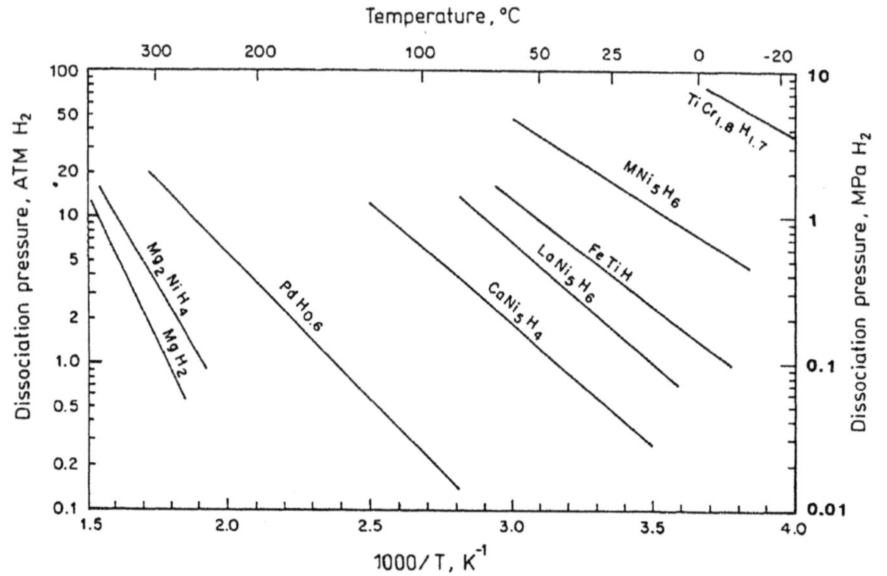

Figure 17 - van't Hoff Plots (Desorption) for Various Intermetallic Hydrides [83]

In addition to the properties discussed above for intermetallic hydrides (e.g. plateau pressures, temperature effects, hysteresis, H-capacity, etc.), there are a few other important properties of interest: activation, decrepitation, reaction kinetics, heat transfer, effects of gaseous impurities, disproportionation and safety. Here I will give only a brief description of each property. Much more detail can be found in [27] (especially [83]) and references cited therein.

Activation is the process required to hydride an IMC sample the first time. The sample usually has a natural oxide coating which must be penetrated by the H_2 gas. Often the process simply involves evacuating the air and holding at some pressure higher than the absorption plateau pressure for some period of time. Sometimes heating is required to open up the oxide film and allow catalytic contact of the H_2 molecules with clean metal atoms to produce active H atoms. There is usually an incubation period after which hydrogen enters the sample and results in particle cracking and new surface generation by the H-induced expansion of the brittle lattice.

For metallic hydrides (including IMCs) there are significant volume increases during hydriding and corresponding decreases during dehydriding. In many cases, this process can result in decrepitation or particle size breakdown during extensive cycling. This must be carefully considered during container design and construction because fine powder can be entrained in the gas stream and can pack in the container leading to gas impedance, poor heat transfer and stressing of the container walls. Similarly, the heats of reaction (exothermic on hydriding, endothermic on dehydriding) must be accounted for in the container design. As mentioned earlier, the heat of reaction is given (more or less) by the slope of the van't Hoff line and depends on the particular IMC. See [48] for compilations of ΔH_f values.

Intrinsic reaction kinetics of H_2 with IMCs are generally very high. Samples can often be hydrided and dehydrided within seconds if care is taken to quickly move the reaction heat out and in. For the most part, practical reaction rates are dominated by heat transfer and, because of this, it is quite difficult to quantify true isothermal kinetics. Powder generally has poor thermal conductivity and special means of thermal conductivity enhancement is required for many practical devices in order to achieve rapid cycling.

The effects of gaseous impurities in the H_2 used can be dramatic in damaging the IMC particle surfaces, resulting in a surface barrier that either reduces the usually good kinetics, lowers capacity or or sometimes even eliminates all H_2 reactivity. Different IMCs vary markedly in their response to given impurities. Given IMC-impurity interactions can be classified as poisoning, retardation, reaction or innocuous, depending on the nature of the cyclic effect [79,83]. Although it is difficult to completely generalize, some of the common damaging gas impurities are the following (in decreasing order of severity): SO_2, H_2S, CO, O_2 , H_2O and CO_2. Relatively harmless gases (at least near room temperature) include NH_3, N_2 and CH_4, although N_2 can be damaging to AB_2 hydrides.

Disproportionation of the AB_y IMC can occur during extensive cycling or long periods of exposure in the hydrided state. Instead of the metastable reversible storage reaction

$$AB_y + \frac{x}{2} H_2 \text{<---->} AB_yH_x \ , \qquad \text{(Eq.6)}$$

the AB_y IMC usually has the thermodynamic tendency to disproportionate to form a stable A-hydride and free B clusters according to the following, much less reversible reaction:

$$AB_y + \frac{x}{2} H_2 \text{---->} AH_x + yB \ . \qquad \text{(Eq.7)}$$

Fortunately, the disproportionation reaction (Eq.7) depends on the diffusion of the A and

B atoms, which is slow near room temperature, so the metastable storage reaction (Eq.6) generally is valid, except perhaps at high temperature. However extensive cycling, even at room temperature, can result in some distortion of isotherms and lost H-capacity as the disproportionation reaction partially occurs. The tendency for disproportion is a strong function of the IMC composition. Disproportionated intermetallics can usually be reproportionated by annealing in a H-free environment [84].

Safety is an important reason why metal hydrides are of interest for many aspects of hydrogen handling, storage and transportation. Because of the low pressures and near-ambient temperatures associated with hydride storage of hydrogen, they are generally viewed as much safer than LH_2 or high pressure gaseous H_2. Because hydrides are endothermic on discharge, container rupture results in a cooled, self-limiting release of gas. Safety advantages led to the extensive use of IMC hydrides for early H_2-fueled vehicles [1,83]. The safety considerations in the use of hydrides centers around their large expansions and the tendency for some IMC alloys and hydrides to be pyrophoric when suddenly exposed to the air in power form. Expansion must be considered in container design [1]. As a general rule, AB_5 hydrides are mildly pyrophoric, AB hydrides are largely non-pyrophoric and AB_2 hydrides are quite pyrophoric if they contain Mn. All in all, I consider the pyrophoricity of IMCs a relatively small disadvantage compared to the positive safety feature of low pressure.

3.3 COMPLEX HYDRIDES

Thus far we have mainly considered the metallic hydrides, materials that are relatively simple, easy to form and highly reversible, given the appropriate combinations of temperatures and pressures. But they have rather low hydrogen storage capacities, as measured in weight percent. There are families of more complex hydrides that offer the potential of higher H-capacities, although they are generally less reversible. Some of these complex hydrides are old and some are new and have considerable R&D potential for future storage applications. They deserve a brief discussion, at least.

3.3.1. *Non-Transition Metal Complex Hydrides.* This family of hydrides can be described with the general formula $M(M'H_4)_n$, where M is a metal (typically Group IA or IIA) with valence n and M' is a trivalent Group IIIB metal (typically B, Al or Ga) [47]. Classic examples include $NaBH_4$, $LiBH_4$ and $LiAlH_4$. They consist of M^+ cations and $M'H_4^-$ anions (e.g., BH_4^- and AlH_4^-), where the H atoms are tetrahedrally coordinated around the M' atom and have a distinct electron-rich character. These materials offer high hydrogen weight percents: $NaBH_4$=10.7 wt.%H; $LiBH_4$=18.5 wt.%H; $LiAlH_4$=10.6 wt.% H. The problem is that these hydrides cannot be easily made by direct reaction of the metal and gaseous H_2, but must be made by various chemical reactions [47]. In addition, they cannot be thermally decomposed at practical temperatures. But they are long standing hydrides of commercial importance and are used as reducing agents (H-donors) for the production of specialty chemicals, pharmaceuticals, precious metals and other products. Their chemistry is well established and it seems unlikely they can be used as reversible hydrides in the same sense as the interstitial hydrides.

Within this class of hydrides are the family A_xMgH_y, where A= Na, Ba or Eu, for example K_2MgH_4 [85]. These hydrides have striking analogies to structurally similar fluorides [86]. An example is the perovskite hydride $KMgH_3$, which is structurally identical to $KMgF_3$.

3.3.2. *Transition Metal Complex Hydrides.* When certain transition metals are combined with a Group IA or IIA element in the presence of hydrogen, a low valence complex of the transition metal and multiple H atoms will form. Such complexes are stabilized by the donation of electrons from the more electropositive IA or IIA elements [87]. A well used example of this is Mg_2NiH_4, where Mg donates electrons to stabilize the $[NiH_4]^{-4}$ complex. In effect 4 hydrogen atoms bond with a single Ni atom and the two Mg atoms donate two electrons each to stabilize that high-H transition metal complex. The structure of such complex hydrides is typically antifluorite.

Although Mg_2NiH_4 has a corresponding Mg_2Ni intermetallic, it is very significant that transition metal complex hydrides can be synthesized from combinations of electropositive elements and transition metals that do not form intermetallic compounds. For example, it is well known that Mg and Fe do not alloy at all in the H-free metallic state [61]. Yet when Ni and Fe powder are sintered in H_2, the high-H complex hydride Mg_2FeH_6 forms [88]. The structure of Mg_2FeH_6 is shown in Figure 18. Note the 6 H-atoms coordinated around the single Fe atom.

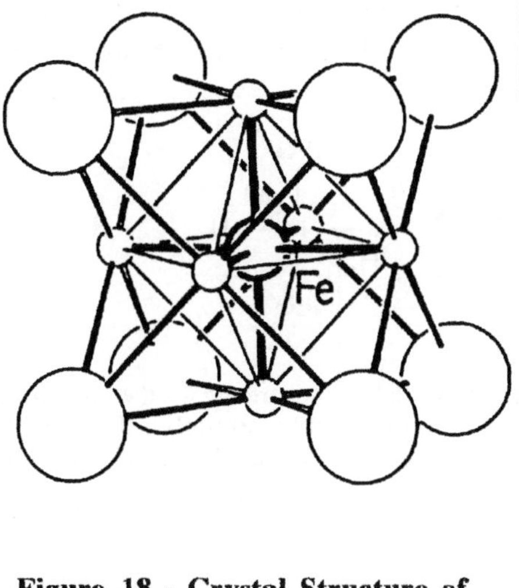

Figure 18 - Crystal Structure of Mg_2FeH_6. Small circles are H.[88]

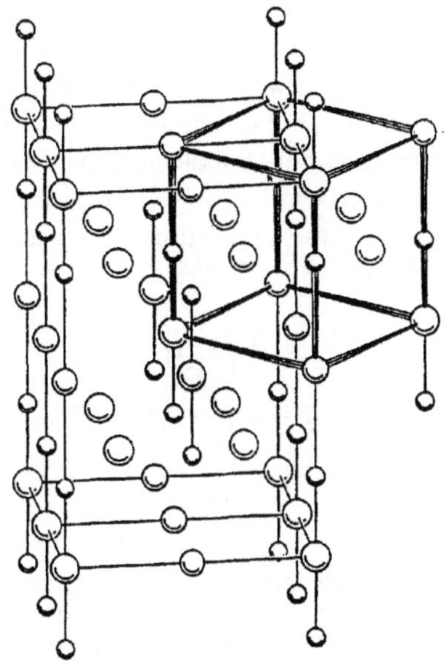

Figure 19 - Crystal Structure of $NaPd_3H_2$. Large circles are Pd. Intermediate and small circles are Na and H, respectively. [85]

The formation and decomposition of transition metal complex hydrides usually require some metal atom diffusion, so kinetics tend to be rather slow compared to the traditional interstitial hydrides. However the high hydrogen contents possible (e.g. Mg_2FeH_6 = 5.5 wt.% H) give promise to these materials as a new approach to storage hydrides.

In recent years Kadir in Stockholm has studied this class of hydrides as a function of transition metal (d-electron content) and has discovered a number of new materials and structures [85]. One is $NaPd_3H_2$, which has characteristics intermediate between interstitial and transition metal complex hydrides. The structure of $NaPd_3H_2$ is shown in Figure 19 in relation to the boldly delineated, and closely related, $PdH_{0.6}$ subcell. Note the linear array of H-atoms coordinated to the Pd atoms.

Work in Geneva on $[ReH_9]^{2-}$ complexes has identified $BaReH_9$, with the largest H/M ratio of any known metal hydride [90]. Made by wet chemistry, the H/M ratio is 4.5 and the hydrogen packing density is twice that of LH_2. However, the H-content in terms of weight percent is only 2.7, because of the very heavy Re atom present. Such extraordinary complex hydrides give hope for future advancements in the area of lighter weight storage materials.

3.3.3. *Nonclassical Polyhydride Metal Complexes.* There is another broad series of organometallic compounds, widely used as chemical and petrolium catalysts, that are termed transition metal hydrides [91]. They usually contain H and CO ligands and have historically not been considered for reversible hydrogen storage. Recently, however, Jensen and coworkers at Hawaii have identified the potential of certain nonclassical metal complex hydrides for H_2 storage [92]. They have concentrated on Group VIIIA based materials such as $IrH_aX_b(H_2)(PR_3)_2$, where X = Cl, Br or I; R = C_6H_{11}, $C(CH_3)_3$ or $CH(CH_3)_2$; a = 1-2; and b= 1-2. In this general formula (H_2) represents a dihydrogen species (a sort of H_2 molecule) that can bound to the Ir metal atom to form a weak ligand. This ligand can be easily broken and reformed during the reversible H_2 desorption / absorption storage reactions. For example, in the X = Cl version, the reversible reaction with H_2 can be written as

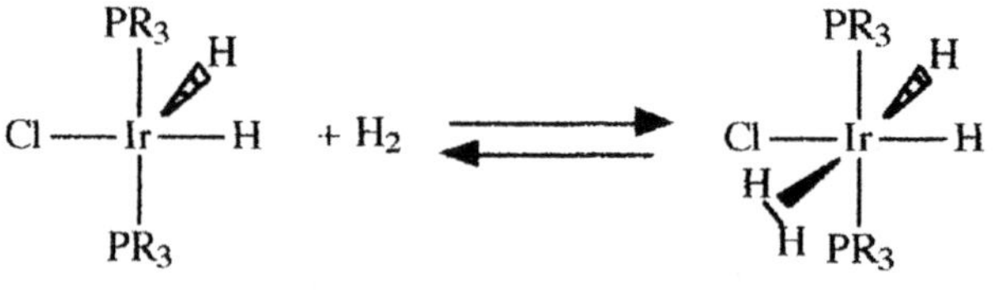

(Eq.8)

The thermodynamics of the storage reaction is controlled by the strength of the dihydrogen ligand with the transition metal atom (Ir) which in turn can be strongly influenced by varying the PR_3 (phosphine alkyl) ligands to that atom. It is not yet clear if this approach will lead to a practical new H_2 storage compound, but it is certainly a different approach that deserves thorough evaluation.

3.3.4. *Hydrides of Metastable and Disordered Alloys.* Various non-equilibrium alloys can be made by special techniques, such as rapid quenching, vapor deposition and mechanical alloying. Using such techniques, a number of attempts have been made to improve the properties of hydriding alloys.

The area of disordered and amorphous (and/or glassy) alloys has been especially active and there has even been an earlier NATO ASI on this subject [37]. The high degree of structural disorder in such materials seems to offer some potential for improved H-capacity and absorption kinetics [37,76,93,94]. However, these effects are certainly not always observed, for example glassy Zr-Ni alloys often show significantly lower capacities than their crystalline counterparts [95]. Because highly disordered alloys do not have a uniform and consistent set of interstitial sites, they do not show absorption / desorption plateaus typical of hydride formation, but instead highly sloping isotherms that are typical of solid solution behavior. The lack of a first-order hydride transformation often prevents the usual embrittlement associated with hydriding, a practical advantage [95,96]. However, the lack of a distinct plateau is usually undesirable for gas applications. A flat plateau is not particularly necessary for Ni-MH batteries, so amorphous and disordered alloys seem to be directed toward that application in particular. At present, the commercial Ovonic (USA) Ni-MH battery is based on multiphase V-Ti-Zr-Ni-Cr alloys that consist of disordered amorphous, microcrystalline and/or polycrystalline phases [76,97].

Another potential technique for making metastable alloys is mechanical alloying, or high energy ball milling of powders. Not only is it possible to make active, high surface area, stable Mg-Ni alloys by this technique [98,99] and conventional AB_5 IMCs [100], composites of metallurgically insoluble species such as Mg-Fe can be made [101]. Mechanical alloying seems to offer the potential for other new and interesting metastable materials.

3.3.5. *Surface Engineered Materials*. Hydriding alloy surfaces can be modified to increase H_2 absorption kinetics and provide a significant degree of protectivity from impurity species. Following are a few examples, among others.

Mg is a reasonably good H_2-storage medium (MgH_2 = 7.6 wt.% H), but the kinetics of H_2 absorption and desorption are very low. Over many years Bogdanovic and associates have developed various Ni surface doping techniques for Mg that greatly increase H_2 reaction kinetics [102-104]. These include Ni in organo-transition-metal solutions (e.g. COD_2Ni, $NiCl_2$ in THF-anthracene and $Niacac_2$-toluene [with and without Al-i-Bu_3 reduction]), as well as dry methods (decomposition of $Ni(CO)_4$ gas and mechanical mixing of Ni powder)[103]. The kinetic improvements are attributed to the catalytically active, fine "fractal-like" surface produced [104].

Imamura and co-workers have also made high activity Mg by evaporation and condensation in a THF atmosphere or high activity Mg_2Ni by treatment with perylene or tetracyanoethylene [105]. The improvements are associated with the formation of electron donor-acceptor complexes formed on the surfaces.

Suda and co-workers have successfully coated AB_5 and Mg_2Ni IMCs with protective rare-earth or complex Mg fluoride coatings by exposure of the alloy particles to an aqueous F^- containing solution [106]. In the case of $La(Ni,Al)_5$, for example, the surface structure shown in Figure 20 is produced. The outer LaF_3 layer easily allows the penetration of H_2 but restricts the passage of impurities such as CO and H_2O (vapor and liquid). Below the LaF_3 layer is a Ni-rich sublayer that is catalytically active for H_2 ---> 2H dissociation, resulting in very rapid activation of treated alloy at low pressure [107]. Mg_2Ni can be similarly treated to result in a surface that is quite active to H_2 absorption at temperatures that approach ambient [108].

Surface engineering remains an active area in the field of metal hydrides and offers a

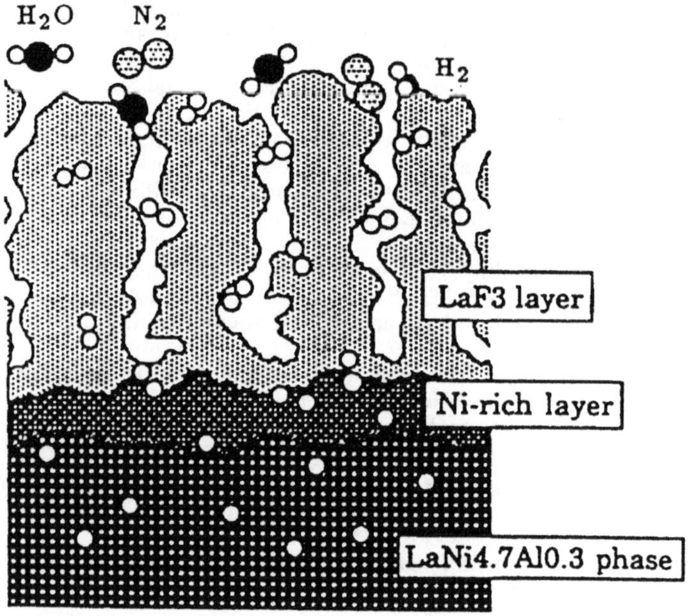

Figure 20 - Protective LaF_3 on the Surface of $LaNi_{4.7}Al_{0.3}$ [105]

good potential for solving some of the gaseous impurity problems that restrict the use of metal hydride use with impure H_2.

4. Hydrogen Degradation of Structural Materials

Our discussions above concentrated on the controlled charging and discharging of metals for useful applications such as hydrogen storage and handling. Unfortunately, there is another side of the Hydrogen-Metals Systems story, the undesirable entry of H into structural materials and the associated degradation of those materials that often results. The degradation is economically distressing, at least, and sometimes catastrophic in nature. Because of the widespread problems associated with hydrogen degradation, there have been thousands of publications over the last century. The subject is ever changing in its terminology, phenomenology and proposed mechanisms, leading to potential confusion even to those well-entrenched in the field of hydrogen or metallurgy. The following is a highly simplified and brief outline, based on my own limited experience and a few books on the subject [38-40,109,110].

There are two broad classes of degradation often described - hydrogen reaction effects and hydrogen embrittlement. The second term is often subdivided into various sub terms associated with the origin of the hydrogen or specific nature of the embrittlement.

4.1 HYDROGEN REACTION EFFECTS

This term, sometimes also called Hydrogen Reaction Embrittlement is associated with internal chemical reactions between hydrogen and impurities (intentional or unintentional) in the metal. It is usually a problem only at high temperature in H_2 atmospheres where H_2 (or atomic H) can readily diffuse into the metal bulk and react with other elements. The two classic examples of reaction effects are carbon steel and O-containing Cu. H_2 reacts with the C in steel (usually existing as cementite Fe_3C) at high temperature to internally form methane:

$$Fe_3C + 2H_2 \text{ ----> } 3Fe + CH_4 \,, \qquad \text{(Eq.9)}$$

which results in blistering, cracking and flaking of the steel. In the case of high-O Cu, the reaction results in H_2O (steam) which also causes blistering:

$$CuO + H_2 \text{ ----> } Cu + H_2O. \qquad \text{(Eq.10)}$$

The materials to be used for any application involving high temperature H_2 must be carefully selected to avoid internal reactions. Fortunately, the long experiences of the chemical and refining industries have gone a long way toward qualifying various materials for high-temperature H_2 use.

4.2 HYDROGEN EMBRITTLEMENT

This term is certainly rather broad and maybe a bit archaic to those deeply involved in this field. It is usually broken down into at least two subterms: (1) Hydrogen Environment Embrittlement, involving phenomena associated with metals strained in gaseous H_2 atmospheres, and (2) Internal Reversible Hydrogen Embrittlement, involving atomic H effects within the metal resulting from processing, welding, electrolytic or corrosion origins at temperatures near ambient [109]. Reversible indicates the H can be removed to at least partly reverse its deleterious effects. The term can be further subdivided. For example, the term Stress Corrosion Cracking is often associated with H-assisted crack propagation, the atomic H resulting from aqueous corrosion reactions:

$$M + xH_2O \text{ ----> } M(OH)_x + xH. \qquad \text{(Eq.11)}$$

It is the general effect of atomic H on crack propagation that most concerns those studying hydrogen embrittlement. It is abundantly clear that most metals show lower macroscopic ductility in the presence of H_2 or loaded with H. The embrittlement H produces is often strain-rate dependent e.g., more pronounced at low strain rates where the H atoms have time to diffuse to dislocations or crack tips. This can result in a particularly distressing phenomenon called Delayed Failure. The phenomenology of H-induced delayed failure is shown in Figure 21, and involves the slow growth under stress of cracks in H-containing metals (e.g. high strength steels). Fig.21 represents a classic schematic stress vs. time curve [111]. The dotted line represents the time required for a crack to start and the solid line represents final failure. At high enough stress (above the Upper Critical Stress, there is nearly immediate failure. At low enough stress (below the Lower Critical Stress), no crack forms and the part effectively never fails. The intermediate stress region is defined by two sigmoidal curves, crack initiation and final part fracture. The band between represents the slow H-assisted growth of a crack or cracks. The problem is that growing cracks are often hard to detect and the end result can be the apparent sudden and catastrophic failure of a major structure.

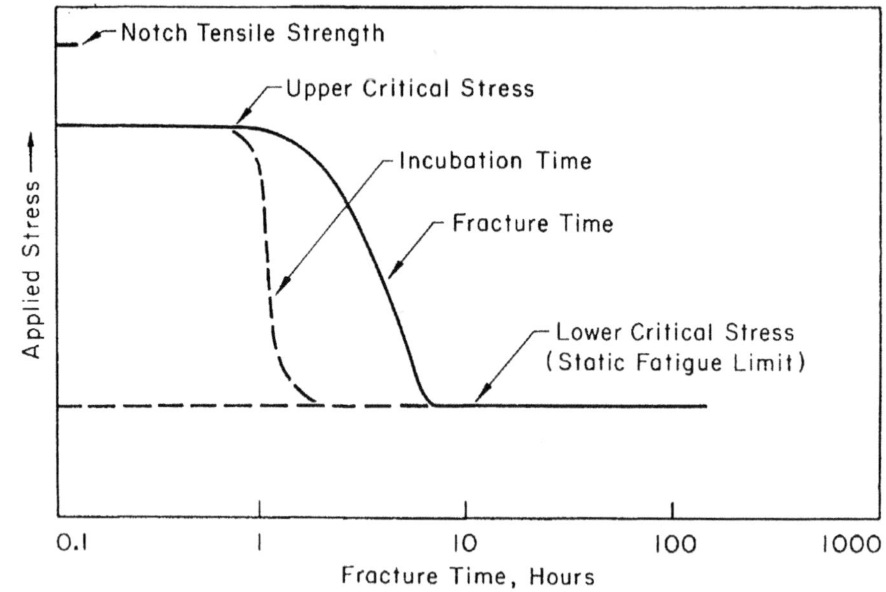

Figure 21 - Schematic Representation of the Failure Characteristics of a Hydrogen-containing High Strength Steel [110]

The mechanisms for hydrogen embrittlement (i.e., the reason(s) H assists crack growth) are varied and the experts are far from unanimous. Birnbaum lists three main possibilities occurring in the stress-field around a crack tip [112]:

1. Hydrogen Related Phase Changes - Here the hydrogen promotes a phase change at the crack tip, the product phase being brittle. One possible phase change is the precipitation of a hydride. This can occur not only in strong hydride formers such as Ti, but also in the high H-fugacity crack tip environments of weak hydride formers such as Ni and Fe. Non hydride brittle phases can also form, such as martensites in normally austenitic stainless steels.

2. Hydrogen Induced Local Plasticity - This theory is based on evidence that the addition of H to a metal actually decreases the flow stress and allows cracks to propagate by a localized ductile mechanism, a sort of "lubrication" of dislocation motion. Even though the macroscopic fracture may appear brittle, careful fractographic examination often gives evidence of highly localized plastic flow.

3. Decohesion - This rather old concept says the H somehow decreases the strength of the metal-metal bond and locally weakens the metal lattice, thus allowing the crack to propagate.

In spite of the many and complicated facets of hydrogen degradation that face the coming of Hydrogen Energy, we should learn from the fact that today's industry has learned to contain and handle hydrogen in gas, liquid and solid form. To be sure, for each new hydrogen device, we must carefully consider possible hydrogen degradation effects and design that device accordingly. There are problems, but they are not insurmountable.

5. Summary

Hydrogen is an almost universal solute in metals. The H in Metals coin exhibits two faces to the coming of the Hydrogen Energy era. In one case the controlled addition of H into metals and its removal on demand offers a practical "solid" alternative to liquid and gaseous H_2 for storage and handling. On the other side of the coin, the unwanted ingress of H_2 into structural metal, and the associated property degradations, must be understood and reckoned with. I have tried to give a broad introduction to both sides of the H in Metals coin.

6. Acknowledgements

I wish to express my thanks to the many colleagues who have contributed to my knowledge of Hydrogen in Metals over the last 20 years. This especially includes the many friends who are regular attendees of the long-standing series of Symposia on this subject. I have attended those represented by [16-23] and the personal interactions therein have been most valuable and rewarding. For this particular chapter, I wish to give special thanks to Dr. George Libowitz (who has long given me much advice on the whole field of metal hydrides and allowed generous use of his library) and Professor Dag Noréus (who introduced me to transition metal complex hydrides).

7. References

1. G. Sandrock: "Applications of Hydrides", this volume
2. T. Graham: Phil. Trans. Royal Soc., **156** (1866) 415
3. T. Graham: Proc. Royal Soc., **16** (1868) 422
4. D.P. Smith: Hydrogen in Metals, U. of Chicago Press, Chicago, 1948
5. L Schlapbach: "Introduction", in Ref.27, Vol.I, p.1
6. G. Taubes: Bad Science: The Short Life and Wierd Times of Cold Fusion, Random House, New York, 1993
7. J.J.G. Willems: Philips J. Research, **39** **(Sup.1)** (1984)
8. F.A. Lewis: Platinum Metals Review, **12** (1968)
9. F.A. Lewis: Platinum Metals Review, **14** (1970)
10. A.S. Darling: Platinum Metals Review, **15** (1971)
11. G. Alefeld, H. Browdowsky, J. Völkl, E. Wicke: Hydrogen in Metals, Rep. Jül Conf-6, KFA Jülich, 1972 (2 Volumes); see also Ber. Bunsengesellschaft Phys. Chem., **76** (No.8)(1972)
12. I.R. Harris, J.P.G. Farr, Eds.: J. Less-Common Met., **49** (1976)
13. E. Wicke, H. Züchner, Eds.: Z. Phys. Chem. NF, **111-116** (1979)
14. B. Baranowski, Ed.: J. Less-Common Met., **101** (1984)
15. F. Lewis, E. Wicke, Eds.: Z. Phys. Chem., **143, 145-147** (1985-86)
16. A.F. Andresen, A.J. Maeland, Eds.: Hydrides for Energy Storage, Pergamon, Oxford, 1978
17. G.G. Libowitz, G.D. Sandrock, Eds.: J. Less-Common Met., **73** & **74** (1980)
18. W.E. Wallace, T. Schober, S. Suda, Eds.: J. Less-Common Met., **88** (1982) & **89** (1983)

19. M. Ron, Y. Josephy, I. Jacob, D. Shaltiel, Eds.: J. Less-Common Met., **103** & **104** (1984)
20. A. Percheron-Guegan, M. Gupta, Eds.: J. Less-Common Met., **129-131** (1987)
21. R. Kirchheim, E. Fromm, E. Wicke, Eds.: Z. Phy. Chem. NF, **163** & **164** (1989)
22. F.D. Manchester, Ed.: J. Less-Common Met., **172-174** (1991)
23. D. Noréus, S. Rundqvist, E. Wicke, Eds.: Z. Phy. Chem., **179**, **181**, **183** (1983)
24. Proc. Miami International Symposium on Metal-Hydrogen Systems, Pergamon, Oxford, 1982
25. Proc. 2nd International Congress on Hydrogen in Metals, Pergamon, Oxford, 1978
26. G. Alefeld, J. Völkl, eds.: Hydrogen in Matals I & II, Topics in Appl. Phys., **28** - **29** (1978)
27. L. Schlapbach, Ed.: Hydrogen in Intermetallic Compounds I & II, Topics in Appl. Phys., **63** & **67** (1988 & 1992)
28. G.G. Libowitz: The Solid State Chemistry of Binary Metal Hydrides, W.A. Benjamin, New York, 1965
29. H. Buchner: Energiespeicherung in Metallhydriden, Springer, Vienna, 1982
30. W.M. Mueller, J.P. Blackledge, G.G. Libowitz, Eds.: Metal Hydrides, Academic, New York, 1968
31. D.T. Hurd: An Introduction to the Chemistry of the Hydrides, John Wiley, New York, 1952
32. K.M. Mackay: Hydrogen Compounds of the Metallic Elements, E & F.N. Spon, London, 1966
33. B.L. Shaw: Inorganic Hydrides, Pergamon, Oxford, 1967
34. Y. Fukai: The Metal-Hydrogen System, Springer, Berlin, 1993
35. G. Bambakidis, Ed.: Metal Hydrides, NATO ASI Series B: Physics, Vol. 76, Plenum, New York, 1981
36. P. Jena, C.B. Satterthwaite, Eds.: Electronic Structure and Properties of Hydrogen in Metals, NATO Conf. Series VI, Plenum, New York, 1983
37. G. Bambakidis, R.C. Bowman: Hydrogen in Disordered and Amorphous Solids, NATO ASI Series B: Physics, Vol. 126, Plenum, New York, 1986
38. R. Gibala, R.F. Hehemann, Eds.: Hydrogen Embrittlement and Stress Corrosion Cracking, ASM International, Materials Park, OH, 1984
39. A.W. Thompson, I.M. Bernstein, Eds.: Effect of Hydrogen on Behavior of Materials (1976) and Hydrogen Effects in Metals (1981); N.R. Moody, A.W. Thompson, Eds.: Hydrogen Effects on Material Behavior (1990), The Minerals, Metals and Materials Society, Warrendale, PA
40. R.A. Oriani, J.P. Hirth, M Smialowski, Eds.: Hydrogen Degradation in Ferrous Alloys, Noyes Publications, Park ridge, NJ, 1985
41. H. Bruning, A. Sieverts: Z. Phys. Chem., **163A** (1933) 409
42. F.A. Lewis: The Palladium Hydrogen System, Academic, New York, 1967
43. E. Wicke, H. Brodowsky: "Hydrogen in Palladium and Palladium Alloys", in Ref.26, Vol.2, p.73
44. R. Speiser: "The Thermodynamics of Metal-Hydrogen Systems" and "Statistical Mechanics of Metal-Hydrogen Systems", in Ref.30, p.51
45. T.B. Flanagan, W.A. Oates: Thermodynamics of Intermetallic Compound-Hydrogen Systems", in Ref.27, Vol.I, p.49
46. D.M. Gruen, M. Mendelsohn: J. Less-Common Met., **55** (1977) 149
47. E.A. Sullivan, R.C. Wade: "Hydrides", Kirk-Othmer Encyclopedia of Chemical Technology, V.12, John Wiley, New york, 1980, p.772
48. R. Griessen, T. Riesterer: "Heat of Formation Models", in Ref.27, p.219
49. C.S. Barrett, T.B. Massalski: Structure of Metals, 3rd Ed., Pergamon, Oxford, 1980

50. A.C. Switendick: Z. Phys. Chem., **117** (1979) 89; "The Change in Electronic Properties on Hydrogen Alloying and Hydride Formation, in Ref.26, Vol.I, p.101
51. K.C. Hoffman, J.J. Reilly, F.J. Salzano, C.H. Waide, R.N. Wiswall, W.E. Winsche: Int. J. Hydrogen Energy, **1** (1976) 133
52. J.J. Reilly, R.H. Wiswall, Jr.: "The Effect of Minor Constituents on the Properties of Vanadium and Niobium Hydrides", Rept. BNL-16546, Brookhaven National Lab, Upton, NY 11973, 1972
53. W.M. Mueller: "Titanium Hydrides", in Ref.30, p. 336
54. B. Baranowski: "Metal Hydrogen Systems at High Hydrogen Pressures", in Ref. 26, Vol.II, p.157
55. G.G. Libowitz, A.J. Maeland: Matl. Sci. Forum, **31** (1988) 177
56. A.F. Maeland G.G. Libowitz, G. Rak: U.S. Pat. 4,440,736, Apr.3, 1984
57. J.H. Westbrook, Ed.: Intermetallic Compounds, Jojhn Wiley, New York, 1967
58. G.G. Libowitz, H.F. Hayes, T.R.P. Gibb: J. Phys. Chem., **62** (1958) 76
59. K.H.J. Buschow, P.C.B. Bouten, A.R. Miedema: Rep. Prog. Phys, **45** (1982) 937
60. J.H.N. van Vucht, F.A. Kuijpers, H.C.A.M. Bruning: Plilips Res. Repts., **25** (1970) 133
61. T.B. Massalski, Ed.: Binary Alloy Phase Diagrams, 2nd Edition, ASM International, Materials Park, OH, 1990
62. C. Lartigue, A. Percheron-Guegan, J.C. Achard, J.L. Soubeyroux: J. Less-Common Met., **113** (1985) 127
63. J.F. Lakner, F.S. Uribe, S.A. Steward: J. Less-Common Met., **72** (1980) 87
64. G.D. Sandrock: "Hydrogen Storage", in Hydrogen Energy, Rept. 199, Royal Swedish Acad. Eng. Sci., Stockholm, 1981
65. W.M. Mueller: "The Rare Earth Hydrides", in Ref. 30, p. 384
66. H.H. van Mal: "Stability of Ternary Hydrides and Some Applications", Doctor's Dissertation, Technische Hogeschool Delft, 1976
67. P.D. Goodell, G.D. Sandrock, E.L. Huston: "Microstructure and Hydriding Studies AB_5 Hydrogen Storage Alloys", Rept. SAND79-7095, Sandia National Lab., 1980
68. G.D. Sandrock: "Development of Low Cost Nickel - Rare Earth Hydrides for Hydrogen Storage", in Proc. 2nd World Hydrogen Energy Conf., V.3 (Pergamon Press, Oxford, 1978), p.1625
69. Y. Osumi, H. Suzuki, A. Kato, K. Oguro, N. Nakane: J. Less-Common Met., **74** (1980) 271
70. A. Pebler and E.A. Gulbransen: Trans. Met. Soc. AIME, **239** (1967) 1593
71. D. Shaltiel, I. Jacob, D. Davidov: J. Less-Common Met., **53** (1977) 117
72. Y. Gamo, Y. Moriwaki, T. Yamashita, M. Fukuda: U.S. Pat. 4,144,103, Mar. 13, 1979
73. D.G. Ivey, D.O. Northwood: Z. Phys. Chem., **147** (1986) 191
74. R.M. van Essen, K.H.J. Buschow: Mat. Res. Bull., **15** (1980) 1149
75. F. Pourarian, H. Fuji, W.E. Wallace, V.K. Sinha, H.K. Smith: J. Phys. Chem., **85** (1981) 3105
76. S.R. Ovshinsky, M.A. Fetcenko, J. Ross: Science, **260** (9 April 1993) 176
77. J.J. Reilly, R.H. Wiswall: Inorg. Chem., **13** (1974) 218
78. J.J. Reilly, J.R. Johnson: "Titanium Alloy Hydrides and Their Applications", Proc. First World Hydrogen Energy Conf., V.II, Int. Assoc. Hydrogen Energy, 1976, p.8B-6
79. G.D. Sandrock, P.D. Goodell: J. Less-Common Met., **104** (1984) 159
80. H. Buchner, M.A. Gutjahr, K.-D. Beccu, H. Säuferer: Z. Metall., **63** (1972) 497
81. J.J. Reilly, R.H. Wiswall: Inorg. Chem., **7** (1968) 2254
82. E.L. Huston, G.D. Sandrock: J. Less-Common Met., **74** (1980) 435

83. G. Sandrock, S. Suda, L. Schlapbach: in Ref.27, V.II, p.197
84. G.D. Sandrock, P.D. Goodell, E.L. Huston, P.M. Golben: Z. Phy. Chem. NF, **164** (1989) 1285
85. A. Kadir: "Studies of New Ternary Hydrides with Different d-Electron Content", Doctoral Dissertation, Stockholm Univ., 1994 (Chem. Commun., 1994, No.2); K. Kadir, D. Noréus: Z. Phys. Chem., **129** (1993) 249
86. A.F. Maeland, W.D. Lahar: Z. Phys. Chem., **179** (1993) 181
87. D. Noréus: Z. Phys. Chem., **163** (1989) 575
88. J.J. Didisheim, P. Zolliker, K. Yvon, P. Fischer, J. Schefer, M. Gubelmann, A.F. Williams: Inorg. Chem., **23** (1984) 1953
89. P. Zolliker, K. Yvon, J.D. Jorgensen, F. Rotella: Inorg. Chem., **25** (1986) 3590
90. N.T. Stetson, K. Yvon, P. Fischer: "On the Structure of the Complex Metal Hydride $BaReH_9$", Inorg. Chem, 1994, in press
91. D.W. Slocum, W.R. Moser: Catalytic Transition Metal Hydrides, Anals N.Y. Acad. Sciences, **415**, New York, 1983
92. M. Mediati, G.N. Tachibana, C.M. Jensen: Inorg. Chem., **31** (1992) 1827
93. A.J. Maeland, L.E. Tanner, G.G. Libowitz: J. Less-Common Met., **74** (1980) 279
94. S.R. Ovshinky, K. Sapru, K. Dec, K. Hong: U.S. Pat. 4,431,561, Feb.14, 1984; K.C. Hong, K. Sapru: Int. J. Hydrogen Energy, **12** (1987) 411
95. F.H.M. Spit: "The Thermodynamics and Kinetics of Hydrogen Solution in some Metallic Glasses", Doctoral Dissertation, U. of Utrect, Jan., 1982
96. D.H. Ryan, F. Dumais, B. Patel, J. Kycia, J.O. Ström-Olsen: J. Less-Common Met., **172-174** (1991) 1246
97. K. Sapru, K. Hong, M.A. Fetcenko, S. Venkatesan: U.S. Pat. 4,551,400, Nov.5, 1985
98. M.Y. Song, E. Ivanov, B. Darriet, M. Pezat: J. Less-Common Met., **131** (1987) 71
99. A. Stepanov, E. Ivanov, I. Konstanchuk, V. Boldyrev: J. Less-Common Met., **131** (1987) 89
100. A. Kr. Singh, A.K. Singh, O.N. Srivastava: Int. J. Hydrogen Energy, **18** (1993) 567
101. I.G. Konstanchuk, E.Y. Ivanov, M. Pezat, B. Darriet, V.V. Boldyrev, P. Hagenmuller: J. Less-Common Met., **131** (1987) 189
102. B. Bogdanovic, B. Spliethoff: Int. J. Hydrogen Energy, **12** (1987) 863
103. B. Bogdanovic, T.H. Hartwig, B. Spliethoff: Int. J. Hydrogen Energy, **18** (1993) 575
104. R. Zidan, D.K. Slattery, J. Burns: Int. J. Hydrogen Energy, **16** (1991) 821
105. H. Imamura, M. Kawahigashi, S. Tsuchiya: J. Less-Common Met., **95** (1983) 157
106. S. Suda: "Mechanism of Hydrogen Uptake on the F-Treated Surface of the $LaNi_{4.7}Al_{0.3}$-alloy", Research Reports of Kogakuin University, **74** (Apr.1993) 29
107. F.-J. Liu, G. Sandrock, S. Suda: J. Alloys and Compounds, **190** (1992) 57
108. F.-J. Liu, G. Sandrock, S. Suda: Z. Phys. Chem., **183** (1994) 163
109. D.A. Mathis: Hydrogen Technology for Energy, Noyes Data, Park ridge, NJ, 1976
110. C.D. Beachem, Ed.: Hydrogen Damage, ASM, Metals Park, OH, 1977
111. A.R. Troiano: Trans. ASM, **52** (1960) (reprinted in [110])
112. H.K. Birnbaum: "Mechanisms of Hydrogen Related Fracture of Metals", in Ref.39 (1990), p.639

MASS STORAGE OF HYDROGEN

J. D. POTTIER and E. BLONDIN
Gaz de France
Etudes et Technique Nouvelles
La Plaine-Saint-Denis
France

ABSTRACT. Hydrogen is stored in bulk quantities in liquid form in 360 and 3800 cubic meter tanks and in gas form underground. The two techniques of underground storage for hydrogen in the future are analogous to those used for natural gas. Cavities created by dissolving salt layers, affording volume capacities of the order of 100,000 m^3, make it possible to store a few million cubic meters of H_2. Storage facilities in aquifers fitted out in the same way as natural gas fields provide capacities of the order of 100 to 1000 million m^3. This paper gives a technical description of both types of underground storage and illustrates the economic case for future hydrogen utilizations with two examples: annual storage in aquifers and daily storage in salt cavities.

Introduction

The hydrogen transportation and storage functions support the increased use of hydrogen in chemical or energy applications. Since large and medium scale applications are as yet limited, the value of a synthetic study lies precisely in bringing out the technical and economic factors involved in the transportation and storage of hydrogen gas that will affect future developments.

Consideration was given to the large scale use of hydrogen following the energy crisis of 1973. Hydrogen's qualities as a future energy carrier or "vector" have been inventoried and they include among other things, obviously, transport and storage possibilities comparable to those known today for natural gas (1, 2). The cost of these operations, which are greater for hydrogen by a factor of 2 to 3, must be included in the overall evaluation of hydrogen as an energy vector, but the potential for a gradual and easy transition from oil and natural gas to hydrogen is a highly advantageous factor. Hydrogen transport and storage techniques are also important for the development of other energies, such as nuclear and solar, since hydrogen is easy to store and transport and can be integrated into an overall energy scheme.

Y. Yürüm (ed.), Hydrogen Energy System, 167–179.

There are various ways to store hydrogen: in pressure vessels, as liquid hydrogen, as metal hydrides and in underground storages. However, for large volumes, the only techniques used are those of liquid hydrogen and underground storage.

1. Liquid Hydrogen Storage

The storage of liquid hydrogen at a temperature of -253°C is practiced on an industrial scale using vacuum-insulated spherical tanks. The largest storage facilities are those set up on rocket launch sites. The National Aeronautics and Space Administration (NASA) in the USA uses a spherical tank with a geometrical capacity of 3,800 cubic meters. On the French launch site in Kourou, used for the Ariane rocket, there are four 360 m^3 tanks (horizontal cylinders 5 m in diameter and 25 m long, Fig. 1). These tanks are semi-mobile. They are brought to the launch site before each launch.

In case of substantial growth in the future, it is certainly possible to build larger capacity tanks, of the order of 12,000 m^3, corresponding to approximately 10 million m^3 of hydrogen gas (at 0°C and 1.013 bar). It should be pointed out that in the plans for sea transport of liquid hydrogen, ships equipped with 4000 m^3 spheroid tanks are being envisaged. As a comparison, liquefied natural gas is currently stored at -160°C in land tanks having capacities up to 80,000 to 120,000 m^3. The unit capacity of gas tanks, spherical or otherwise, on methane carriers is as great as 30,000 m^3.

Evaporation losses from large tanks are small, in fact less than 0.1% per day. Actually, the limitations on storage of liquid hydrogen stem rather from the high cost of the tank and from the economic and energy cost of liquefaction (about 20 to 25% energy loss). This technique is used presently only when the availability of liquid hydrogen presents a synergy with the downstream utilization, ie. when direct loading as rocket fuel or filling of distribution tanks follows. It can also be used to transport hydrogen by ship.

For higher capacities, and for applications that do not derive an advantage from liquefaction, underground storage of gaseous hydrogen is a more worthwhile option. This can be achieved by two alternatives: storage in cavities and storage in porous reservoir-rocks (1)(2)(3)(4). In both cases, favorable geological sites must be available, which puts a constraint on location of the storage.

2. Storage in Salt Cavities

Cavities created deep underground by controlled dissolution of salt formations are commonly used to store methane gas and occasionally other products (ethylene, propylene, propane), including hydrogen gas.

Use of other types of underground cavities, such as cavities obtained by nuclear blasts, natural caverns of karstic origin and abandoned mine galleries, has been envisaged. Until now, none of these techniques have been developed for gas storage.

Other galleries excavated to shallow depths, in compact media, for storages associated with industrial complexes are commonly reserved for storing liquids, such as ammonia, ethylene and LPG, sometimes under pressure.

We will therefore limit the scope of this technical evaluation to cavities dissolved in salt formations several hundred meters below grade (2)(3)(4).

Salt (sodium chloride) is quite frequently present in sedimentary basins, in the form of strata tens, even hundreds of meters thick, or of domes or diapirs. As a material, it is practically impermeable and fluid tight. Each cavity is leached out from an oil or gas type well that is lined and cemented from the surface to a depthwise dimension beneath the salt dome. Figure 2 shows diagrammatically the leaching methods used. In direct leaching, the well being equipped with two concentric columns of tubing, fresh water is injected through the central tube and the brine produced by the leaching action is discharged up the center annulus. A cushion of a product inert with respect to salt is placed in the outside annulus from grade down to a few meters above the base of the column of larger diameter tubing; it serves to control the progress of dissolution in the top part of the cavity being created. The positions of the bases of the two tubing columns are changed as the cavity grows. The procedure is assisted by reference to a model describing the salt dissolution and cavity formation processes.

This technique is accurate enough to give the cavity a regular shape with a conical roof sufficiently slanted to ensure good stability for gas storage. The shape of each cavity is appreciated by making sonic measurements in the well. In favorable cases, the cavities can reach a height of about 100 meters and a radius of 20 to 30 meters, giving a resulting free volume of a few hundred thousand cubic meters (Fig. 3).

After installing the operating equipment, the gas is injected to replace the brine and the gas pressure can be increased up to high values compatible with the mechanical behavior of the salt and with the strength of the production well casing cementation. The cavity is then exploited by expanding the gas to the lowest pressure compatible with the stability of the cavity roof.

The most extreme working pressures are generally of the order of 1.8 and 0.6 times the hydrostatic pressure at the cavity roof, which means, for a cavity located 1000 m deep for example, a pressure varying between 180 and 60 bar.

Draw-off from the cavities must be carried out in such manner as to ensure good mechanical stability of the cavity and avoid creeping of the salt layer as would occur if the pressure of the gas in the cavity became too small. It is therefore necessary to perform tests of the mechanical properties of the salt, which are specific to each site, and models of the mechanical behavior of salt cavities. It is normal during exploitation of the cavities to find a slight reduction in volume, due to creep flow. Measuring this volume reduction precisely makes it possible, subsequently, to validate the mechanical behavior models. A volume reduction of 5 to 10% after five years can be considered normal.

For natural gas storages, several adjacent cavities are made in the same salt layer to provide a storage site served by a surface gathering network, a compression station and processing units.

Storage facilities of this type have been made for more than 20 years in various industrialized countries. There is therefore a long experience with operation of natural gas storages. Gaz de France has developed, based upon the first storages completed back in 1969, numerical models allowing their performance to be predicted in the following areas: total storable stock, active volume, residual stock, output rate and stock available at any given time.

These models can be applied to hydrogen storage by substituting the thermodynamic and physical data of hydrogen for those of natural gas. It deserves to be pointed out on this subject that hydrogen, unlike natural gas, has a compressibility factor greater than one and that its compressibility increases with pressure.

By way of example, a simulation made to compare the operation with natural gas and with hydrogen of a 220,000 m^3, 1430 meters deep cavity from which stock is drawn off continuously at a rate of 1 million cubic meters (n) per day gave the following results:

– for a given maximum permissible pressure in the cavity and a given temperature of the gas mass (55°C), the cavity contains approximately $47 \cdot 10^6$ m^3 (n) of natural gas and only $37 \cdot 10^6$ m^3 (n) of hydrogen,

– the pressure drop in the cavity is faster in the case of hydrogen than it is for methane; the speed at which the salt block is mechanically stressed is therefore slightly greater,

– the cooling of the gas phase is not as fast and the temperature difference observed at the end of draw-off, at the minimum permissible pressure, between natural gas and hydrogen is of the order of 7°C,

– respiration (the difference between the maximum stock and the stock when the working volume has been drawn off) is, in volume terms, less for hydrogen than that seen for natural gas; the same is true of the cushion (the stock remaining when the working volume has been drawn off),

– the well head pressures are about 24 bar greater for hydrogen (difference in pressure losses and especially in the weight of the gas column).

In respect of the fitting out and operation of hydrogen storage cavities, the carry-over of technologies proven with natural gas to hydrogen storage cavities raises two types of problem :

– the qualification of the well equipment: embrittlement of certain metals, sensitivity of plastic and elastomer materials. Adaptation of the equipment shall have to be envisaged,

– the hydrogen produced after storage contains water vapor and sometimes other impurities in very small concentrations.

One application of salt cavities to the storage of hydrogen has been in existence for more than ten years in Teeside, U.K. The subsoil of a petrochemical plant has a layer of salt about 35 m thick at a depth of 350 m. Flat cavities about 30 m thick and 70 m in diameter were made in the layer, giving a volume of approximately 55,000 m^3. The cavities are operated at a constant pressure of about 44 bar by injection or withdrawal of salt water stored at the surface. The storage capacity is of the order of 2 million cubic meters of hydrogen per cavity.

3. Aquifer Storage

Deep aquifer strata are located in porous and permeable geological layers. The principle of gas storage consists in injecting gas in the rock pores initially filled with water. This creates a kind of gas accumulation which is similar to certain natural hydrocarbon deposits. The use of this technique requires special geological conditions (2)(3)(4).

The porous reservoir-rock layer to be used for gas storage must be located under a cap rock which prevents gas from moving upwards because it is impermeable or because its very fine interstitial structure prevents gas from entering below a certain capillary pressure. To make gas accumulate upwards round the production wells, in the form of a bubble and without risk of leakage, cap rock and reservoir layers must be bell or anticline shaped, creating a trap in which the gas remains enclosed.

Figure 4 show a cross section of an aquifer storage. The gas stock is limited by the anticline threshold which determines the reservoir closure or useful height of the anticline. When the stock is drawn off, the gas rises until water fills the production wells and makes it impossible to keep up the output gas flow rate. The gas quantity which can be injected and recovered during normal annual production cycles is called the "working volume". It is about half the total volume.

The storage facility is equipped with production wells (often twenty or more), used for injection and draw-off, and with control wells (fig. 5-6). At surface level it includes a gathering system and a processing and compression station (fig. 7).

The initial aquifer pressure is generally close to hydrostatic pressure (around 1 bar per 10 m depth). During injection this pressure must remain within values ranging from 1.2 to 1.5 bar per 10 meter depth. Conversely, during production cycles, it drops below the original equilibrium pressure.

Natural gas storages normally have an annual injection and draw-off (input and output) cycle. In order to forecast pressure trends and production well parameter trends, models have been developed which are similar to those used for production from hydrocarbon fields. Such models describe the change in mean pressure of the stored gas on the basis of the overall behavior of the surrounding aquifer and also show relative water and gas movements during gas production.

This storage technique, long applied to natural gas, can also be used for hydrogen. Obviously, the physical properties of hydrogen, which are different from those of natural gas, need to be taken into account.

The first factor to consider is the tightness of the cap rock and this involves two problems:

– Hydrogen in solution in the connate water of the cap rock spreads faster than methane, but at such a slow rate nevertheless that the effect on the reservoir cap rock is not noticeable on a human timescale.

– Gas penetration into the cap rock by capillary pressure depends on the interfacial tension and the penetration rates of methane and hydrogen are comparable. As their viscosities are of the same order of magnitude, the capillary tightness of the cap rock is not a particular problem for hydrogen.

These considerations are corroborated by industrial examples:

– The operation in Beynes, France, from 1957 to 1974, of a town gas reservoir with a volume of 330 10^6 m^3 (n). This reservoir, whose gas contained about 50% hydrogen, has a very thin cap rock (7 m). So a high concentration gradient between the reservoir and the upper aquifer was able to be experimented. Yet no trace of hydrogen was ever found in the gas extracted under vacuum from the water taken from the upper level.

– The operation, without detectable loss, since 1963 in the United States, near Amarillo, Texas (USA), of a helium storage of more than 1 billion m^3 (n) beneath a limestone cap rock; calculations based on laboratory data showed a theoretical diffusion rate of 3 m^3 (n) per year and per square kilometer, through a 100 meter thick overburden. This result, which would have to be at most doubled for hydrogen, shows that regardless of the theoretical contribution that an in-depth study would provide, there is little risk that diffusion might have an industrial bearing.

The other problems concern production technology and economy, which differ for hydrogen and natural gas because of their different compressibility, viscosity, density and calorific value.

As in the case of salt cavities, plans will have to be made to:

– adapt the equipment and piping used in the storage wells,

– purify the gas produced from storage. Prior experience with natural gas storage teaches that each site is a specific case, but that frequently the drawn-off gas contains, besides water vapor, an appreciable amount of hydrogen.sulfide. A processing station is always required to eliminate this compound.

4. Constraints on Underground Storage Sites

Like all industrial installations, underground storage surface installations must meet environmental requirements. But the geographical siting of the storage is based first of all on the possibility of finding favorable geological conditions. Moreover, the construction of the underground equipments must be compatible with other utilizations of the subsoil. This requires, in particular, strict protection of underground water. Subsoil-related siting constraints can result in a storage site being located far from the place where hydrogen is used, sometimes at a distance of one hundred kilometers or more. The role of a large-scale storage and its economical aspects are therefore in most cases associated with hydrogen transmission, which may be limited to a single line from the storage to an industrial site but will often be part of a hydrogen transmission network. For comparison, a map of natural gas storages and transmission network in France is shown in fig. 8.

It must also be emphasized that underground storages take a long time to complete. The exploration phases are spread over several years and face a certain probability of success. If they fail, a work phase needs to be repeated, resulting in additional delays. When the quality of the site is ascertained, the actual construction work – drilling and equipping wells for cavity leaching, erecting surface installations, etc. – must be done. So, the search for and construction of an underground storage take from 5 to 10 years.

Salt cavity storages are well suited for a working volume of 50 to $100 \cdot 10^6$ m^3 (n), since every cavity can contribute tens of millions of cubic meters of working stock.
Aquifer storages are better suited for higher working volumes, greater than 200 to $300 \cdot 10^6$ m^3 (n).

5. The Economy of Storages

The large-scale storage of hydrogen obviously fits into a broader industrial system. The structure and cost of the storage facility depend directly on its function within that system. Storage can become necessary either because of variation of the production function (production from solar energy or from tidal power) or because of end-use requirements (weather-related annual or daily load modulations and/or safety reserves).

By way of example, two very different cases will now be described in detail.

5.1 ANNUAL STORAGE IN AQUIFERS

In France, we have considered the case of an electrolysis plant using nuclear electricity during the summer season (3.500 h) and associated with regular utilization of hydrogen by a chemical plant. From the total 360 MW electrolysis power (72.000 m3/h of hydrogen), 200 MW is used for storage of hydrogen (40.000 m3/h). The plant is connected to an aquifer storage site located 100 km from the chemical/electrolytic plant complex and providing a working capacity of $140 \cdot 10^6$ m^3 (n). The overall cost of equipping the storage facility for this system was estimated to be between 2 to 3 French francs per working m^3. This cost estimate includes the cost of drilling, of surface installation and of the cushion gas. In normal operation, the cost of storing and draw-off includes operating and debt service costs. It is estimated to be 0,30 to 0,40 FFr per m^3 (n) for storage and 0.03 to 0.04 FFr per m^3 (n) for transmission (in 1993 French francs, 1 FFr roughly equal to US$ 0.17).

Presently, the cost of producing hydrogen by natural gas reforming is about 0.8 to 1 FFr/m^3, but over the long term a very significant increase in the price of natural gas must be envisaged, and stricter environmental constraints in the form, for example, of a tax on CO_2. Assuming long term hydrogen prices of approximately 1.5 to 2 FFr/m^3, it can be seen that for the transmission and storage system associated with the above project the transmission cost is small (a few percent) and the storage cost is moderate.

This situation would be sufficient to allow electrolysis to be commercialized in France during off-peak periods of nuclear plant electricity production if it is assumed that fossil fuels will see a marked price increase.

On the long term, this scheme could be applied to energy systems supported by a large production of nuclear electricity, with season variable energy needs.

5.2 DAILY STORAGE IN SALT CAVITIES (5)

The example taken is that of a tidal power station supplying 1000 MW to an electrolysis plant during 4.5 hours over each 12.5 h tidal cycle. This facility is associated with a salt cavern storage (affording a working stock of 2 Mm^3) and the hydrogen is used to meet the daily peak load of electricity.

The hydrogen-from-electrolysis output rate is 220,000 m^3/h, which corresponds to a total hydrogen production of $\cdot 10^6$ m^3 per tide cycle. The storage equipment is adapted for peak load electricity generation of 500 MW over 3.6 hours (8 energy recuperating turbines of 60 MW each).

Two storage scenarios are being considered using salt caverns located about 500 m deep.

Type of storage	Dry	Brine compensated
Maximum pressure (atm)	25	57
Minimum pressure (atm)	17	-
Diameter of spherical cavity (m)	79	40
Compressors	7 x 6000 HP	3 x 3500 HP

The economic evaluation of the two scenarios gives very comparable costs. The author J.B. TAYLOR (5) estimates that the storage function adds 10% to the cost of the production of hydrogen by the tidal power plant and its associated electrolysis unit.

5.3 DISTINGUISHING BETWEEN RESERVOIR AND STORAGE-FUNCTION COSTS

It can be seen by analyzing these two examples that the storage equipment or reservoir in the narrow sense (tank, cavity or aquifer) is associated with operating equipment (compressors, liquefaction) adapted to the service expected from the storage. Because of this, the cost of storage can be assessed at different levels as follows:

the reservoir itself, without its auxiliaries: the cost per working cubic meter has limited significance;

the reservoir together with its auxiliaries (compressors, supply system, monitoring system): this set of things is related to the storage function. The compression power in particular will be very different depending on whether an annual storage, a daily storage or a peak-shaving storage is involved.

It is possible to define a cost of equipment per working cubic meter, as well as an operating cost per cubic meter of draw-off, but in these terms the different types of storage are not comparable.

The storage function together with the system as a whole

In the case of a storage associated with a fluctuating production of hydrogen, it is possible to assess the financial impact of the storage on total production over a full operating cycle, in terms of cost per overall cubic meter produced.

Thus, in the first example, the summertime storage of 140 Mm^3 is associated with an annual consumption of 250 Mm^3 of hydrogen (110 Mm^3 in summer without storage and 140 Mm^3 used after storage). The storage cost can therefore be related to the 250 Mm^3 annual output.

The article by J.B. Taylor presents detailed examples of various large-scale storage modes and of systems that may be associated with them. Although it dates from 1985, it remains entirely relevant and the second example above is taken from that publication :

The cost of the salt cavity (total volume 4.5 Mm^3, working volume 2 Mm^3) is evaluated as \$6m at 1985 prices. The auxiliaries and especially daily compression cost \$25m. The total cost of the storage (including expenses during construction, startup and working capital) is about \$50m. Referred to the working volume, this equipment is very costly (\$25 per cubic meter), but it allows twice-daily storage and draw-off.

If one wishes to use salt cavities on the basis of modulation over a year (as is the case with natural gas), the installation will be very different:

– the compression unit, designed to fill the reservoir over 100 to 200 days, will cost much less,
– the cavities, which will be deeper, will have a higher unit capacity (10 to 20 Mm^3),
– a group of 5 to 20 cavities can be served by a single compression and monitoring/control plant.

Under these conditions, the cost per cubic meter of working volume will be much lower and could fall in the range of \$2 to \$5.

6. Large-Scale Hydrogen Storage and the Environment

Underground storages are installations that respect the environment. They are generally located outside of urban or industrial areas. Their surface installations occupy a limited area; they include, among other things, noise-proofed compression stations, purification equipment and pipe network manifolds. The pipes making the connection to the wellheads are buried. Once drilling and commissioning are completed, the well platforms can be arranged to integrate harmoniously with the landscape.

Underground storages are quite safe. No hydrogen is stored above ground. The wells are equipped with safety devices, notably with automatic downhole valves which prevent accidental blow-out of the production wells. As regards aquifer storages, systematic monitoring is provided of both the storage aquifer and those above it. The law often requires an environmental impact study to be made at the start of a storage project. Such studies provide an independent, objective guarantee that the environment will be preserved and that there will be no nuisances for the surroundings.

Storages of liquid hydrogen in surface reservoirs are subject to special regulations (regulated industrial facilities) as are storages of liquefied natural gas, propane, butane and so on.

7. Conclusion : Large-scale hydrogen storage technologies will be gradually implemented in the future when hydrogen-based energy systems are developed. Storage techniques, for both cryogenic storage of liquid hydrogen and underground storage of gaseous hydrogen, are already available on a significant scale, or can be adapted from those used for natural gas.

REFERENCES

(1) Large Scale Storage and Transmission of Gaseous Hydrogen
J. Pottier, E. Blondin and A. Garat
Hydrogen Energy Progress VII (7th WHEC) pp. 1095-1108
Moscow, USSR 25-29 Sept., 1988 - Pergamon Press
(2) Underground gas storage facilities in France
E. Blondin-Hydrocarbon and Petroleum Geology of France (A. MASCLE)
EAPG n°4 June 1994 p 51/60 Springer Verlag
(3) Les stockages souterrains de gaz naturel
E. Blondin, J. Ferrier
Gaz d'Aujourd'hui N°5 (1994) p 319/327
(4) Les possibilités offertes par le sous-sol pour le stockage de grands volumes de gaz
A. Lecoanet, E. Blondin
Annales des Mines Oct. 1987 Vol 10 p 50/63
(5) Technical and Economic Assessment of Methods for the Storage of Large Quantities of Hydrogen
J.B. Taylor et al.
International Journal of Hydrogen Energy, Vol. 11 No. 1, pp. 5-22, 1986.

Fig : 1 Manufacturing of large size liquid hydrogen tank (Air Liquide)

Solution mining principle
1 Brine extraction. **2** Fresh water injection. **3** Fuel oil injection.
4 Salt caprock. **5** Rock salt. **6** Insoluble residue.

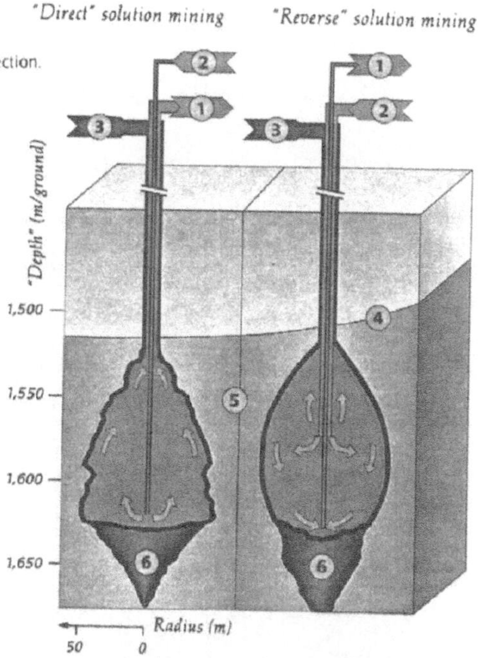

figure 2

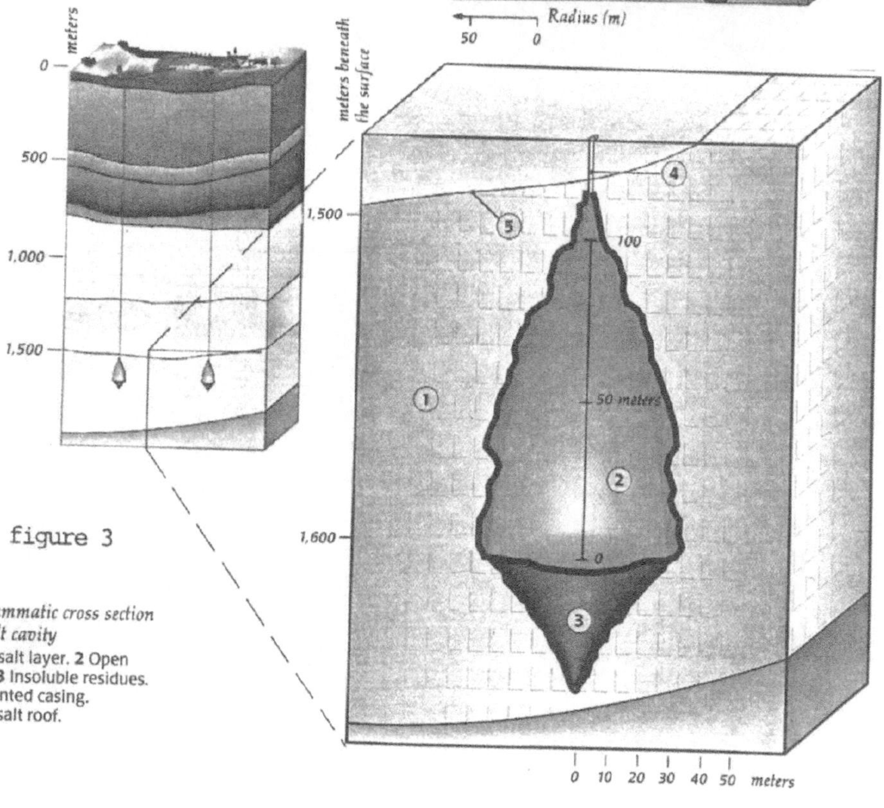

figure 3

Diagrammatic cross section of a salt cavity
1 Rock salt layer. **2** Open cavity. **3** Insoluble residues.
4 Cemented casing.
5 Rock salt roof.

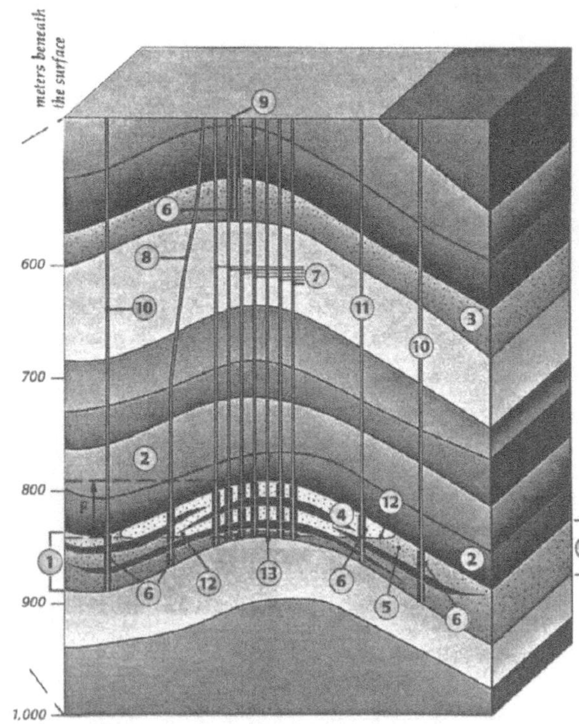

Diagrammatic cross section of an aquifer storage reservoir
1 Aquifer. Porous, permeable layer (reservoir). **2** Impermeable cap rock. **3** Upper control aquifer. **4** Gas. **5** Water. **6** Strainers. **7** Operating well. **8** Deviated well. **9** Upper aquifer observation well. **10** Peripheral observation well. **11** Water level monitoring well. **12** Water level gas/water interface. **13** Neutron logging well. **F** Closure.

figure 4

Cross section of a storage well
1 Wellhead. **2** Safety valve control. **3** Connection pipe. **4** Flow tubing string. **5** Automatic safety valve. **6** Cemented casing. **7** Packer. **8** Caprock. **9** Strainers (– 500 to – 1,200 m). **10** Reservoir.

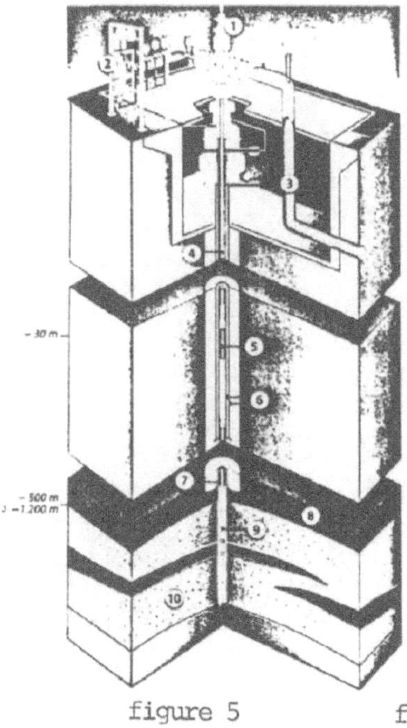

figure 5

figure6

Drilling rig.

figure 7 *Underground storage facility at Saint-Illiers (Yvelines). Above-ground installation.*

GAS TRANSMISSION AND UNDERGROUND STORAGE SYSTEMS

figure 8

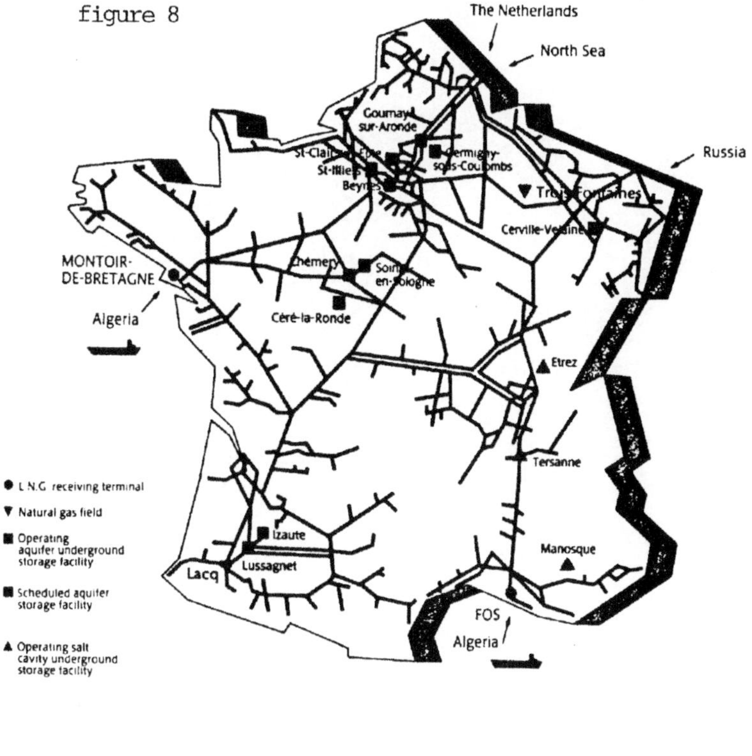

HYDROGEN TRANSMISSION FOR FUTURE ENERGY SYSTEMS

J. D. POTTIER
Gaz de France
Etudes et Technique Nouvelles
La Plaine-Saint-Denis
France

ABSTRACT. In future, hydrogen-based energy systems, hydrogen will be moved mainly in gas form, in pipelines under high pressure, based on the techniques currently employed for natural gas. Some technical adaptations will be required concerning the steels used, the pipeline auxiliary equipment and the compression stations. The energy transportation cost is estimated to be 1.5 to 2 times greater for hydrogen than for natural gas, which makes it quite acceptable in the long run. This paper also describes the transportation of merchant hydrogen (gas or liquid) in moderate quantities. The sea transport of liquid hydrogen, which is very costly, might be used in special systems in the very long term.

1. Introduction

With the view of developing hydrogen as an energy carrier in the future, the main transportation mode that suggests itself is obviously transmission of hydrogen in gas form, in pipes, under pressure (1).

Presently, this transmission mode is implemented either for links between nearby production and utilization sites (a few kilometers or tens of kilometers apart), or in more extensive networks (roughly 200 km). Future developments will certainly entail greater flow rates and distances.

There are also other types of transportation used for specific cases:
-transportation of compressed or liquid hydrogen in trucks,
-transportation of liquid hydrogen associated especially with space vehicle launching sites.

Also worth mentioning is the transportation of liquid hydrogen by sea. These more specific applications will be examined in the second part of this paper. As stated, hydrogen piping is already done on routine basis. Table 1 lists the principal hydrogen pipelines around the world (2, 3). The most highly developed are the networks constructed in the Ruhr in Germany, in northern France and in Belgium. The Ruhr hydrogen transmission network (4) built in 1938 links 17 companies, four of which produce hydrogen by cryogenic separation after electric arc plasma reforming of

Y. Yürüm (ed.), Hydrogen Energy System, 181–193.

natural gas, by recovery from electrolyzed sodium chloride and by PSA separation of gases obtained by various dehydrogenation processes. This branched network, totalling approximately 220 km in length, is made up of pipes 100 to 300 mm in diameter operated at 25 bar from a central control room. The network, managed since the beginning by the Huls company, was recently taken over by BOC Gase GMBH. Its capacity is about 1000 million cubic meters of pure hydrogen per annum. As this network is installed in a heavily built-up area and in mineland, the pipelines are equipped with special safety systems such as valves tripped in response to rapid pressure drop.

The L'Air Liquide company, also, has installed in the North and East of France and in Belgium several pipelines, mainly of 100 mm diameter, working at a pressure of 100 bar and about 340 km in overall length (4). The quantity of high purity (99.995%) hydrogen that can be moved through this system is 175 million m^3 annually.

The transmission of very pure hydrogen as described in the two above examples, involving flow rates of a few thousand m^3/h, is based on conventional techniques but it requires special precautions to guarantee that users will receive a product stable in quality and quantity throughout the year. This can be done on the one hand by using equipment with outstanding fluid tightness and built with special steels and on the other hand by implementing exact and carefully controlled operating procedures.

2. Properties of Hydrogen

The important properties affecting the flow of hydrogen and for pipeline and compression station engineering are its composition, density, specific volume, specific heat, conductivity, viscosity, enthalpy and entropy, its Joule-Thomson coefficient and the velocity of sound in the gas. These properties are a function of pressure and temperature; the pressure and temperature ranges concerned in hydrogen transmission are 1 to 100 atmospheres (10^2 to 10^4 kPa) and -40 to +50°C. For any given composition, it is necessary to know the relevant equation of state. Several formulations of this equation and for predicting various properties of hydrogen, in particular the viscosity of hydrogen gas, have been proposed. Those of McCarty (1979) and Mortveev (1984) are worthy of mention.

To compute gas flow, the general formula is given below. Other presentations have been given in the literature.

The pressure drop for gas flowing in a horizontal pipe can be expressed by the following equation:

$$P_1^2 - P_2^2 = K \cdot s \cdot Zm \cdot Tm \cdot (Q_0^2 / D^5) \cdot L \cdot \lambda$$

where P_1 and P_2 are the upstream and downstream pressures respectively,
s is the density of the gas,
T_m is the mean absolute temperature of the gas in the pipe,
Z_m is the compressibility factor of the gas under the average temperature and pressure conditions of the flow,
L is the length of the pipeline section,
D is the inside diameter of the pipe,

Q_0 is the volume flow rate of gas transmitted, expressed in the reference temperature and pressure conditions,

λ is the friction factor which depends upon the Reynolds number characteristic of the flow and the relative roughness (referred to the diameter) of the pipes,

K is a constant the value of which depends on the system of units and the reference conditions adopted.

If we overlook the variation of the friction coefficient as a function of the flow rate, the diameter and the temperature and pressure conditions, as well as the variation of the compressibility factor as a function of the temperature and pressure conditions, the pressure drop formula can be simplified as follows:

$$P_1^2 - P_2^2 = k \cdot s\,(Q_0^2 / D^5) \cdot L$$

where k is a constant equal to K • s • Zm • Tm • λ

The above formula can also be written $Q_0 = D^{\frac{5}{2}} \cdot (P_1^2 - P_2^2)^{\frac{1}{2}}$

Hydrogen presents the particularity of having a Joule-Thomson effect inversion temperature of -43°C – lower than room temperature. Contrarily to natural gas and other gases, its temperature therefore increases during expansion.

The safety-related properties of hydrogen result from its small molecular size and low viscosity. These have consequences from two standpoints:

- Leakage and inflammation hazard: the flammability limits are very broad (4 to 75%). The ignition temperature (570°C) is rather high but the ignition energy (0.02 mJ at a concentration of 30% H_2 in air) is 10 to 15 times smaller than for natural gas. A pure hydrogen flame is invisible and not very radiative. Moreover, hydrogen disperses upwardly and diffuses rapidly. The risks of explosion in an unconfined environment are very small. These properties bring constraints on the operation of compression stations as well as on the tightness and operation of pipelines.

- In metallic materials, hydrogen has damaging effects, particularly due to diffusion in the grain boundaries, and can cause cracking.

Gaz de France in 1983-84 carried out a general study for the Commission of the European Communities of the safety problems associated with hydrogen (5), some aspects of which are relevant to piping and compression techniques described below.

3. Hydrogen Gas Pipelines

The composition of the steels used and the engineering design of pipelines for hydrogen gas are similar to those for natural gas. It is necessary, obviously, to choose steels less prone to embrittlement by hydrogen under pressure, but this phenomenon only becomes critical in the transmission of very pure hydrogen (higher than 99.5%), which would not be the case where transmission of energy in hydrogen form, rather than of the pure hydrogen required in certain chemical processes, is concerned.

As concerns hydrogen transmission pipelines, the investigations of the selection of materials made partly with the support of the General Directorate XII of the EC Commission, came to the following conclusions (6):

- Extended static pressure tests (10,000 hours) under stress conditions close to the elastic limit show that the steel grades and welds generally used for natural gas transmission are not adversely affected by hydrogen as long as its purity is not greater than 99.5%.

- These results were confirmed by comparative burst tests of steel rupture disks in pipes used for pressurized hydrogen or helium transmission.

- Additional tests have shown that these steel grades, when subjected to severe cyclic stresses, appear to be rather sensitive to hydrogen embrittlement but that this relative sensitivity can be lessened by taking certain precautions such as limiting the stress rate, using steel with a low sulphur content and applying a normalizing heat treatment to it.

This data must therefore be taken into account for hydrogen transmission even under operating conditions resembling those of natural gas transmission, where pressure variations are usually of medium magnitude and rather slow.

This can be expected to increase pipe prices by 30% to 50% over the amounts budgetted for natural gas.

For welding these pipes on site, it is best to use a high-energy process which prevents hardening of the metal around the welds, which entails a cost increase of the same order for pipe joining and laying operations.

Some general guidelines for gas pipelines can be found in the publication "Manuel pour le transport et la distribution du gaz" by the French gas industry association, Association Technique de l'Industrie du Gaz en France (ATG) (7). The publications by M. Mohitpour et al (2) (3) provide information about steels used in hydrogen pipelines, pipeline construction and pipeline auxiliary equipment.

4. Hydrogen Compression Stations

There are two compression technologies available for hydrogen: reciprocating compressors and centrifugal compressors. The latter are used extensively in natural gas compression stations for medium and large flows.

For hydrogen, use of centrifugal compressors certainly creates more problems than the use of reciprocating compressors. In fact, given the lightness of hydrogen, its recompression ratio is four times less than that which can be obtained for natural gas, for the same given tangential speed of the rotor. This requries a large number of stages. Also because of its low specific gravity, hydrogen tends to return to the inlet, thus decreasing centrifugal compressor efficiency.

To avoid having to build huge machines, it is best to limit the use of centrifugal compressors to small outputs, which reduces the worthwhileness of the technique.

Besides that, it is difficult to estimate the cost of such equipment in advance, without a heavy detailed study.

Reciprocating compressors used for natural gas can be used for hydrogen without major design modifications. However, since hydrogen diffuses more readily, special attention must be given to sealing. Furthermore, steel used for parts subjected to fatigue stress (valves in particular) should be adapted for hydrogen.

Reciprocating compressors for use with hydrogen should in principle be driven by electric motors but they could also be powered by hydrogen engines, although the latter's efficiency would be lower than that of natural gas engines.

These differences would lead to higher equipment costs, of roughly twice those incurred for natural gas for the same compression capacity.

The gross driving power for a compression station can be written in the following simplified form:

$$W = K' \cdot Qo \cdot Z'm \cdot T_1 \cdot Ln(r)$$

where W is the gross compressive driving power,
Qo is the compressed volume flow rate (under reference conditions),
Z'm is the mean compressibility factor (between intake and discharge),
T1 is the absolute temperature at the intake,
r is the compression ratio

This formula can be further simplified by overlooking the variations in the compressibility factor, to:

$$W = k' \cdot Qo \cdot Ln(r)$$

where k' is a constant equal to $K' \cdot Z'm \cdot T_1$

5. Transmission Cost of Hydrogen Gas

Estimations of the transportation cost of hydrogen gas and comparisons with that of natural gas are indicated in various litterature published over the last twenty years. Comparisons are difficult because the conditions of the studies – length, flow rate, optimization mode – differ.

Considering the matter from the standpoint of energy transport, the structure of the optimum installations for an equivalent energy output is different for hydrogen and natural gas.

To give an idea of the respective size of natural gas and hydrogen transmission pipelines, a comparison was made of the transmission of a thermal output of 1000 MW at the same operating pressure of 68 bar. Hydrogen has a calorific value 3.6 times lower and pressure losses 5 times smaller than natural gas. In these circumstances, 1000 MW requires 330,000 m^3 (n)/h of hydrogen to be transmitted down a 500 mm diameter pipeline or 92,000 m^3 (n)/h of natural gas to be transmitted down a 400 mm pipeline. For a given recompression ratio, the recompression stations need to be spaced twice as far apart for hydrogen as for natural gas, but the total recompression power required for a given thermal output is about twice as great for hydrogen as for natural gas.

As a rule, hydrogen transmission requires larger diameter piping and more compression power than natural gas for the same energy throughput. In economic terms, most of the studies found that the cost of large-scale transmission of hydrogen is about 1.5 to 1.8 times that of natural gas transmission. The comparative study made by the Gaz de France transportation production department, reviewed in the appendix, effectively leads to a factor of 1.8.

Taking a long-term view, that would certainly imply a substantial increase in the cost of energy products, such a difference in transmission cost would not have a significant effect. Hence future energy systems will be able, in due time, to implement hydrogen transmission systems on a scale of hundreds, even thousands of kilometers.

Estimates of the real cost are few. The work of Williamson and Edeskuty, "Recent Developments in Hydrogen Technology" (8), dating from 1986, evaluates the cost of transmission over a distance of 100 miles (30 inch pipeline, 19,000 hp of compression) at a nominal flow rate of $3.24 \cdot 10^{11}$ Btu/day to be \$0.225/10^6 Btu/100 miles.

Various studies focus on comparing the transmission of hydrogen with that of medium calorific value (medium-Btu) gas or that of substitute natural gas (SNG) (9) both of which can be obtained by coal gasification.

Hydrogen transmission is envisaged in the medium-term to transport the hydroelectrical energy available in sites remote from the utilization zones: beyond 1000 km, it is in fact more economical to transmit hydrogen than electricity. Projects have been researched for, among other places, the Canadian Far North, Zaire and Brazil (10).

6. Use of Natural Gas Networks for Hydrogen Transmission in the Long Term

In the previous section we compared the costs of installation of new pipelines for carrying natural gas and hydrogen: the cost for hydrogen transmission, although higher than for natural gas, is nevertheless quite acceptable in future energy systems.

A further possibility for hydrogen is that it can be transported in the same networks as natural gas, something which is very worthwhile in the period of transition between natural gas and hydrogen. As a first step, hydrogen can be mixed at low concentrations (10% maximum) with the natural gas without any noticeable disturbance to the usual utilizations of natural gas. Next, conversion of the network to hydrogen transmission can be envisaged. If the hydrogen contains a small proportion (1 to 5%) of methane or inert gas the steel pipelines could still be used, except in certain sections sensitized by metal fatigue where the risks of stress-corrosion cracking must be considered.

Gas distribution networks will likewise be reusable (5). For steel pipelines, the pressure level, under 20 bar, avoids stress-corrosion cracking. For the polyethylene piping now in widespread use in distribution networks, research done at the IGT (11) has shown that at the same pressure level hydrogen permeation through the polyethylene was 4 to 6 times that of methane by volume. Hydrogen permeation is therefore still acceptable.

On this issue of converting pipelines from methane to hydrogen, two important points deserve emphasis:

- In terms of technology, the actual pipelines are reusable but the auxiliary equipment (fittings, seals, valves, pressure reducing valves) must be qualified and, possibly, modified or replaced. The compression stations also must be converted.

- In economic terms, the network flow capacity, evaluated in energy, is lessened by the conversion from natural gas to hydrogen. This capacity reduction is difficult to estimate accurately and its incidence in an evolving urban context is also difficult to evaluate. It would be necessary to take into account such factors as better insulation of premises, and the improved efficiency of hydrogen-burning appliances, which reduce specific energy consumptions.

7. Transportation of Gaseous Merchant Hydrogen

The transportation of hydrogen in a discontinuous mode, whether in the gaseous or liquid state, is currently used by isolated or occasional users, or low-volume users. The cost of discontinuous transport is very high. The cost of the hydrogen to the user may be as high as 2 to 5 times the production cost. In far-term energy systems based on hydrogen this type of transportation would see little use.

Hydrogen in the gas phase is generally transported in pressurized cylindrical vessels (at 200 bar) arranged in frames adapted to road transport. The unit capacity of these frames or skids can be as great as 3000 m^3. Hydrogen gas distribution companies also install such frames at the user site to serve as a stationary storage. Gaseous (and liquid) hydrogen carriage is subjected to strict regulations ensuring public safety, which in some countries are very constraining.

8. Transportation of Liquid Merchant Hydrogen

Liquid hydrogen is transported in double-skin vacuum tanks. The quality of the insulation of these tanks is such that hydrogen can be kept in them for several days. The tanks, which exist in capacities of up to 60 m^3 (5000 m^3 of hydrogen gas), can be placed on semitrailers or railroad wagons (12).

This type of transportation equipment is employed by specialized companies which have the technology and the personnel to ensure safety in transport and handling. The tanks are emptied by pressurization through controlled heating, avoiding the use of discharge pumps.

Liquid hydrogen transportation is suitable for medium-scale utilizations (of a few thousand cubic meters per day), for which delivery of frames of gaseous hydrogen would not be very practical. The liquid hydrogen is regasified at the delivery site.

The distribution companies supply the cryogenic tanks from a central station that includes a liquefaction unit. In Europe these units are located in France (Waziers 10 t/day), in the Netherlands (5t/d) and in Germany (3t/d).

9. Transportation of Liquid Hydrogen by Sea

Sea transport of liquefied gas began with natural gas, with the startup in 1965 of the Arzew liquefaction plant in Algeria and of the Arzew-Le Havre shipping link to France served by 40,000 m^3 methane tankers.

Since that time, sea transport of LNG has increased to reach a volume of 60 million tonnes (metric) in 1993 (24% of the volume of international trade). Methane tankers with a capacity of 130,000 m^3 are now common.

Transport by ship of liquid hydrogen in a cryogenic tank was put into practice to supply the French space station in Kourou, Guiana.

Development in the long-term of hydrogen-based energy systems may involve intercontinental movements of hydrogen. The Euro-Canadian project, "Euro Quebec Hydro-Hydrogen Pilot Project " (EQHHPP), has produced a technical and economic study of a system comprising a leg of oversea transport of a significant amount – 100 MW – of liquid hydrogen (13).

The hydrogen would be transported in a ship 180 m long equipped with five 3000 m^3 cryogenic tanks, for a total of 15,000 m^3.

The hydrogen is produced by electrolysis in Canada and liquefied before shiploading. Within the scope of a classic economic calculation, the breakdown of the overall cost per Kwh is as follows, in European currency units (Ecus):

Hydroelectric energy	(Canada)	0.0206
Electrolysis	(Canada)	0.0251
Liquefaction	(Canada)	0.0464
Storage and harbor	(Canada)	0.0146
Ship		0.0128
Receiving and storage	(Hamburg)	0.0190
Distribution		0.0020
	TOTAL	**0.1405**

As can be seen, in this project the cost of hydrogen production represents 30% of the cost of the distributed product, the remainder of the cost being spread over liquefaction (30%) and storage/transportation (40%). The price of the hydrogen sold in Hamburg – about $0.20 per Kwh – is much higher than the price of industrial hydrogen gas. However, certain specific utilizations can benefit from a supply of hydrogen in liquid form. The cost figures indeed underscore the experimental nature of this project. For larger shipments, the cost would be substantially reduced. Large shipments of hydrogen in cryogenic ships are nevertheless unlikely to be considered in the first stages of development of hydrogen-based energy systems.

References

1 Pottier, J. Blondin, E. and Garat, A., "Large Scale Storage and Transmission of Gaseous Hydrogen". *Proc. 7th World Hydrogen Energy Conf.* Moscow (25-29 Sept., 1988) Vol. 2 pp. 1095-1108, Pergamon Press (1988)

2 Mohitpour, M., Pierce, C.L.and Hooper R., "The design engineering of cross-country hydrogen pipelines". *Journal of Energy Resources Technology,* Vol. 110 pp. 203-207 Dec 1988.

3 Mohitpour, M., Pierce, C.L.and Graham, P., "Design basis developed for H2 pipeline". *Oil and Gas Journal*, Vol. 88 No? 22 pp 83-94, 1990.

4 Kaske, G. and Plenard, F.J., "High purity hydrogen distribution network for industrial use in Western Europe". *Int. J. Hydrogen Energy* Vol. 10 No. 7-8 pp 479-482, 1985.

5 Jungbluth, C., Langrognat, B., Saunière, J., Siegler, F., Tellier, C., Thierry, C. "Eléments pour un guide de sécurité Hydrogène". *Contract EHC 32004F, European Communities Commission* EUR 9689 FR, 1985.

6 Bezin, S.and Garat, A., "Etude des possibilités de conversion à l'hydrogène des réseaux de transport de gaz". *Proc. Int. Seminar "Hydrogen as an Energy Vector"*. Brussels 12-14 Feb. 1980, A.A. Strub and G. Imarisio eds pp 603-62 - D. Riedel Pub. Cop.

7 Association Technique du Gaz, "Manuel pour le transport et la distribution du gaz tome 17, Exploitation et Entretien des Réseaux de Transport", 1993 - ATG 62, rue de Courcelles 75008 Paris, France.

8 Williamson, K.D., Frederick, Jr., Edeskuty, J., "Recent developments in hydrogen technology", Vol. 2 Boca Raton, Fla. CRC Press 1986, pp 14-18.

9 Lang, P., "Comparison of hydrogen and SNG transmission", *Energy World* No. 167 pp 7-8, March 1989.

10 McLanagan, SRB, "Towards a hydrogen economy – the storage and transmission of hydrogen – Technical note", *Int. J. of Ambient Energy,* Vol 13 No 1 pp 49-52, Ambient Press, Jan 1992.

11 Jasionowski, W.J., Ding Huang, H., "Gas distribution equipment in hydrogen service - Phase II" *Am. Inst. of Aeronautics and Astronautics.* 1980 pp 2295-2300.

12 Savostianoff, D., "Hydrogène: l'Air Liquide affirme ses ambitions" *Informations chimie* No 320 pp 123-126 Jan 1989.

13 Gretz, J. et al, "Status of the hydro-hydrogen pilot project (EQHHPP)" *Int Journal of Hydrogen Energy* Vol. 19 No. 2 pp 169-174, 1994.

WORLD'S HYDROGEN PIPELINES - D'après MOHITPOUR (4)

Location	Pipeline Material	Years of Operation	Diameter, mm	Length km	Pressure (kPa) & gas purity
AGEC, Alberta, Canada	Gr, 290 (5LX X42)	since 1987	273x4.8 W.T.	3.7	3,790 kPa - 99,9
Air Liquide France	Carbon steel, seam-less	since 1966	Various sizes	290	6,484 - 10,000 kPa : pure and raw
Air Products, Houston area, USA		since 1969	114.3 - 324	100	345 - 5,516 (Pure H_2)
Air Products, Louisana, USA	ASTM 106	since 1990	101.6 - 304.8	48.3	3,447
Air Products, Texas, USA	Converted natural gas line (steel)	before 1980	114.3	8	5,500 - Pure H_2
Air Products, Texas USA	Steel, schedule 40	before 1980	219.0	19	1,400 - Pure H_2
Chemische Werke Huls AG-Marl. Germany	Seamless equipment to SAE 1016 Steel	since 1938	168.3-273	215	to 2,500 raw gas
Cominco B.C., Canada	Carbon Steel (ASTM 210 seamless)	since 1964	5 x 0.8125 W.T.	06	>30,000 62-100% H_2
Gulf Petroleum Canada Petromont to Varennes Montreal east Canada	Carbon Steel, Seamless, schedule 40		168.3	16	93,5% H_2 7.5% methane
Hawkeye Chemical, lowa USA	ASTM A53 Gr. B	since 1987	152.4	3.2	2,757
ICI, Bilingham, U.K.	Carbon Steel			15	30,000 kPa: pure
LASL, N.M. USA	ASME A 357-Gr.5		25.4	6.4	13,788
Los Alamos, N.M. USA	5 Cr. - Mo (ASME A357 Gr.5)	before 1980	30	6	13,790 pure
NASA-KSC, Fla USA	316 SS (austinitic)	before 1975	50	1.6-2	42,000 kPa
Philips Petroleum	ASTM A524	since 1986	203,2	20.9	12,000 kPa
Rockwell international S. Los Angeles, USA	SS-116	before 1980	250		>100,000kPa ultra pure
South Africa				80	

APPENDIX

TECHNICAL AND ECONOMIC COMPARISON OF HYDROGEN AND NATURAL GAS TRANSPORTATION

Economic Assumptions – The following economic assumptions were made:

- the price of pipelines by unit length is in the form
  $a \cdot D^c$ where D is the pipe diameter (with c on the order of 1.3 (1))
- the price of compression stations is in the form
  $b \cdot W^d$ where W is the compression power (with d on the order of 0.65(1))
- for natural gas, in the optimum, the cost of transmission is distributed in the proportion 3/4 for pipelines and 1/4 for compression;
- a hydrogen transmission pipeline costs 30% more than a natural gas transmission pipeline of the same diameter (choice of a suitable steel grade, heat treatment);
- a hydrogen compression station costs twice as much as a natural gas compression station of equal capacity (need to provide motor or engine compressor sets instead of turbocompressors, modifications, choice of materials).

Results

1. For a given transmitted energy flow, the volume flow rate is in the ratio of the higher heat rates (11.5/3.6), ie. about 3.2 times greater than the flow rate of natural gas.

2. Optimum pipeline diameter

   The optimum diameter is proportional to:

$$D = \left[\frac{1}{a} \cdot k \cdot (k')^d \cdot Qo^{2+d} \cdot b \right]^{\frac{1}{c+5}}$$

where a, k, k', d, c and Qo are as defined in the body of the paper.

For same given conditions of relative roughness of the pipes, of average temperature and pressure, assuming furthermore that the friction factor varies as the 0.039th power of the kinematic viscosity (Panhandle formula), we can calculate the ratio of the constant K for hydrogen to the constant K for a natural gas, on the basis of the physical characteristics below.

	Hydrogen/Natural gas (NG) ratio
Density	0.12
Zm (60 bar, 15°C)	1.17
Viscosity v	7

hence $[k_{H2} / k_{NG}] = [s_{H2} / s_{NG}] \cdot [Zm_{H2} / Zm_{NG}] \cdot [v_{H2} / v_{NG}]^{0.039} =$
0.15

Based on the compressibility ratios Zm for the inlet conditions (15°C, 45 bar) and outlet conditions (60°C, 68 bar), we calculate the ratio of the constant k′ for hydrogen to the constant k′ for natural gas, ie. 1.1.

We then deduce the ratio of the optimum diameter for hydrogen to the optimum diameter for natural gas.

$$D_{H2} / D_{NG} = \Big[\underbrace{\left(\frac{1}{1.3}\right)}_{a\text{-factor}} \cdot \underbrace{0.15}_{k\text{-factor}} \cdot \underbrace{(1.1)^{0.65}}_{k'\text{-factor}} \cdot \underbrace{(3.2)^{2.65}}_{Qo\text{-factor}} \cdot \underbrace{2}_{b\text{-factor}} \Big]^{\frac{1}{6,3}} = 1.30$$

For a given transmitted energy flow and at optimum cost, the diameter of a hydrogen transmission pipeline is approximately 30% greater than that of a natural gas pipeline.

3. Compression Optimum

At the optimum cost and for a given energy flow the stations are spaced approximately 2.4 times further apart for hydrogen transmission than for natural gas transmission, but the total recompression power is 50% greater for hydrogen.

For a given compression ratio, the station spacing, based upon the pressure drop formula, is proportional to:

$L / n = D^5 / Qo^2 / k$ *where D, Qo, k and L are as defined above and n is the number of stations*

$$[L / n]_{H2} / [L / n]_{NG} = \left[\underset{D\text{-factor}}{\underset{\uparrow}{(1.30)^5}} \quad \underset{Qo\text{-factor}}{\underset{\uparrow}{(3.2)^{-2}}} \quad \underset{k\text{-factor}}{\underset{\uparrow}{(0.15)^{-1}}} \right] = 2.4$$

The compression power of each station is proportional to k′Qo, whence

$$W_{H2} / W_{NG} = \underset{spacing}{\frac{1}{2.4}} \quad x \quad \underset{k'\text{-factor}}{\underset{\uparrow}{1.1}} \quad x \quad \underset{Qo\text{-factor}}{\underset{\uparrow}{3.2}} = 1.5$$

4. Transmission Cost Optimum

We will use C to denote the optimum cost of natural gas transmission referred to an energy unit.

By hypothesis, the cost attributable to the piping, for natural gas, is 0.75 C. To obtain the corresponding cost for hydrogen, it is necessary to correct for the diameter effect ($1.3^{1.3}$) and the pipeline cost effect at the given diameter (1.3). The cost attributable to the piping for hydrogen is thus 1.35 C.

The cost attributable to compression for natural gas is 0.25 C. To obtain the corresponding cost for hydrogen, it is necessary to correct for the spacing effect (1/2.4), the unit power of each station effect ($(2.4 \times 1.5)^{0.65}$) and the cost at given power effect (2). The cost attributable to compression for hydrogen is thus 0.45 C.

We find that both components of the optimum cost are 80% greater. Consequently, at the optimum, the cost of hydrogen transmission per energy unit and per unit length is roughly 80% more than that of natural gas.

FUEL CELLS

A. YILDIZ and K. PEKMEZ
Department of Chemistry
Hacettepe University
Ankara 06532
Turkey

ABSTRACT. Electrode reactions, characteristic properties, types of cell constructions, advantages and disadvantages and application areas of various types of fuel cell systems such as alkaline fuel cell, molten carbonate fuel cell, phosphoric acid fuel cell, solid polymer electrolyte or ion exchange membrane fuel cells and solid oxide fuel cell will be presented.

1. Introduction

In a fuel cell cold form of combustion or oxidation of fuel gases occurs and chemical energy contained in fuel gases is converted into electrical energy in an electrochemical process that is almost non - pollutant with low noise and with minimal moving parts. Hydrogen is oxidized at an anode with a concurrent reduction of oxygen at a cathode. Methanol can also be converted to water and carbon dioxide or carbon monoxide becomes carbon dioxide in a fuel cell. Some heat is also produced along with the generation of electricity due to the exothermic nature of electrode reactions and Joule heating.

The fuel gas, for example molecular hydrogen releases electrons as the oxidation takes place at an anode. These electrons are conveyed to a cathode through an external conductor and the products of hydrogen oxidation, hydrogen ions, migrate through the electrolyte. The electrons reaching a cathode reduce molecular oxygen producing hydroxide ions. The ions produced at both electrodes then combine to form water. The anode and cathode used in fuel cell systems are permeable to gases and conduct electricity well. The electrolyte, on the other hand, must be impermeable to gases and conduct the ions well. As long as the cell is supplied with a suitable fuel and oxidant electrical power can be obtained. An electrical voltage of about one volt drives the current. Higher voltages can be achieved by connecting several cells in series. Gas diffusion electrodes are either made of fine carbon powder bonded with polytetrafluoroethylene (PTFE) with deposited catalyst or sintered Ni powders. The cells are constructed, either as unipolar or as bipolar type.

The most important application of fuel cells have been their use as chief electricity source for space vehicles. The use of hydrogen - oxygen fuel cells in manned spaceships

Y. Yürüm (ed.), Hydrogen Energy System, 195–202.

provides also pure water as product for drinking and cooling purposes. Fuel cells work at high efficiencies with very low emission of pollutant products. Power plants based on fuel cells can be installed in even densely populated areas. It is predicted that fuel cells will play an important role in near future because of their low emission and noise levels and high efficiencies especially if hydrogen becomes main energy source. High purity hydrogen for fuel cells utilizing hydrogen as a fuel can be produced by the electrolysis of water where cheap electricity is available from hydropower stations or during off - peak hours of electricity demand. The storage of high purity hydrogen can be accomplished in three different ways. It can be stored as gas in high pressure cylinders or as liquid in cyrogenic vessels or as reversible solid metal hydrides. The storage of hydrogen introduces important problems with respect to weight and volume of the systems.

Fuel cells using pure H_2 as a fuel are called direct fuel cells. In indirect fuel cells the fuel is converted (cracking of hydrocarbons or steam reforming of alcohols) into hydrogen containing mixture of gases. This process is usually carried out with the aid of catalysts at high temperatures. Low temperature fuel cell systems require a reformer that is coupled with the fuel cell. Reforming reactions are usually endothermic. Since fuel cells need cooling due to the evolved heat the reformers can be coupled with the fuel cell systems. The fuel cells operating at high temperatures can process fuels such as hydrocarbons inside the cell stack. If hydrazine is used as a fuel catalytic decomposition of this material occurs at the electrode surface, so it can be added to the electrolyte directly. NH_3 is an easily stored and relatively inexpensive fuel that is free of problems encountered with petroleum based fuels. Feeding ammonia to a catalytic reactor where it is decomposed to hydrogen and nitrogen gases is the preferred way of using it as a fuel although direct feeding of ammonia to the fuel cell anode is also possible. Coal can also be used as solid fuel from which hydrogen containing gases are produced. Oxygen, air and in alkaline hydrazine fuel cell hydrogen peroxide are used as oxidant.

Fuel cell systems are usually classified according to electrolyte used. Alkaline fuel cells (AFC), molten carbonate fuel cells (MCFC), solid oxide fuel cells (SOFC) and proton exchange membrane fuel cells (PEMFC) or solid polymer fuel cells (SPFC) are various types systems that were developed during the past few decades. The fuel cell systems are also classified according to their working temperatures. Low temperature fuel cells operate at 80 - 200°C, high temperature cells at 600 - 1000°C. Low temperature cells are of particular importance for space and vehicular applications. High temperature cells are more suitable for large - scale production of electricity and heat. Waste heat energy in high temperature cells can be used to supply hot water and heat pumps and for space heating and cooling to provide maximum overall efficiency. At high operating temperatures electrode reaction rates are enhanced considerably, so that no expensive noble metal catalysts are needed. Carbon monoxide does not poison the electrodes at high temperatures and it can therefore be used as fuel. Other low cost fuels can also be used in high temperature systems.

2. Alkaline fuel cells (AFC)

Alkaline fuel cells use KOH as electrolyte (30 - 45 % by weight). These cells operate well at room temperature and yield the highest voltage at comparable current densities of the other fuel cell systems. The reactions are given as follows:

$$2\,H_2\,(g) \;+\; 4\,OH^-\,(l) \longrightarrow 2\,H_2O\,(l) \;+\; 4\,e^- \qquad \text{Anode reaction}$$

$$O_2\,(g) \;+\; 2\,H_2O\,(l) \;+\; 4\,e^- \longrightarrow 4\,OH^-\,(aq) \qquad \text{Cathode reaction}$$

$$2\,H_2\,(g) \;+\; O_2\,(g) \longrightarrow 2\,H_2O\,(l) \qquad \text{Cell reaction}$$

No CO_2 is allowed in the gas inlet since CO_2 forms carbonates in alkaline medium that black the pores. CO_2 is removed from the air by absorption with soda lime. Lithiated porous nickel cathodes and porous nickel anodes are used. In some applications Pt catalyst is deposited on the electrodes. Some cells operate at around 200°C and higher pressures (5 mPa). Figure 1 shows the general features of AFC system. AFC systems have been used in space crafts, submarines and electric vehicles. Long operating life times (10 000 - 15 000 h) have been achieved. High initial costs of using noble catalysts, consumption of extra energy for compression and liquefaction of hydrogen, high price of pure hydrogen and lack of infrastructure for hydrogen distribution are drawbacks encountered in large scale use of these fuel cells.

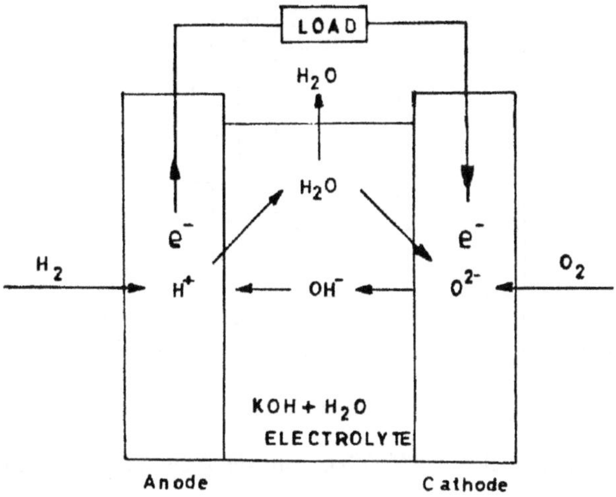

Figure 1. Principle of the alkaline hydrogen - oxygen fuel cell.

3. Molten carbonate fuel cells (MCFC)

Mixtures of molten lithium - potassium carbonates immobilized in $LiAlO_2$ matrix are used as electrolyte in MCFC systems. The reactions occurring in these systems are;

$$H_2\,(g) \;+\; CO_2^{2-} \longrightarrow CO_2\,(g) \;+\; H_2O \;+\; 2\,e^- \qquad \text{Anode reaction}$$

$$1/2\,O_2\,(g) \;+\; CO_2\,(g) \;+\; 2\,e^- \longrightarrow CO_3^{2-} \qquad \text{Cathode reaction}$$

$$H_2\,(g) \;+\; 1/2\,O_2\,(g) \longrightarrow H_2O \qquad \text{Cell reaction}$$

As seen from electrode reactions recirculation of CO_2 from anode exhaust is essential for the operation of the cell. Operating temperatures of these fuel cell systems are between 500 to 700°C. No expensive catalysts such as Pt to activate the electrodes are needed at these temperatures. Internal steam reforming is possible. Hydrocarbons can be led directly to the cell where they are converted into hydrogen containing gases. Coke oven gas or coal and air are used as fuel and oxidant, respectively. CO does not poison the anodes while H_2S content of the incoming fuel is a serious problem as far as poisoning the electrode surfaces is concerned. Bipolar stack configuration is usually used. Cell endurance is a problem. Only about 5 000 hours of lifetimes have been achieved. Finding low cost materials in constructing these cells presents difficulties. Small power plants, based on MCFC systems have been developed. MCFC is shown schematically in Figure 2.

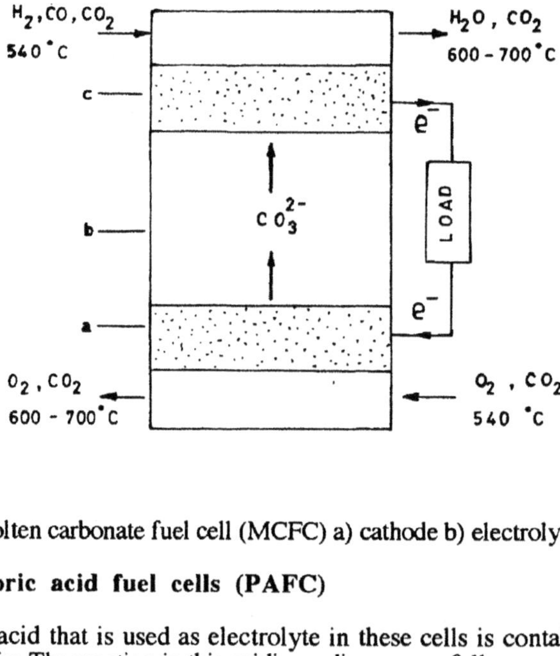

Figure 2. Molten carbonate fuel cell (MCFC) a) cathode b) electrolyte c) anode

4. Phosphoric acid fuel cells (PAFC)

Phosphoric acid that is used as electrolyte in these cells is contained in a thin silicon carbide matrix. The reaction in this acidic medium are as follows:

$$H_2(g) \longrightarrow 2H^+(aq) + 2e^- \qquad \text{Anode reaction}$$

$$1/2\, O_2(g) + 2H^+(aq) + 2e^- \longrightarrow H_2O \qquad \text{Cathode reaction}$$

$$H_2(g) + 1/2\, O_2(g) \longrightarrow H_2O \qquad \text{Cell reaction}$$

Electrolyte is stable and the conductivity is improved in the temperature ranges encountered in operating conditions. These cells operate with ambient air containing carbon dioxide. PAFC system can be used with reformed hydrocarbons and alcohols. Reforming is carried out in a separate system. Operating temperature of the system is around 170 - 200°C with reformed fuels. Fairly high operating temperatures minimizes catalyst poisoning by carbon monoxide. Stacks are of bipolar type. The lifetimes of about 40 000 hours have been achieved and the system is relatively low cost. PAFC system using diesel oil as fuel for military purposes with 1 - 5 kW power output and power plants based on PAFC using natural gas with 40 kW - 5 MW power output have been used. Japan has recently constructed the world's largest phosphoric acid based demonstration plant with 11 MW power output. Figure 3 shows the PAFC system schematically.

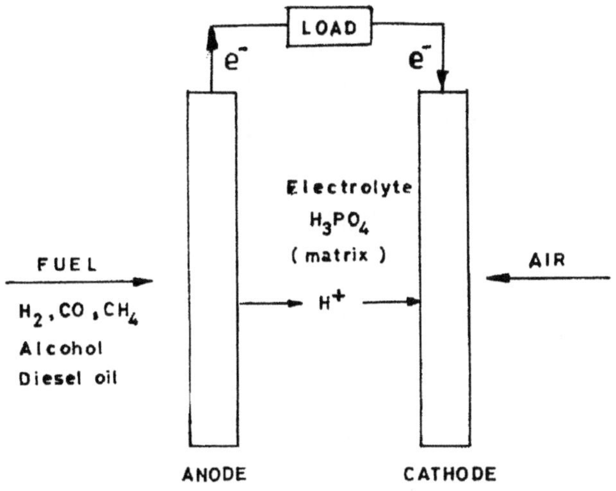

Figure 3. Phosphoric acid fuel cell (PAFC)

5. Proton exchange membrane fuel cell (PEMFC) or solid polymer fuel cells (SPFC)

These systems use sulphonated polytetrafluoroethylene (Nafion) as electrolyte. Since the utilized membranes are about 50 μm thin, significant size reduction is achieved. The ohmic losses are thus reduced and very high current densities can be produced consequently. The electrode reactions correspond to those discussed above for acid fuel cell systems. Carbon electrodes containing Pt as catalyst are pressed onto both sides of the proton exchange membrane. Hydrocarbon reformates can be used as fuel. Fuel must be free of carbon monoxide. Poisoning of Pt catalyst by carbon monoxide can be reduced by the addition of small quantities of air at the SPFC anode or by alloying Pt catalyst with various metals to increase its tolerance to poisoning. Another approach is to use thin palladium or palladium - silver membranes which can diffuse hydrogen through and is impermeable to other gases. The operating temperature of SPFC is about 100ºC. SPFC systems have also considerable commercial potential for transport applications as replacement for the internal combustion engines. The search for better anode catalysts with increased tolerance to carbon monoxide poisoning and cathode catalysts with better oxygen reduction performance and cheaper membrane electrolyte which can operate at higher temperatures continues. The schematic description of SPFC system is shown in Figure 4.

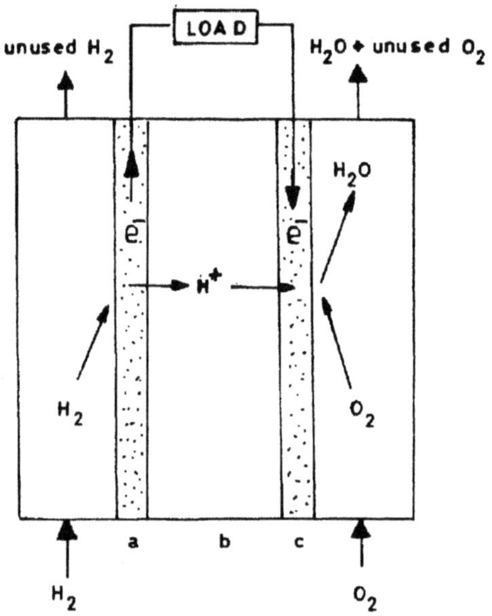

Figure 4. Proton exchange membrane (PEMFC) or solid polymer fuel cell (SPFC) a) anode (H_2 electrode) b) ion exchange membrane or solid electrolyte, c) cathode (O_2 electrode)

6. Solid oxide fuel cells (SOFC)

In SOFC system both electrodes and the electrolyte between them are made of solid oxide ceramic materials such as stabilized yttria or zirconia. These materials become O^{2-} ion conductor at temperatures above 800ºC. If the cell is operated at around 900ºC the following reactions take place;

$$a\,H_2\,(g) + b\,CO\,(g) + (a+b)\,O^{2-} \longrightarrow a\,H_2O\,(g) + b\,CO_2\,(g) + 2\,(a+b)\,e^- \qquad \text{Anode reactions}$$

$$1/2\,(a+b)\,O_2\,(g) + 2\,(a+b)\,e^- \longrightarrow (a+b)\,O^{2-} \qquad \text{Cathode reaction}$$

$$1/2\,(a+b)\,O_2\,(g) \;+\; a\,H_2\,(g) \;+\; b\,CO\,(g) \longrightarrow a\,H_2O\,(g) \;+\; b\,CO_2\,(g)$$ Cell reaction

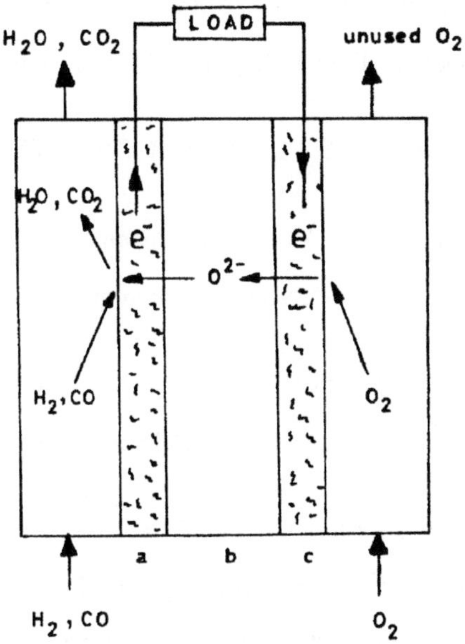

Figure 5. Solid oxide fuel cell (SOFC) a) anode (H_2 electrode) b) electrolyte c) cathode (O_2 electrode)

The cell is shown in Figure 5. Recirculation of CO_2 from anode exhaust is not necessary. The commercial goal is to develop a SOFC based power plants of 100 kW to 100 MW which are suitable for decentralized energy distribution. Natural gas, coal gas or vaporized liquid fuels can be reformed within the cell stack. Waste heat is also available. SOFC system offers the best compatibility with natural gas and therefore have the highest market potential in the near future.

7. References

Austin, L. 1967 'Fuel cells SA SP - 120, Office of Technology Utilization' Washington

Baker, B. S., 1965, 'Hydrocarbon Fuel Cell Technology' Academic Press., New York

Berger, C. 1968, 'Handbook of Fuel Cell Technology' Prentice - Hall, Englewood Cliffs. N.J.

Bockris, J. O'M. and Srinivasan, S., 1969, 'Fuel Cells, Their Electrochemistry' McGraw - Hill., New York

Hart, A. B. and Wormak, G. J., 1967, 'Fuel Cells' Chapman and Hall London

Justi, E. W. and Winsel, A. W., 1962, 'Kalte Verbrennung. Fuel Cells' Franz Steiner Verlag, Wiesbaden

Kordesch, K., 1985, 'Brennstoffbatterien' Springer Verlag. Wien - New York

Liebhafsky, H. A. and Cairns, E. J., 1968, 'Fuel Cells and Fuel Batteries' J. Wiley and Sons, New York

Linden, D., 1983, 'Handbook of Batteries and Fuel Cells' McGraw - Hill, New York

Mitchell, W., 1963, 'Fuel Cells' Academic Press, New York

Ullmann's Encylopedia of Industrial Chemistry, 1985, **A 12**, 55

Vielstich, W., 1965, 'Brennstoffelemente - Fuel Cells, Verlag Chemic' Weinheim

PROGRESS IN PEM FUEL CELL SYSTEMS DEVELOPMENT

F. BARBIR
Energy Partners, Inc.
1501 Northpoint Parkway, Suite 102
West Palm Beach, FL 33407, U.S.A.

Abstract. Proton Exchange Membrane (PEM) fuel cell stacks, consisting of up to 60 cells having an active area of 780 cm^2, have been designed, built, tested, and installed in fuel cell power systems for mobile and stationary applications. The differences between operational parameters in a single cell, stack of cells and fuel cell systems are discussed. The paper addresses the research, design, manufacturing and marketing issues pertinent to commercialization of PEM fuel cell technology.

1. PEM Fuel Cell

Proton exchange membrane fuel cells (PEMFC), also known as solid polymer electrolyte fuel cells (SPEFC), are very attractive as a power source for various transportation, marine and stationary applications. PEM fuel cells have no moving parts, they are quiet in operation, do not produce any emissions, are very efficient (usually about 50%), operate at relatively mild operating conditions (temperatures up to 90°C, and pressures up to 600 kPa), and are very responsive to load changes. The technology, however, needs further development from the standpoints of reliability, performance, manufacturability, and cost.

High energy efficiency and high power density are vitally needed for commercial applications of PEM fuel cells, in order to minimize fuel consumption, weight, volume and capital costs. These two requirements almost contradict each other. The maximum efficiency of a PEM fuel cell is achieved at very low power levels, and efficiency decreases almost linearly as power increases. A PEM fuel cell can operate at practically any efficiency below its maximum efficiency, which is usually about 65% (based on hydrogen lower heating value). Maximum power density in PEM fuel cells is obtained at cell voltages of 0.5 V or lower, corresponding to an efficiency of 33% and below. Hence, in this context, maximum power density has little practical value. A better measure of fuel cell performance would be power density at some practical efficiency level, for example at the efficiency of 50%, which corresponds to cell voltage of 0.64 V. Table 1 shows some reported single cell power densities at 50% efficiency [1-4].

The performance of a PEM fuel cell depends on its design characteristics and operational conditions. Design characteristics of a PEM fuel cell are:

Y. Yürüm (ed.), Hydrogen Energy System, 203–213.

- Cell size and shape. The PEM fuel cell can theoretically be made in any size or shape. However, because of technological reasons the size of the PEM fuel cells built to date vary from 5 cm^2 to 1,180 cm^2, and the shape is usually rectangular.

- Membrane type and thickness. The membrane material is perfluorosulfonic polymer. Nafion™ by DuPont, although not originally developed for fuel cell applications, is the only available commercial product. It comes in different molecular weights (1000 and 1100) and different thicknesses (50, 125 and 175 μm). Dow Chemical Company has produced a 100 μm thick experimental membrane that has a higher water retention capability and specific conductivity than those of Nafion. Very high power densities have been obtained with this experimental membrane [5].

- Electrode properties (such as catalyst layer thickness and composition, diffusion layer porosity, platinum loading, etc.). This is practically the most important information about a fuel cell, and usually it is considered proprietary. Different methods of catalyst layer preparation and its deposition either on porous and conductive support surface or directly on the membrane have been and are being developed, with a goal of improving the performance and reducing the cost. "Standard" platinum loading is considered to be 4 mg/cm^2, but extensive research has resulted in electrodes with as low as 0.05 mg/cm^2 [1,2,6-8].

Table 1 Some Reported Hydrogen/Oxygen Single Cell Performances

Reference	Power density @ 50% efficiency W/cm^2	Maximum power density W/cm^2	Membrane	Operational conditions
[1]	0.9	1.0	"C"	300/500 kPa; 80°C
[1]	1.0	1.4	Dow	300/500 kPa; 80°C
[2]	1.3	n.a.	Dow	400/500 kPa; 95°C
[3]	1.8	2.3	Dow	510/580 kPa; 95°C
[3]	0.9	1.0	Nafion 115	510/580 kPa; 95°C
[4]	2.2	2.3	Dow	780 kPa; 100°C

Operational conditions that affect the performance of a PEM fuel cell include:

- Cathode and anode pressure. The increased pressure, particularly on oxidant side results in a voltage gain, approximately given by the following equation [9]:

$$\Delta V = (2RT/F) \ln (P''/P')$$

where:
ΔV = voltage gain, V
R = gas constant = 8.31 J/mol °K
T = absolute temperature, °K
F = Faraday's constant = 96,500 As/mol

P'' = final pressure
P' = initial pressure

It is therefore advantageous to run a fuel cell at higher pressures. Usual operating pressures are between 300 and 500 kPa for hydrogen and oxygen, which are readily available as compressed gases. However, if air is used as the oxidant, an additional amount of energy is needed for its compression, and in that case an optimization study is required to find out the optimum operational pressure. Successful operations at nearly atmospheric pressures have been reported [10].

- Temperature. Higher temperatures result in lower values of Tafel slope and conductive resistance, higher values of exchange current density, and better overall performance of the fuel cell. Usual operating temperatures are about 80°C, although operation at 95°C has been reported [2-4].

- Humidity of reactant gases. Proton exchange membranes must constantly be maintained wet to retain their conductive properties. Hence, the reactant gases must be humidified before entering the cell. Ticianelli, *et al.*, have concluded that the optimum conditions for humidification are achieved at a temperature 5-15°C above the cell operating temperature [11]. However, Mosdale, *et al.*, reported PEM performance without any humidification [12].

- Purity of reactant gases. PEM fuel cells are not sensitive to CO_2, which only acts as a diluent. However, CO acts as a catalyst poisoning agent and may significantly affect the fuel cell performance, particularly at lower temperatures and pressures. A small amount of CO (< 10 ppm) in the anode gas may be tolerable. The PEM fuel cell can use air as oxidant. Since oxygen's content in air is only 21%, its partial pressure is accordingly lower, which affects the performance of the fuel cell (reduces cell voltage). In addition, nitrogen blocks the passage of oxygen to the electrode-membrane interface, thus creating significant mass transport problems and reducing maximum operational current densities.

2. Fuel Cell Stack

Although, most of the research is based on single cell experiments, a single cell has no practical value due to its low voltage (0.5-1.0 V). For any practical application a stack of cells connected in series is required. The number of cells in a stack is determined by the required voltage. Power, which is a product of voltage and current, determines the size of a cell, i.e, its active area. Although a PEM fuel cell is a very simple device with no moving parts, the processes inside the fuel cell stack are extremely complex. They include:

- flow of reactant gases through the stack and through the individual cells,
- diffusion of the reactant gases through the porous electrodes and membrane,
- electrochemical reactions at the electrode-membrane interface:
  $2H_2 \rightarrow 4H^+ + 4e^-$ at the anode, and
  $O_2 + 4H^+ + 4e^- \rightarrow 2H_2O$ at the cathode,
- hydrogen proton transfer through the membrane,
- electrical current collection and conduction,
- water vapor transport through the membrane (electro-osmotic drag and diffusion),
- water transport through the porous electrode,

- water and oxygen (or oxygen depleted air) flow out of the cell and stack,
- heat generation at the site of the electrochemical reaction and heat transfer throughout the stack.

A design of the fuel cell stack and its components must provide appropriate conditions for each of these processes, because they are all interrelated and equally important for efficient and reliable operation of the stack.

The factors that may affect the performance of a fuel cell stack, beside those that apply to a single cell, are:

- Voltage losses in stack. Although stack material is electrically conductive, some voltage losses result in long stacks mainly due to ohmic resistance.

- Reactant gases supply. Careful stack flow distribution design is required to ensure equal amount of reactant gases in each cell, and also uniform distribution of reactant gases over a large electrode area.

- Temperature distribution. In larger and longer fuel cell stacks it becomes more difficult to maintain uniform spatial and longitudinal temperature than in a small size single cell.

- Water management. Water removal from a cell in a multi-cell stack is more difficult than from a small single cell. Concentration of water near the membrane/cathode interface increases in the direction of water flow, which may cause flooding of the lower parts in large area cells.

A few companies have published limited information on building PEM fuel cell stacks and their performances. General Electric pioneered the early work during 1959-1982. In the mid-1980s, GE transferred the technology to United Technologies, and before this a licence for the basic know-how was transferred to Siemens AG in Germany.

Ballard Power Systems, Inc. has incorporated the latest version of the Dow experimental membrane into a production fuel cell stack, which consists of 35 active cells and an integral gas humidification section [13]. The active electrode area is 232 cm^2. At an air pressure of 440 kPa the stack produces 5 kW, and if oxygen at 300 kPa is used it produces over 10 kW of power (that corresponds to a power density of 1.2 W/cm^2). Ballard has demonstrated stack operation in excess of 2000 hours with the Dow membrane without significant degradation in performance [13].

Siemens has built a 42-cell stack, with a large cell active area of 1,180 cm^2, using Nafion 117 membranes. At rated load of 650 A, the cell reaches a voltage of 30 V (that corresponds to a power density of 0.4 W/cm^2), operating on oxygen (230 kPa) and hydrogen (200 kPa) at 75°C. By using a Dow membrane, much higher power densities were obtained at the same operational conditions (> 1 W/cm^2) [14].

Energy Partners has developed the PEM fuel cell stacks consisting of up to 60 cells, with an active area of 780 cm^2. PEM fuel cell technology transfer occured as a result of the acquisition of Treadwell Fuel Cell Division in 1992.

The following is a description of a typical PEM fuel cell stack. The cells are sandwiched between the so-called collector plates. The plates have a dual role: they connect the cells electrically in series and they physically separate the cells (and the gases in the cells) from each other. Each plate has the embedded hydrogen and oxygen flow fields on opposite sides. The plate accommodates a membrane electrode assembly with an active area of 780 cm^2 made of Nafion™ membranes (117, 115 and most recently 105) and in-house prepared electrodes. The stack contains an internal humidification section for the hydrogen and oxidant inlet streams. Cooling of the stack is provided by water which circulates through cooling cells, strategically distributed through the active section of the stack. The stack contains up to 60 active cells, plus an adequate number of cooling and humidification cells, and it is held together by two aluminum end plates bolted together with tie-rods. All the connections for hydrogen, oxidant and water are provided on one of the end plates. Electrical current is drawn from the two gold-plated bus plates placed at both ends of the stack's active section. Figure 1 shows a fuel cell stack.

The main feature of the EP fuel cell stack is a molded graphite collector plate. The plate is made out of graphite and binder molded under pressure in a two stage process. The molded plate has about 90% of its features already embedded on its surfaces. The same molded collector plate is used for twelve different functions within a multi-cell stack; the functions are obtained by positioning the manifolds, which is made by machining. This process has reduced the manufacturing cost from \$5,000 to \$100 per plate, and it is very well suitable for mass production.

Several stacks with 5 to 60 active cells have been assembled and successfully tested in hydrogen/oxygen and hydrogen/air operational modes. Figure 2 shows a typical I-V curve for a full size cell operating in a multi-cell stack. Operational power densities of 0.45 W/cm^2 in hydrogen/oxygen operation and 0.2 W/cm^2 in hydrogen/air operation have been obtained. Table 2 provides basic information on the latest EP 25-cell stack.

When a fuel cell stack operates on air it needs 5-8 times more oxidant flow than in the case when pure oxygen is used. Large flows require high pressures to push that amount of air through the stack. It has become apparent that a drastically different new design is required for air breathing fuel cell stacks, which would allow for much more air flow at much lower pressures.

Table 2 PEM Fuel Cell Stack Specifications (Hydrogen/Oxygen)

Number of active cells	25
Cell active area	780 cm^2
Membrane	Nafion 115
Rated power	6 kW
Efficiency at rated power (LHV)	0.55
Power density at 50% efficiency	0.45 W/cm^2
Operating voltage range	25 V - 16 V
Operating pressure range	300-500 kPa
Operating temperature	70°C
Dimensions	36×36×30 cm
Weight	65 kg

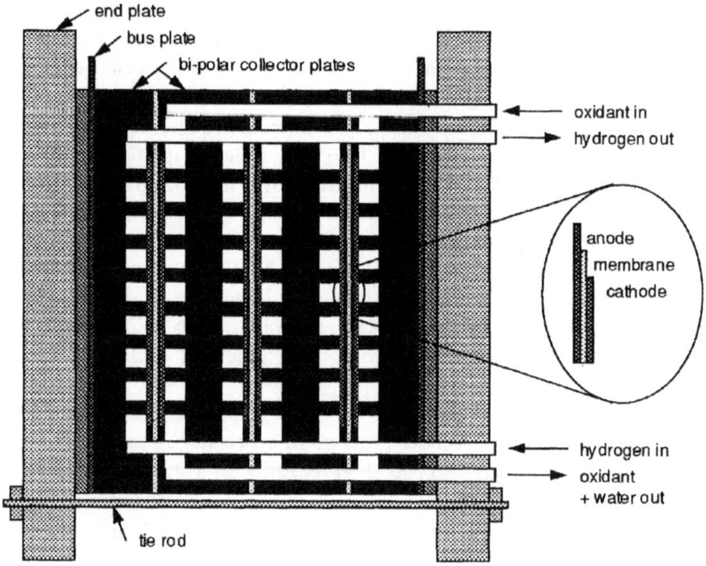

Figure 1. Schematic of a Typical PEM Fuel Cell Stack

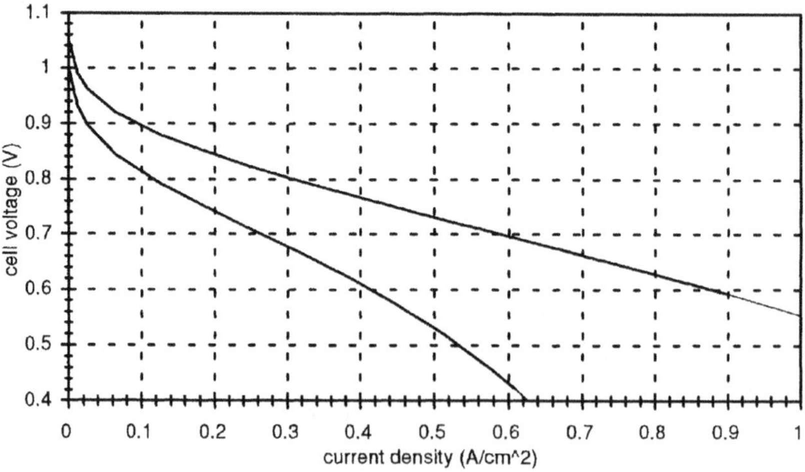

Figure 2 Polarization curves for O_2 and air operation of a typical fuel cell with active area of 780 cm^2 in multi-cell stack (at 400 kPa and 70°C)

3. Fuel Cell System

A fuel cell stack needs a supporting system for fuel and oxidant handling, water management, power conditioning, monitoring, control, safe operation and emergency shut-down. EP has developed and patented a closed loop reactant/product management system for PEM fuel cells, which was applied in a first commercial fuel cell powered submarine PC1401 in 1989 [15]. The main feature of this system is closed loop operation. Both, unused fuel and oxidant are recirculated from the fuel cell outlet back to the inlet by means of recirculating pumps. This feature provides fuel cell operation at higher stoichiometric ratios, but the system operates at exact stoichiometric consumption of hydrogen and oxygen. One modification of this system is hydrogen operation in dead-end mode without recirculation and without exhaust (with only intermittent purging). Water also circulates in a closed loop, and it has dual function: (i) cooling of the stack, and (ii) humidification of reactant gases. Product water and condensate from exhaust streams are collected and used for these purposes. Power requirement to run the pumps, solenoid valves, sensors, and controller is less than 10% of the fuel cell output.

The system has recently been improved and adapted for various mobile and stationary applications. In an automobile, called EP Green Car, the system was modified to operate with ambient air as oxidant [16]. Hydrogen operates in dead-end mode with intermittent purging of inerts. Air flows through the fuel cell stacks pushed by a compressor. A compressor significantly increases parasitic power requirements (up to 25% of the fuel cell power output).

Most recently, several sophisticated fuel cell systems have been designed and built for laboratory research. Figure 3 shows a functional diagram of a fuel cell system. It can operate using either oxygen or air (or any combination of oxygen and nitrogen) in various operational modes:

- hydrogen: closed loop, dead end and flow through
- oxygen: closed loop and flow through
- air: flow through only.

The system is fully instrumented with measurements of reactants and water inlet and outlet pressures and temperatures, reactants flows, stack current and voltage, and individual cell voltage. Design operating pressure is 250-400 kPa, and operating temperature is 70°C (measured at cooling water outlet). The system is equipped for automatic operation. The start-up of the system, monitoring of its performance, and shut down in the event that any of the monitored parameters exceeds a prescribed envelope, is provided by a programmable controller. The system successfully supported a fuel cell operation at 8.0 kW power level. Total electrical power requirement for the fuel cell support systems does not exceed 0.6 kW. The efficiency of the system at maximum power, defined as net electricity generation (fuel cell output minus auxiliary power) divided by hydrogen energy input (lower heating value) was 48% .

4. Further Development and Prospects for Commercialization

PEM fuel cell technology is already sufficiently advanced to go into the commercial phase. Potential markets include stationary power generation (remote stand-alone power systems, dispersed power generation and co-generation), transportation (automobiles, fleet vehicles, buses, trains, ships, submarines) and special - military applications.

The main obstacles for widespread applications are the cost of fuel cells and fuel cell systems, and lack of hydrogen infrastructure. Today, the production cost of a fuel cell stack is about

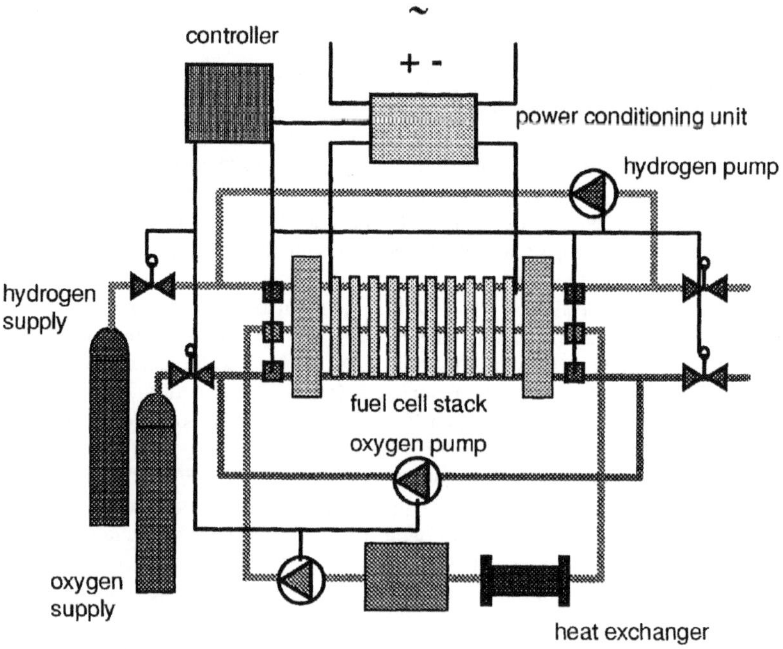

Figure 3 Functional Block Diagram of a Fuel Cell System

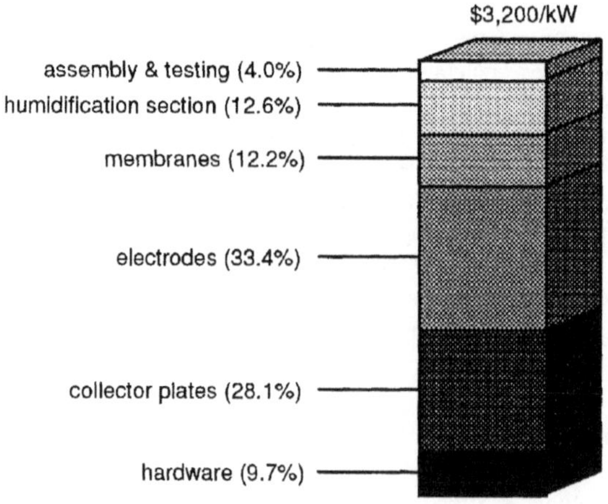

Figure 4 Production Cost Breakdown of a 10 kW PEM Fuel Cell Stack

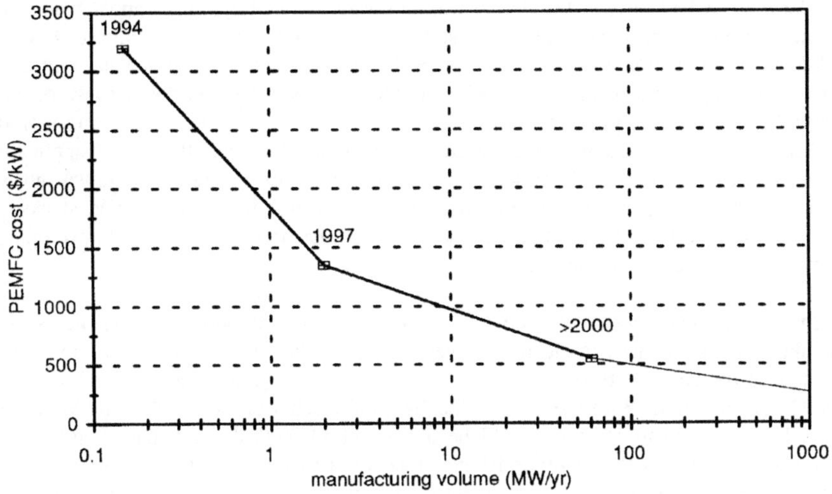

Figure 5 Projections of PEM Fuel Cell Stacks Costs

$3,200/kW for a 10 kW unit, but the cost of an entire fuel cell system is about twice as much. Current Energy Partners' production capabilities are about 150 kW/yr. Figure 4 shows the breakdown of production costs of a typical PEM fuel cell stack. Collector plates and electrodes are the most expensive parts of the fuel cell stack. Energy Partners has already developed a process suitable for mass production of collector plates, and another process for mass production of membrane-electrode assemblies is currently under development (the patent application has already been filed). Hence, it is reasonably to expect that with increased manufacturing volume and improvement in performance (increased power density) the cost of PEM stacks will be reduced, as shown in Figure 5. Figure 5 also shows a time-frame in which this development and commercialization can be achieved.

In order to achieve those cost projections, further research and development is required, particularly in membrane and electrode assembly. Low platinum loading (< 0.1 mg/cm^2), high performance (power density of > 1 W/cm^2 at 50% efficiency), and simple membrane-electrode assembly are required. With the cost of the fuel cell stack well below $1,000/kW, the cost of fuel cell systems up to 50 kW should be at around $1,000/kW, and that would make these systems competitive with conventional small power generation systems, particularly if other advantages of PEM fuel cells are taken into account, such as high efficiency, zero emissions, quiet operation, quick start-up and fast response to change of load. However, for automobile applications, even lower costs would be required.

The PEM fuel cell runs on hydrogen and oxygen or air. Hydrogen is not (yet) readily available as a commercial fuel (with space applications being the sole exemption). Proponents of hydrogen energy envision large scale hydrogen production from direct and indirect forms of solar energy and its widespread applications as a clean and inexhaustible fuel. If such an energy future is inevitable, in a transition period hydrogen could be produced from natural gas, either by steam reforming - an already commercially available process, or by thermal decomposition [17]. Many energy specialists forecast that natural gas will become the most used energy source in the near term future [18]. The cost of hydrogen from natural gas, produced in large quantity, should not exceed $10/GJ - a price already competitive with other synthetic fuels. Taking into account its high energy conversion efficiency and cleanliness in fuel cells, hydrogen may soon become a fuel of choice, and fuel cells will be the technology that will enable hydrogen's widespread applications.

References

1. Wilson, M.S., and S. Gottesfeld, High Performance Catalyzed Membranes of Ultra-Low Pt Loadings for Polymer Electrolyte Fuel Cells, *J. Electrochem. Soc.* Vol. **139**, No. 2, pp. L28-L30, 1992

2. Srinivasan, S., O.A. Velev, A. Parthasarthy, D.J. Manko, and A.J. Appleby, High Energy Efficiency and High Power Density Proton Exchange Membrane Fuel Cells - Electrode Kinetics and Mass Transport, *NASA Conf. Publ.* 3125 (Space Electrochem. Res. Technol.), pp. 101-122, 1991

3. Murphy, O.J., Lyntech, Inc., private communication, December 1992

4. Watkins, D.S., Research, Development, and Demonstration of Solid Polymer Fuel Cell Systems, in L.J.M.J. Blomen and M.N. Mugerwa (eds.) *Fuel Cell Systems*, Plenum Press, New York, 1993

5. Ballard Power Systems, Inc., *The Ballard Fuel Cell*, brochure #700.306.008

6. Taylor, E.J., E.B. Anderson, and N.R.K. Vilanbi, Preparation of High-Platinum-Utilization Gas Diffusion Electrodes for Proton Exchange Membrane Fuel Cells, *J. Electrochem. Soc.*, Vol. **139**, No. 5, pp. L45-L46, 1992

7. Wilson, M.S., and S. Gottesfeld, Thin-Film Catalyst Layers for Polymer Electrolyte Fuel Cells, *Journal of Applied Electrochemistry*, Vol. **22**, pp. 1-7, 1992

8. S. Srinivasan, D.J. Manko, H. Koch, M.A. Enayetullah, and A.J. Appleby, Recent Advances in Solid Polymer Electrolyte Fuel Cell Technology with Low Platinum Loading Electrodes, *Journal of Power Sources*, Vol. **29**, pp. 367-387, 1990

9. Dhar, H.P., On Solid Polymer Fuel Cells, *J. Electroanal. Chem.* Vol. **357**, pp. 237-250, 1993

10. Lehman, P.A., C.E. Chamberlin, T.G. Herron, and R.M. Reid, Use of a Proton Exchange Membrane Fuel Cell for Standby Power in a Stand-Alone Energy System, presented at the International Conference on Fuel Cells, Long Beach, CA, February 22-25, 1994

11. E.A. Ticianelli, C.R. Derouin, A. Redondo and S. Srinivasan, Methods to Advance Technology of Proton Exchange Membrane Fuel Cells, *J. Electrochem. Soc.* Vol. **135**, No. 9, pp. 2209-2214, 1988

12. Mosdale, R., P. Stevens, F. Novel-Cattin, B. Loppinet, G. Gebel, P. Aldebert, and m. Pineri, Water Management in a Solid Polymer Electrolyte Fuel Cell, Proceedings of the European Space Power Conference, Florence, September 1991, ESA-SP-320, pp. 479-483, 1991

13. Prater, K.B., Solid Polymer Fuel Cell Developments at Ballard, *Journal of Power Sources*, Vol. **37**, pp. 181-188, 1992

14. Strasser, K., Mobile Fuel Cell Development at Siemens, *Journal of Power Sources*, Vol. **37**, pp. 209-219, 1992

15. Perry, J.H., Jr., A. Person, S.M. Misiaszek, D.P. Alessi, Closed Loop Reactant/Product Management System for Electrochemical Galvanic Energy Devices, U.S. Patent No. 5,047,298, 1991

16. M. Nadal and F. Barbir, Development of a Hybrid Fuel Cell/Battery Powered Electric Vehicle, prepared for 10th World Hydrogen Energy Conference, Cocoa Beach, FL, June 20-24, 1994

17. Muradov, N.Z., How to Produce Hydrogen from Fossil Fuels without CO_2 Emission, *Int. J. Hydrogen Energy*, Vol. **18**, No. 3, pp. 211-216, 1993

18. Howell, D.G., (ed.), *The Future of Energy Gases*, U.S. Geological Survey Professional Paper 1570, U.S. Government Printing Office, Washington, 1993

HYDROGEN FUELED TRANSPORTATION

WM. D. VAN VORST
Professor Emeritus, Department of Chemical Engineering
University of California, Los Angeles
Los Angeles, CA, 90024-1592
USA

ABSTRACT. The literature reveals periodic interest in hydrogen as an engine fuel since it was used in the first self sustaining combustion engine in 1820. Current interest is driven by the need for drastic reduction in the noxious pollutants emitted with fossil fuels, as well as the fact that fossil fuels are non renewable and must, inevitably, become unavailable. There are, of course, severe problems to be overcome prior to general acceptance of hydrogen fueled engines. These include: the economics of large scale production and distribution of hydrogen, the storage of hydrogen onboard the vehicle, and the development of the infrastructure necessary for the services for its vehicular usage. Safety is not the problem it is perceived to be by the public. Hydrogen may well play a highly significant role in surface transportation (including rail), and is destined to play a dominant one in high speed air travel.

1. Introduction

1.1 THE ADVANTAGES OF HYDROGEN AS AN ENGINE FUEL

Hydrogen is generally recognized as a non-polluting fuel. Clearly, its emissions will not contain significant amounts of CO_2, CO, or unburned hydrocarbons; the only possible source of such contaminants is the lubricating oil used, which has been assumed responsible for trace amounts found on occasion. Since hydrogen will be burned with air in the great majority of applications, oxides of nitrogen will be generated in various amounts, depending on the flame temperature utilized. This, however, may prove not to be as serious as in the case of a carbon-base fuel, as other components of the photo-chemical reactions will not be generated simultaneously. At the same time, it must be recognized that biogenic hydrocarbons will still contribute to the photochemical synthesis of undesirable pollutants, and that such production may be limited only by the presence of oxides of nitrogen. It must also be mentioned that hydrogen's properties make it an excellent choice for use in fuel-cells, and that no oxides of nitrogen would be produced at the low operating temperatures possible with hydrogen.

In addition to its non-polluting nature, hydrogen also offers advantageous combustion properties. Its flammability limits offer a wide range of fuel/air ratios over which combustion may take place so that operation at both very lean and rich conditions may take place, though there seems to be no practical incentive for rich operation. The energy required for its ignition is quite low, so that starting under a variety of conditions is not a problem. Its flame velocity

Y. Yürüm (ed.), Hydrogen Energy System, 215–239.

is high, as is its diffusivity in air, so that combustion may take place quickly, allowing a closer approximation to the ideal, instant combustion of the Otto cycle,-and others utilizing the concept of constant volume combustion. Finally, hydrogen offers the highest heat of combustion per unit of mass of all fuels.

1.2 THE PROBLEMS WITH HYDROGEN AS A FUEL

Certain of the advantages of hydrogen as an engine fuel set forth above also imply potential problems. The ease of ignition, for example, may result in an undesirable hydrogen combustion being initiated at an inopportune time! Further, the extremely low density of hydrogen either as a gas or liquid makes its heat of combustion on a volume basis very low; thus excessive space is required for its storage onboard a vehicle. And, of course, hydrogen does not appear in its elementary state (in any useful quantity) in nature, so that it must be produced in great amounts if it is to be used to any significant degree in transportation. Thus, three major categories of problems in connection with the use of hydrogen as a fuel for transportation emerge: safety, production, and storage.

There follows a general discussion of the factors noted above, after a review of earlier work and that currently in progress devoted to improving the performance of hydrogen as a fuel, and to solving the problems associated therewith.

2. Review of Earlier Work

2.1 THE NINETEENTH CENTURY

Although our subject matter centers about the use of hydrogen as a fuel for transportation vehicles, it is of interest to dwell briefly on the history of the use of hydrogen as an engine fuel in general. Surprisingly enough, it was the first fuel used in a combustion engine,- or, at least, the first so recorded. This was the work of Reverend William Cecil in England in the early 1800's, described in his paper presented to the Cambridge Philosophical Society in 1820.
Cummins (1976) refers to it as the first "self - sufficient" internal combustion engine. To appreciate fully the true novelty and potential significance of Cecil's contribution, it need only be realized that the "engines" of the day were the water-wheel, and the still - primitive steam engine. Cecil noted the severe limitations these devices: one could only be located near a naturally abundant supply of water, the other required such an excessive time to warm up to produce useful power that the overall economy of its use was adversely effected. Thus, in Cecil's view, the incentive to devise an engine free of these drawbacks. We are not in an entirely dissimilar position today. We are all too dependent on hydrocarbon fuels, and critically concerned with their ultimate exhaustion, as well as the violence their usage has done, and will continue to do, to our atmosphere. Hence our advocacy of the most serious consideration of the use of hydrogen.

Cummins (1976) notes that Barsanti and Mateucci, in Italy, developed a free - piston engine in the 1850's using hydrogen as the fuel. But this, too, predates the internal combustion engine that made possible the transportation system that we know today. The distinction of pioneering the forerunner of today's automotive engine must go to Nicholas Otto, who, with his partner, Langen, in Paris in 1867 demonstrated the first engine to use the Otto cycle, as it is now known, which lead to the automotive engine that has had such a profound effect on transportation. While Otto did not use pure hydrogen as his fuel, he did use a "town gas" of high hydrogen content-probably greater than 50%. Incidentally, in later years, Otto attempted to use a liquid fuel; results were explosively disastrous, and he resumed work with gaseous fuels, considering liquid fuels too dangerous for use. Around the turn of the century, however, the invention of the carburetor made the use of liquid fuels practical, and benzine, or gasoline became the fuel of choice for such engines. It is worthy of note that a fuel once thought intolerably dangerous is used routinely today. There would seem to be no reason to doubt that hydrogen could be used routinely and safely when it becomes a common fuel for use in transportation vehicles.

Though there was further work on internal combustion engines in general during the nineteenth century, it was not with hydrogen as the fuel, and we move into the twentieth century to continue chronicling the early contributions to the use of hydrogen in internal combustion engines

2.2 THE PERIOD 1900 - 1950

The first half of the twentieth century saw great developments in the internal combustion engine. In the 1920's, just a century after the Reverend Cecil's endeavors, considerable work was done with hydrogen as the fuel. Ricardo, in England, seems to have initiated the interest in hydrogen, and Burstall of Cambridge University reported investigations in 1925. Both reported indicated thermal efficiencies of over 40%, and both noted the problem of backflash, or firing back, or as Ricardo put it, "popping back into the carburetor", which, of course, limited performance. About the same time, Rudolph Erren, in Germany, was converting engines to operate with hydrogen, and devising methods of fuel injection to overcome the problem of backflash. Weil (1972) mentions that considerable, serious thought was given to the use of lighter-than-air craft under construction in England and Germany. Since hydrogen was to provide the necessary buoyant force, there would be no additional storage problem. It seemed an appealing and promising suggestion. Ricardo (1968) also alludes to it in his autobiography, and allows the inference that political considerations rather than technical problems overruled it. Erren continued his efforts in England during the 1930's, working with Hastings-Campbell. Their 1933 paper remains good reading today. In addition to its technical content, it stresses the non-polluting nature of hydrogen, as well as the need to become independent of foreign sources of fuel. In particular, it suggested the conversion of buses to run on hydrogen, which would be stored in banks of cylindrical tanks at a relatively high pressure. Erren later developed such a bus, as he continued to convert cars and trucks to run on hydrogen. Hoffman (1981) reports that Erren estimated that he had made over 1,000 such conversions.

Interest in hydrogen as an internal combustion engine fuel continued through the decade of the 1940's, much of it stimulated by military needs arising in World War II. A hydrogen-oxygen system, for example offered the possibility of submarine and torpedo propulsion free of a traceable wake. The idea, apparently due to Erren, seems to have been rejected by the British Admiralty, but pursued by the German navy. The work was done in secrecy, of course, and technical reports have not appeared in the open literature. Hoffman (1981) tells of an interview with Erren (who was interned by the British during the war) in which the latter spoke of German submarines captured by the allies, which were using the system with considerable success. De Boer (1976) has called attention to the efforts of Oemichen (1942) in Germany. The major thrust of this work was directed toward solving the backflash problem via direct cylinder injection of the hydrogen.

2.3 FROM 1950 TO 1990

Interest in hydrogen seems to have waned considerably after World War II, with the return of plentiful, and comparatively cheap gasoline. R.O.King, then at the University of Toronto, carried out an extensive investigation of hydrogen performance, especially on the problems of preignition and backfire. King had been present at the 1933 presentation of the Erren-Campbell paper, and had participated actively in the discussion of it. He served as a Captain in the British Air Ministry during the war, moving to Canada afterward.

The United States space program of the 1960"s led to the development of considerable hydrogen technology in rocket engines, particularly with the Apollo program. Incidentally, the remarkable safety record of this program should be noted here, and serve to reassure those fearful of any use of hydrogen. During the '60s, a key contribution to the potential use of hydrogen was made by Reilly and Wiswall (1968) with their work on the use of metallic hydrides as a storage medium for gaseous hydrogen.

A marked increase in research and development directed toward the potential use of hydrogen as an engine fuel for vehicular use took place in the 1970's. Indeed, MacCarley (1975) tabulated approximately fifty major research centers so involved, either with direct or supporting activity. This was probably due to two factors: increasing concern over air pollution, and the verification of the major role of automotive emissions in generating it; and, the great reduction in petroleum shipments to the United States. The latter was responsible for much of the financial support then available. And, it may be argued that 1972 was a turning point, - or the stimulus for one. The Urban Vehicle Design Contest was held in that year, attracted many entries, and publicly demonstrated the viability of the hydrogen car concept. The entries of UCLA, largely the work of Lynch, Finegold, and Baker and of Brigham Young University, largely reflecting the efforts of Roger Billings were notably successful. In addition to the novelty of cars actually running on hydrogen and performing the standard road tests, the contest offered an appreciation of the reduction in air pollution possible with the use of hydrogen. At a time when the proposed automobile emission standards for the State of California (probably the most severe in the world at the time) were 1 gram per mile for unburned hydrocarbons, 2.4 for carbon monoxide, and 1.5 for oxides of nitrogen the actual figures for the UCLA car were 0.09, 0.15, and 0.04 respectively. Ironically, it was later recognized that these figures were probably higher that the actual, since sampling was done in an enclosed facility and reflected background concentrations. Recognizing the billions of dollars annual damage done to agriculture and forest products by air pollution due in large measure to automotive emissions, one can not resist speculating about the cost of **not** using hydrogen as an engine fuel.

The 7th Intersociety Energy Conversion Engineering Conference was held in San Diego, California in 1972, and also contributed to stimulating interest in the hydrogen fueled engine, devoting a session to the subject. The late Professor Kurt Weil reacquainted the technological community with hydrogen engines and speculated on promises for the future in recognition of the emerging need for improvement, nationally and internationally, in the energy/transportation/environment system. The (modern) pioneering work of Murray and Schoeppel of Oklahoma State University and that of Swain and Adt of Miami University was presented.

UCLA's work with hydrogen engines continued with emphasis on the greater efficiency obtainable with hydrogen, the recycle of exhaust gas to reduce emissions of the oxides of nitrogen, and the development of a cryogenic, liquid hydrogen system. Buchner and Saufferer of Daimler-Benz also made significant contributions, as did Peschka and Carpetitis, working with BMW. The former emphasized the use of metallic hydride storage systems, and demonstrated their promise by modifying several Mercedes vehicles to utilize them. The latter devised a cryogenic system to hold liquid hydrogen, and modified a BMW automobile to operate in this mode; they also developed a liquid hydrogen "filling-station" pump to demonstrate the potential ease of filling the automobile"s fuel tank with liquid hydrogen. Not only were these contributions important in themselves, but the fact that Daimler-Benz and BMW supported them gave credibility to the ultimate practicality of hydrogen fueled vehicles.

Yet another event took place in the '70s that is worthy of note here. This was a rally held in 1975, designed to demonstrate the fuel economy of various engine-fuel systems. It started at the Canadian border on the West Coast of the U.S., and terminated in Los Angeles, covering a distance of 2800 km. What makes it noteworthy from our viewpoint is that one entry was fueled with liquid hydrogen. This was a vehicle developed by Professor Furuhama and his group at the Musashi Institute of Technology in Tokyo. As Furahama stated, the event was not without problems, but the fact that his car completed the course satisfactorily gave much needed evidence that the hydrogen car concept was much more than a laboratory curiosity. It was an accomplishment that should have been much more widely recognized, but unfortunately, the rally was not widely publicized. Fortunately, Furuhama's group has been able to continue work on liquid hydrogen fueled engines and cryogenic storage systems; they have also developed a liquid hydrogen fuel pump, with the objective of injecting hydrogen still in the liquid state into the engine. Other work in Japan includes that of Ikegami at Kyoto University on the operation of Diesel engines with hydrogen. There had been some question about the suitability of hydrogen as a fuel for the Diesel cycle due largely to uncertainties about its autoignition characteristics and the higher pressures involved. De Boer (1976) of Cornell University investigated the autoignition phenomenon specifically using a single cylinder, variable compression engine. He reported no autoignition up to compression ratios as high as 29, but found this no deterrent to the use of hydrogen in the Diesel cycle, as he reported that ignition took place readily with a "glowplug" device providing the ignition source. Ikegami's work further demonstrated the feasibility of Diesel engine operation with hydrogen.

About the middle of the '70s, the availability and cost of gasoline in the U.S. returned to normal, and financial support for work on hydrogen fueled engines diminished significantly. There were, fortunately, notable exceptions. Swain (1983) conducted an extensive study of the performance of some nineteen engines that had been modified to run on hydrogen. The overall objective was to develop a broad data base relative to the engine of the future, designed to operate with hydrogen - in contrast to gasoline engines modified to use hydrogen. Daimler - Benz also made a major contribution during this period. Five passenger cars and five delivery vans, each powered with hydrogen fueled engines, were put into regular service, in Berlin, with two commercial fleet- operations. While the vehicles were serviced by the Mercedes-Benz Berlin branch, and hydrogen supplied at a central fueling station, the vehicles were put on regular duty. The results of one and one-half years testing were quite positive, as reported by Feucht and co-workers (1988), and has served to stimulate interest in hydrogen powered vehicles globally. Mention must also be made of the efforts of Bockris in establishing a hydrogen research center at Texas A&M University. Now the Center for Electrochemical Systems and Hydrogen Research, it has been particularly active, and successful, in fuel cell research.

2.4 NON-AUTOMOTIVE APPLICATIONS OF HYDROGEN AS A FUEL

Although section 2.3, next above, might leave the impression that only automotive utilization of hydrogen has been considered, this is, in fact, not the case. Attention has been devoted to the use of hydrogen fueled engines for aircraft and for trains. Indeed, the first issue of the first volume of the *International Journal for Hydrogen Energy* contains an article by Brewer (1976) on the prospects and problems relative to aviation usage of liquid hydrogen fuel. He concluded that the potential gains significantly offset the problems (primarily modifications in aircraft design and structure, and development of airport facilities to handle liquid hydrogen). Long range, supersonic aircraft would benefit greatly from liquid hydrogen usage, while the gain for subsonic aircraft would be more modest. Momenthy (1977), however, found the prospects more promising for the subsonic case, and set forth requirements for a fuel subsystem for liquid hydrogen aircraft. On the other hand, Mikolowsky and Noggle (1978) concluded that liquid hydrogen would not appear promising for use in subsonic military aircraft, largely on the basis of cost-effective and energy-effective analyses. They noted that the opposite conclusion might be drawn in the case of supersonic, or hypersonic aircraft for military transport, but felt such usage too far in the future to allow realistic analysis at the time.

Since weight considerations dictate the use of liquid hydrogen as an aircraft fuel, the servicing of such aircraft would be a significant problem - the solution of which would most likely involve extensive modification of airport facilities. Brewer (1978) and Korycinski (1978) each addressed this, and concluded that present day technology afforded an acceptable solution. Estimated costs were not out-of-line with those expected for major modifications at large air terminals. Studies were made based on the San Francisco and the Chicago O'Hare airports, and cost estimates were $304 and $469 million, respectively. Adler (1987) reported the findings of the "Swiss Group" of the Ad Hoc Executive Group (AHEG) on the possibility of the Zurich airport accommodating hydrogen fueled aircraft The AHEG was formed in 1979, largely by executives of aircraft companies, international airlines, government agencies, industry and research establishments to investigate the feasibility of liquid hydrogen fueled aircraft. The study was based on the refueling of one cargo aircraft per day with 15 to 30 tons of liquid hydrogen. The study was extensive, centering about the operation of a hydrogen liquification plant at the airport, and modifications necessary to pump the liquid into the plane. It was concluded that the Zurich airport would be highly suitable for a pilot plant operation at this level, but that commercial operation would demand innovations to reduce the economic penalty of the liquefaction expense; limitations of the Zurich airport would be difficult to overcome for a large scale operation, as well.

The propulsion unit for aircraft would not, of course be the engines referred to in the foregoing sections. They would be of the gas turbine type, in all probability supplying gas for expansion through a jet propulsion nozzle. The question then arises as to hydrogen's suitability for fuel in the continuous flow combustion process characteristic of gas turbine, or jet propulsion engines. Earlier work would imply a resounding endorsement for hydrogen in such an environment. Mulready (1964) in contributing a chapter on Liquid Hydrogen Engines to Scott's "Technology and Uses of Liquid Hydrogen" reveals that a jet propulsion engine was run on liquid hydrogen as a project of the U.S. Air Force in 1957, with highly successful results. Tsujikawa and Sawada (1982,1984) studied the use of hydrogen in gas turbine combustion throughout the 1980s, and found it an excellent fuel. Tsujikawa and Hirano (1988) suggested the use of liquid hydrogen to precool incoming air in order to reduce the compressor work required, thereby increasing the cycle efficiency. Their thermodynamic analysis indicated this to be a practical and promising scheme.

The utilization of hydrogen for locomotive engines has not generated a degree of interest comparable to that in automobiles (including trucks and busses) and aircraft. Still, some attention has been paid the subject, perhaps the first being that of Geir (197?). BLAH- Steinberg and Scott (1984) have made a computer simulation study of a diesel-electric locomotive engine powered by hydrogen, and a locomotive powered by an alkaline fuel cell. The results were used primarily for economic studies, and were considered preliminary. Simulation studies were continued by Jones, et al (1985), focusing more on the potential alkaline fuel cell usage. The study makes an impressive case for consideration of the hydrogen fuel cell, should political, economic or environmental factors result in the elimination of diesel fuel usage or electrification. The possibility is especially interesting in that the storage problem alluded to earlier (section 1.2) is largely alleviated since railroad operation is not limited by weight or volume to the degree found in other modes of transportation.

With regard to the possible use of hydrogen in marine vessels, the military usage for submarine propulsion has been discussed earlier. There has been little incentive to study surface ship usage of hydrogen; the current increase in legal emission standards for such vessels, however, may generate more in the near future.

2.5 A NOTE ON THE REVIEW OF EARLIER WORK.

The foregoing review has been intended to convey an appreciation of the extent and scope of the work that has been done on hydrogen fueled internal combustion engines. For a more specific review of engine performance *per se*, reference is made to DeLuchi (1989) and Das (1990).

3. Hydrogen as a Fuel

The foregoing review has alluded to benefits from the use of hydrogen as an internal combustion engine fuel, as well as some problems encountered in doing so. It is appropriate at this point, then, to discuss each of these in a more quantitative manner.

3.1 SOME PROPERTIES OF HYDROGEN, AND THEIR SIGNIFICANCE.

Table 1 shows the properties of particular interest of several fuels, although we are, perhaps most interested in the comparison of hydrogen with gasoline, simply because of our greater familiarity with the latter. Consider first the autoignition temperature. For the ordinary, carbureted, spark ignition engine, the lower this value the more limiting it is. For the engine efficiency is dependent on the compression ratio, which in turn dictates the temperature of the fuel-air mixture. The implication is that a spark ignition engine could operate at a compression ratio with hydrogen approximately 50% greater than that with gasoline, with the consequent increase in efficiency. Incidentally, there is some question about the autoignition characteristic of hydrogen. De Boer (1976), of Cornell University has investigated this phenomenon specifically using a single cylinder, variable compression engine. He reported no such occurrence (with hydrogen) up to compression ratios as high as 29. This in itself is not a deterrent to the use of hydrogen in diesel engines, however, as De Boer reported ignition took place readily with a "glowplug" device.

The energy required for ignition appears to be an order of magnitude less for hydrogen than for most other fuels. This we must acknowledge to be a mixed blessing. It certainly makes for prompt ignition, but is also probably responsible for the preignition problem for which hydrogen is notorious.

The flammability limit, considered together with the stoichiometric mixture, is an exceptionally important property. It is desirable to combust the fuel with an excess of air. This should encourage complete combustion at a temperature below the maximum so that pollutants are generated in reduced amounts. We note that the "theoretical" mixture for gasoline is 1.7, while its flammability (ignition) limits are 1.3 to 7.1. But we do not wish to operate on the high side of the stoichiometric, and "lean" operation is severely inhibited. With hydrogen, however, the range is 4 to 75, while the theoretical is 19.6. Thus we should be able to operate well below the stoichiometric, substantially reducing the nitrogen oxides emitted.

Table 1. Combustion Properties of Several Fuels
(Generally Accepted Values from the Literature)

	H_2	CH_4	C_3H_8	Gasolin e	CH_3OH
Autoignition Temp., °C	585	540	510	440	385
Min. Ignition Energy, MJ	0.02	0.28	0.25	0.25	--
Flammability Limit, Vol. % in air	4-75	5-15	2.2-9.5	1.3-1.7	6.7-13.6
Stoichiometric Mixture, Vol. % in air	29.6	9.5	4	1.7	12.3
Max. Flame Velocity Laminar, cm/s	270	38	40	30	--
Diffusivity in air, cm^2/s	0.63	0.2	0.09	0.08	0.13
Lower Heat of Combustion					
1000 Btu/lb	52.6	21.5	20	20	9.1
1000 kJ/kg	68.8	50.0	46.5	46.5	21.1
1000 Btu/gal	31.0*	76.1*	96.6*	120	60
10^6 kJ/m^3	123.6	303.5	285.2	478.5	239.3

*as saturated liquid at atmospheric pressure

The flame velocity is an index of how well we might approach the constant volume combustion called for in the Otto cycle. Note again, the order of magnitude difference for hydrogen over gasoline. Some caution is advisable here, however, for these are maximum values obtained in stoichiometric conditions IF we operate much leaner, as discussed above, we might expect the flame velocity of hydrogen to be much lower. De Boer has suggested that the figures for hydrogen might, in fact, be about the same as for gasoline if the hydrogen is operated at approximately 40% the fuel called for at stoichiometric conditions.

The diffusivity of the fuel is important in that is relates to mixing; the higher the value the more readily natural mixing should take place. Moreover, it has direct bearing on potential safety considerations, in that it indicates an "escaping tendency" highly desirable in the event of spills for whatever reason. Thus the much greater diffusivity of hydrogen than that of gasoline--or even methane--should be viewed as a plus for hydrogen.

The values of the lower heat of combustion are almost self explanatory, and indicate another mixed blessing aspect of hydrogen as an engine fuel. Per unit mass, it releases more energy on burning than any other fuel, and 2 1/2 times that of gasoline. But hydrogen is also the lightest of all fuels, so a great deal more volume is required to contain it. A vehicle designer must generally be more concerned about volume allowance than weight, so this becomes a very important consideration. It is an unfavorable one for hydrogen, for a specified volume of gasoline would release about 3.5 times the energy as that of an equal volume of liquid hydrogen.

We acknowledged earlier the relation of hydrogen's ease of ignition and its tendency to backfire, or flash back through the ignition. The wide flammability limits also contribute to the overall problem. Much attention has been devoted to this problem, and while one hesitates to claim complete solution, some confidence may be expressed that it is solvable. The accumulated evidence reinforces the analysis advanced by King in the 1950's: "hot spots" developed in the system, or any sort of particulate matter--as from pyrolysis or decomposition of oil--serve as ignition sources. Other causes have been suggested: valve overlap permitting contact of the fuel intake charge with hot, perhaps still-burning exhaust, and possible "voltage-crossover" resulting in spark plug activation before the desired time. The work of Swain, however, supports King's contention. King reported being able to operate carbureted engines without backflash as long as he was able to keep the system scrupulously clean, which required he most painstaking efforts of several graduate students, as one of them has reported in personal conversation. Swain's findings are consistent with this; after breaking in a new engine with gasoline, for example, it is necessary to disassemble it and remove completely all residues and deposits before operating on hydrogen. The effectiveness of (cooled) exhaust gas recycle (EGR) and water injection in minimizing backflash with hydrogen is also consistent with King's hot-spot hypothesis. These act as thermal diluents, tending to smooth out the hot spots and reduce their effectiveness as ignition sources. Still, this preignition induced backflash occurs despite EGR and water injection. The spark plug electrodes themselves may serve as hot spots with time, though the use of a narrower gap and "cold" plug minimizes this. The exhaust valve also becomes heated during operation, especially at higher loads, as the flow during the exhaust valve operation promotes heat transfer to it. Thus, backflash seems almost axiomatic with a carbureted, or premixed charge of hydrogen and air. Cold spark plugs, special, cooler operating exhaust valves and operation with EGR or water injection certainly minimize the possibility, but it is still there, and the latter two must reduce power output. Swain suggests that improved cooling passage design in the engine block should alleviate the problem and has supporting evidence--but this invites extensive redesign and substantial engine modification.

The solution would seem to be in preventing the presence of a combustible mixture until combustion is desired. That is, some form of injection system. The ideal injection would be directly into the cylinder (DCI) just as ignition is desired. This is not simple operationally, however. The thermal environment is severe, of course, so there is a basic materials problem. Furthermore, injection at this point would require a comparatively high pressure hydrogen delivery system which would imply an energy penalty, in addition to raising further questions of general safety. In keeping with the theme of simply keeping the hydrogen and air separated for as long as possible, however, several "intermediate" schemes have been, and are currently being investigated. Swain has devised a system of hydrogen induction via separate venturi tubes to each cylinder. He reports considerable success, though noting EGR of water injection still desirable to minimize backflash as the mixture ratio approaches the stoichiometric. MacCarley (1978) has devised an electronic-microprocessor controlled system allowing the injection of hydrogen just at the intake valve. Experience in the laboratory with a prototype was very promising--even without water injection. It offers great promise and would be directly adaptable to direct cylinder injector when the "materials" problem is solved. Lynch has suggested the use of a "third valve" with mechanical timing to allow the induction of hydrogen. A Dodge Colt D-50 pickup truck with the Mitsubishi engine has been adapted to test this concept. It was driven from Denver, Colorado to Riverside, California (approximately 1000 miles) where it is being used for test and demonstration purposes by the Clean Fuels Institute. Results are extremely encouraging--indeed the 1000 mile drive to deliver the vehicle is another milestone in credibility for the hydrogen fueled vehicle. Favoring direct injection, Furuhama is developing a liquid hydrogen pump so that hydrogen may be injected in, or near, the liquid state. This has great implication for more feasible operation of diesel engines as well as higher compression spark ignition engines.

There is, then, documentation that allows the prediction of a solution to the backflash problem.

3.2 SOME ADVANTAGES OF HYDROGEN

Having pointed out a severe problem or two (backfire and storage) with hydrogen, it is in order to stress one of its greater advantages. This is its benign effect on the environment. The carbon dioxide, carbon monoxide and unburned hydrocarbons typical of gasoline and most other fuels are simply not present as there is no source for them. Lubricating oil may generate some but to a trivial degree and of course no more than with other fuels. The elimination of carbon dioxide, incidentally, may be of extreme importance even though it may not be considered toxic. Much more of an effort should be directed toward ascertaining how serious the carbon dioxide "atmospheric blanket" effect really promises to be. If there is indeed a clearly predictable imbalance in radiation to and from the earth due to the accumulation of carbon dioxide in the atmosphere, renewed emphasis on non-fossil fuels is certainly mandatory. The immediate concern, however, is with the production of oxides of nitrogen. This is rather well defined. As long as operation is at low equivalence ratios, there is little problem. As higher ratios are approached, however, the NO_x presence increases markedly. An interesting and favorable by-product to the use of water injection to reduce backflash is a consequent reduction in NO_x, as reported by Billings and Lynch (1973). Overall then, hydrogen fueled vehicles offer great relief to our atmosphere, now so threatened by fossil-fueled engines.

Mention should also be made of MacCarley's suggestion that the electronic injection control system could permit literally skipping the region of maximum NO_x production in going from an equivalence ratio of, say 0.65 to 0.95. The argument is based on the model of McLean and De Boer which indicates maximum NO_x at equivalence ratios between 0.7 and 0.9--and validated to a degree with experimental data due to MacCarley. The idea is that with a multicylinder engine it would be possible to jump through this region, one cylinder at a time or in pairs of opposing cylinders as the throttle is advanced from about two thirds to full power (equivalence ratio, 0.65 to 1). This must be left as somewhat speculative, however, as Swain's more recent data indicate a direct relation between maximum NO_x and maximum power.

The recent World Hydrogen Energy Conference (10th, June 21-24, 1994) offered further documentation of the possibilities for reduced NO_x emissions with hydrogen fueled engines. For example, Jorach et al. (1994) reported on the development of a heavy-duty hydrogen engine, and compared its performance with that of a standard diesel engine; torque, power output and fuel consumption correspond closely. Exhaust emissions were only about 20% of the diesel--including NO_x! Those particularly interested in this aspect of hydrogen fueled transportation might wish to consult the proceedings of the conference for more recent detailed information.

All in all, the hydrogen powered vehicle has much to offer the atmosphere by way of cleanliness--as is verified by ample documentation.

We discuss next the issues of thermal efficiency and power limit. They should be discussed together as several studies of thermal efficiency versus brake horsepower indicate a cutoff point for hydrogen implying a decrease in power available. Figure 1 shows this; it occurs when operating the engine on hydrogen in the same basic manner as with gasoline. That is, by "throttling' or controlling power output through the amount of premixed fuel-air charged to the engine at a constant fuel-air mixture ratio. Appropriate means of governing by controlling the fuel-air ratio itself, however--quality governing--alleviates this problem. We emphasize that this is possible with hydrogen due to the wide flammability limits it exhibits, particularly on the lean side of stoichiometric. Figure 2 shows some results of Swain's work, using his scheme of unthrottled operation. Note that very nearly the full power range may be achieved, and that the thermal efficiency remains noticeably high with hydrogen throughout the full range.

The significance of the remarkably higher efficiency at low power (Figure 1) must not escape attention. One has but to recall the high percentage of stop-and-go, low power operation in the normal city driving cycle to appreciate the potential contribution this would make. We may conclude again, that hydrogen should be a preferred fuel--this time from the standpoint of thermal efficiency--potentially without significant sacrifice of power generation.

The question of uniformity of the mixture admitted to the cylinders is of great importance. Significant cylinder-to-cylinder variation is not uncommon with carbureted gasoline engines, with a consequent reduction in efficiency. While the higher diffusivity of hydrogen in air would seem to permit more uniform distribution, it does not seem to do so. Billings Energy Corpora-

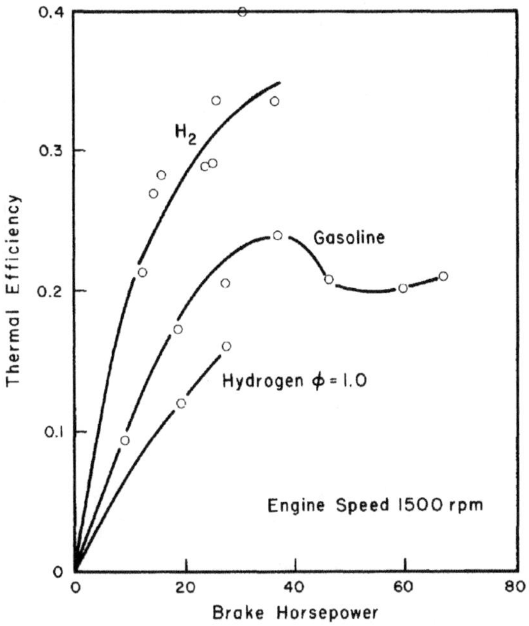

Fig. 1. Thermal efficiency vs. power.

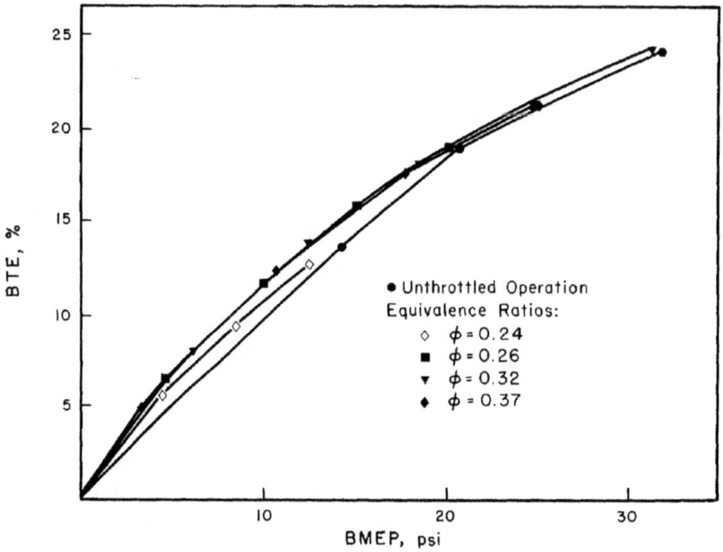

Fig. 2. BTE as a function of BMEP for various values of Φ.

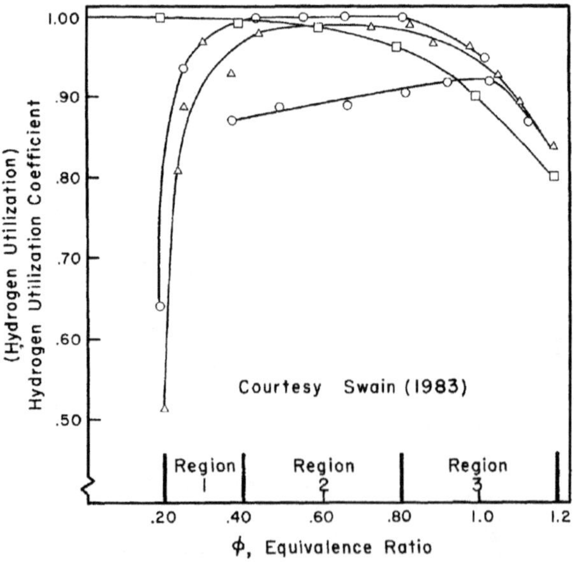

Fig. 3. Illustrates effect of Φ on hydrogen utilization coefficient.

Table 1, however, that the lower limit of hydrogen-air flammability range is above the corresponding point for gasoline. Furthermore, the higher diffusivity of hydrogen (nearly an order of magnitude) and the great buoyancy of hydrogen imply a rapid dispersal into the atmosphere. The heavier-than-air vapor of gasoline tends to concentrate until an ignitable mixture develops. This does not, of course, change the fact that hydrogen systems are more vulnerable to leakage than gasoline. Hydrogen systems (again, as in the case of any high-energy fuel) must be made leak proof. This has not proved to be a problem in the author's experience, or that of others judging from the literature.

In the case of liquid spillage, hydrogen is far less hazardous than gasoline. The extremely low boiling point (approximately -424°F or -253°C and great buoyancy of hydrogen combine to assure almost instantaneous evaporation and dispersion. This has been documented by several spillage tests and recorded on film. Liquid gasoline, on the other hand, lingers on, practically inviting ignition from a variety of possible sources.

There is some evidence to support the implications of comparable or superior safety with hydrogen as compared to gasoline. The Billings Energy Corporation has conducted tests firing armor-piercing incendiary ammunition into one of their hydride tanks, filled to capacity with hydrogen and into tanks filled with gasoline. The comparison is dramatic. The result with the hydrogen is a simple, straightforward burning; with gasoline, a destructive rupture and conflagration results.

Further evidence of the basic safety of a hydrogen fueled vehicle is afforded by the result of an unplanned "crash test" of a liquid hydrogen vehicle, as reported by Finegold (1976). The vehicle had been prepared for a long-distance rally, and its tank was full of liquid hydrogen. It was being towed on a trailer on the highway at 55 miles per hour when the tow car suffered a tire blowout. The net result was that the trailer and hydrogen car came to rest upside down after considerable interaction with the road and a sign post or two. There was no fire despite consequent leakage of hydrogen. In fact, the car was separated from the trailer (the latter being damaged beyond repair), righted, and driven back as far as possible before running out of hydrogen.

These arguments and statements must not, of course, be taken to minimize the need for proper handling of hydrogen--or any other fuel. It is argued, however, that with proper handling, hydrogen is no more inherently hazardous than gasoline. It may, indeed, be less so.

3.3.3 *Onboard Storage of Hydrogen.* Hydrogen's low mass density makes its storage a severe problem. The same volume of air would weigh over 14 times as much as hydrogen, while the factor is about 8 for methane. In the liquid state, its density at most is approximately one-tenth that of gasoline. This may prove to be a dominant constraint, limiting the ultimate acceptance of hydrogen as an internal combustion engine fuel. At the present time, three means of storage are employed on hydrogen powered vehicles: as a gas at high pressure, as a metal hydride, and in the liquid state.

3.3.4. *Onboard Storage of Hydrogen as a Gas.* In evaluating onboard storage systems a decade ago, it was almost automatic to eliminate consideration of gaseous hydrogen as impractical. And with good reason. The first UCLA hydrogen car as described by Finegold and coworkers (1973) used a gaseous system consisting of two steel tanks, with a loading pressure of approximately 400 atm; the overall volume was about 35 gal (0.13 m^3). The range of the vehicle was 60 miles (96 km)! In order to increase this to a more practical distance, say 300 miles (480 km) while keeping the storage volume the same, a rough estimate indicates the loading pressure must approach 800 atm and the tank weight 1360 kg--and both figures might well be higher. The compressibility of hydrogen at high pressure casts some doubt on the issue. Gordon (1984) has re-opened the case for gaseous hydrogen, however, alluding to modern pressure vessel technology developed largely for the space exploration program. The technology is based on composite materials, for example fibers of S-2 glass or graphite imbedded in an epoxy resin filament wound over seamless metal liners. Such a tank is installed in a small hydrogen powered pickup truck operated by the Clean Fuels Institute[1] and has been described and demonstrated by Allen (1982). It is emphasized that vessels approved by the U.S. Department of Transportation for duty at 300 atm are now in serial production and in everyday use. It is predicted that vessels operating at 650-700 atm will be available and that hydrogen may be stored with an installed density of 20.8 kg/m^3, at 5.5% the weight of the pressure vessel. For the 300 mile (480 km) range mentioned earlier, a rough estimate of tank weight is 250 kg; while volume appears excessive (well over one half a cubic meter), the low weight is attractive and suggests a tradeoff with range to make the volume reasonable. More recently, Fulton, Lynch and Marmaro (1994) state that recent advances in compressed gas containers makes high pressure gas storage feasible, and discuss a 3600 psi (about 250 bar) system, displacing just under 1 cubic foot of volume for each pound of hydrogen stored. The desirability of a gaseous system is greatly enhanced by the basic simplicity of control of the flow to the engines, and refilling of the storage tank. A note of caution must be expressed regarding the latter, however; the Joule-Thomson coefficient of hydrogen is such that hydrogen will heat on irreversible expansion (as in flow from high-to-low pressure during filling) as long as the initial temperature remains above 200 K. Thus its density will decrease during filling, negating the initial high density. Considerable cooling might be required to regain it. One must agree with Gordon (1984) that with modern technology gaseous storage must no longer be dismissed as impractical but considered a serious candidate. Yet mass and volume would seem to remain excessive for driving ranges of 300 miles (480 km).

3.3.5. Onboard Storage of Hydrogen as a Metal Hydride. Reilly and Wiswall (1968) reported the property exhibited by some metals and metallic compounds to adsorb hydrogen with the release of heat and to regenerate it on the addition of heat. Hoffman et al. (1969) were, apparently the first to suggest this phenomenon as the basis of a storage system potentially suitable for automotive usage. The magnitude of the heat effect, the temperature-pressure equilibrium relation, and the rate at which each process (adsorption and regeneration) occurs are all critically important properties of the system. Indeed, the desirable properties of a hydride may be summarized as follows:

1. low heat of dissociation
2. adequate pressure discharge level (several atmospheres) at comparatively low temperature (15 to 40°C)
3. high percentage (by weight) of hydrogen carried
4. fast release rate
5. minimal pulverization on continued use.

Unfortunately, no hydride known at the present time has an entirely acceptable "mix" of these properties. A few examples will illustrate, the problem.

---

[1]Now the Clean Air Now organization.

tion studies indicate a 10% cylinder-to-cylinder variation in equivalence ratio for the premixed, carburetion-type operation. This would be a difficult problem to overcome in a single mixing chamber operation, and in some sense is an additional argument for the more direct cylinder induction or injection of the fuel into the intake port, or ultimately into the cylinder itself.

One of the greater surprises of the nature of hydrogen as a vehicular fuel is its significantly less than 100% combustion efficiency, or utilization. With its low ignition energy and high flame speed, one expects almost intuitively that it should burn to completion (100%). MacCarley challenged this assumption, reporting as low as 60% combustion completeness at low speeds, while the number climbed toward 100% at higher speeds. These data were the by-product of testing injector designs, however, and not validated by extensive repetition. Swain's data confirm the less than 100% utilization, however, if not to the same degree. Swain correlates hydrogen utilization with the equivalence ratio for a timed induction hydrogen intake and obtains the results of Figure 3. Note the three identifiable regions. In region 1, the flame speed may be so low that combustion is not propagated across the entire chamber before the charge is exhausted. A "bulk quenching" results. Region 2 shows things pretty much as might be expected, i.e. 98 to 100%. These results are consistent with those of Homan (1978), incidentally, who used a gas chromatograph to analyze the exhaust gas. In region 3, the utilization falls off--probably due primarily to dissociation, though incomplete mixing of the more concentrated hydrogen mixture may contribute.

It is difficult to compare this phenomenon directly with an analogous term for gasoline. Perhaps it is best to acknowledge the problem as such--and to note that it is, fortunately, most serious outside the normally expected range of operation.

3.3 SOME PROBLEMS WITH HYDROGEN.

The problems with the use of hydrogen may be put into three categories: economics, safety and storage. Combustion, with its ancillary problems, is simply not a primary concern at this point in time. This is not to suggest that there should be any lessening of the efforts to improve combustion, but rather to define problems that remain in desperate need of solution before hydrogen can be considered for practical use as a fuel.

3.3.1 *Economics.* The cost of hydrogen is treated by others in this volume, so will not be discussed in detail here. This author would emphasize, however, that comparison with the cost of other fuels must, somehow, be made on a normalized basis. If the use of gasoline, leads to billions of US dollars in agricultural damage, to say nothing of its effect on the public health, it simply is not realistic to compare its cost at the pump with that of hydrogen, the use of which causes no such damage! Clearly, a total, or life-cycle cost analysis must be made the basis for comparison. We look to the economist to generate such a meaningful cost estimates.

3.3.2 *Safety Considerations.* The public perception of hydrogen as a dangerous substance is, perhaps, more difficult to deal with than hydrogen itself. The first reaction seems to be to recall the Hindenberg accident; but it seems little known that the Hindenberg's sister ship operated safely for over a million miles. Of course, any substance with properties that make it a good engine fuel is potentially dangerous and must be properly treated. The point is that hydrogen may indeed be handled safely, as the following discussion implies. It has been noted that hydrogen ignites readily and has a wide range of flammability limits. Thus, in the event of a leak, it might be feared that hydrogen is much more apt to ignite than, say, gasoline. Note from

Iron-titanium ($FeTiH_{16}$) has a relatively low heat of dissociation (16,948 kJ per kg of hydrogen), and dissociation occurs at acceptably low temperatures (35-40°C) at a tolerable pressure (15 atm). Experimental results indicate a satisfactorily high release rate. This combination would permit supplying all the heat for releasing the hydrogen required from the engine exhaust, or from the engine cooling water if a high pressure cooling system were used. The "hydrogen capacity" of iron-titanium is low, however, and the bulk density high. Based on the data of an experimental storage reservoir developed by the Brookhaven group (1974), approximately 1900 pounds (863 kg) of the hydride would be required to fuel an automobile with a range of 300 miles (480 km); the containment volume would be approximately 12 ft^3, or 90 gallons (0.34 Lm^3).

Magnesium based hydrides offer the advantage of a higher hydrogen storage capacity--6 to 7% have been reported. This would imply a total weight of approximately 425 pounds (193 kg), based on the cruising range postulated above. The necessary volume would be in the neighborhood of 7.5 ft^3 or 57 gallons (0.215 m^3). But the heat requirement is high, as is the dissociation temperature at reasonable pressures. In fact, neither the engine cooling system nor the exhaust gases could supply sufficient energy for hydrogen release. Thus, auxiliary energy would have to be used, in an amount earlier estimated by the author to be equivalent to 15% of the energy supplied to the engine. Magnesium based hydrides may also tend to fractionate more than the iron-titanium type on continued usage though definitive evidence is not yet available on this phenomenon. It is, of course, potentially a serious problem as hydrogen flow passages would tend to be restricted by the powdery particulate matter formed.

Reilly and Sandrock (1980) provide an excellent review of the subject, and discuss use for vehicular hydrogen storage in some detail. They show that the hydrides in general have an energy density greater than pure hydrogen even in the liquid state--though it is not clear how much of the containment configuration is included. In any case, the figures make a persuasive case for hydrides.

Ten-You and coworkers (1982) have reported research in the People's Republic of China exploring the effects of additional elements to the basic iron-titanium hydride. They prepared alloys that included manganese and mischmetal, manganese and aluminum, mischmetal and niobium, chromium and nickel, manganese and chromium, and manganese only in addition to iron-titanium. Hydrogen capacities of up to 1.8% were reported--essentially twice that of untreated iron-titanium--and were apparently unaffected by recycle, at least up to 40 times.

Watson and coworkers (1984) also report experience with an Australian hydrogen car, using a hydride storage system--they employed a titanium-iron-manganese alloy with good results.

As implied above, system weight seems to be an inherent problem with hydrides. It must be noted, then that there are situations in which weight is an advantage. Turillon (1982) in a very good treatment of the design aspects of hydride containers discussed a forklift truck developed by Hydrogen Energy Specialists as such an application. Escher (1981) also reported discussions with railroad officials in which the desirability of a heavy system was noted.

All-in-all, while problems remain with the hydride system, it is an entirely viable candidate for automotive systems, and a favored one for certain other applications.

*3.3.6. **Onboard Storage of Hydrogen as a Liquid.*** The critical temperature of hydrogen is 33 K (-240°C, 400°F) and its critical pressure is 12.8 atm. Clearly, then, the problem of storing liquid hydrogen is one of maintaining extremely low temperatures rather than high pressures. Modern cryogenic technology and equipment makes this problem amenable to solution. The critical problems are those of "boil off" and "lock-up" time. The former is the "unintentional" rate at which hydrogen may be lost due to heat leakage into the, tank. It is usually expressed as a percentage of the volumetric capacity of the tank lost per 24 hours. The latter refers to the time required for the pressure in the tank to build up from the filling pressure to the relief pressure, at which time boil-off begins. While it is dependent upon the relief pressure, it is primarily a function of the rate at which heat leaks into the hydrogen. While it is axiomatic that the boil-off should be as low, and lock-up time as great as possible, the importance of these properties should not be misinterpreted. A boil-off rate of even 5%, say, need not imply that 5% of the tankful of liquid hydrogen would be continually evaporating to the atmosphere every 24 hours. The fuel flow system may be modified readily to feed vapor from the tank to the engine as long as the pressure in the tank remains above the design control pressure. This will draw down the tank pressure until the desired pressure is reached. At this point the liquid feed system would automatically become operative as the vapor line to the engine would be closed off. During operation, then there is no boil off, nor will there be any after the engine is shut off until the lock-up period is expended. As a practical matter, therefore, with operation of the engine at least everyday or two, the loss of hydrogen due to boil off would be negligible.

Emphasis then has been on the development of a container-tank with suitable properties and sufficient volume to permit a reasonable vehicle range. Successful efforts have been reported by Stewart and coworkers as early at 1975, Peschka et al. (1980,1982) and Stewart (1982). The latter includes road testing and the refueling of the vehicle from a "service-station" type pump; results seem to justify the claim of the basic simplicity of this operation with liquid hydrogen and a filling time of about five minutes seems realistic.

An implicit consideration with the use of a liquid system is the added expense of hydrogen in the liquid state. Unfortunately, the thermodynamic properties of hydrogen are such that its liquefaction is inherently expensive and the cost of the liquid seems destined to be about 30% greater than that of the gas at moderate (150 atm) pressures. This cost may be offset by the advantages of using liquid hydrogen, particularly those associated with the possibility of the injection of hydrogen as a liquid being investigated by Furuhama (1982). It should also be noted that liquid storage would seem to be the only practical system for aircraft usage.

The liquid system seems to be well established as a strong candidate for fuel storage, suffering primarily from the added cost of hydrogen in the liquid state as well as that of "service equipment" that would have to be developed for its large usage.

3.3.7. Onboard Storage of Hydrogen by Other Means. Several other possibilities have been considered, and some continue under development. Taube et al. (1985) utilize a "liquid hydride"; essentially, toluene is the storage medium, and is converted to methyl cyclohexane when charged with hydrogen. Subsequently, this is thermally and catalytically decomposed, yielding the hydrogen to fuel the engine. The heat required is furnished by the exhaust gas, so that hydrogen is not available for start-up; toluene, or an auxiliary fuel may be used until sufficient exhaust heat is available. The system is bulky, and might be restricted to buses and trucks.

At one time, it seemed feasible to store hydrogen in a *chemically-bonded* form; Ullman (1976) suggested that methyl alcohol be stored in the fuel tank, then catalytically decomposed to carbon monoxide and hydrogen prior to combustion. Since the decomposition reaction is endothermic, the heat of combustion is increased over that of methyl alcohol, and the process is cleaner. Extensive tests, however, indicated that the system could not perform as well as anticipated, and the idea has been abandoned. MacCarley (1976) suggested the hydrolysis of sodium borohydride to generate hydrogen:

$$NaBH_4 + 4H_2O \rightarrow NaB(OH)_4 + 4H_2$$

If the water formed during the subsequent combustion of hydrogen could be condensed and recycled to hydrolyse the sodium borohydride, one would have quite a compact storage system. While laboratory tests were satisfactory, the expense, of the, borohydride would seem prohibitive, and work was not continued.

Hydrogen may also be encapsulated in glass microspheres, as discussed by Yan and coworkers (1985). It is analogous to the metal hydride system, with the significant advantage of lighter weight. At the same time, relatively high pressure is required, which would call for heavier tank construction. The capacity is comparable to that of the iron-titanium system, about 2% by weight. The possible use of glass microspheres for the storage of hydrogen is attractive, and further research is desirable.

"Slush" hydrogen has also been mentioned as a means of obtaining a higher storage density of hydrogen, and Smith (1989) has discussed its production. This is a state between the solid and liquid, and has a density approximately 15% greater than liquid hydrogen. It must be contained at a temperature about 7 K lower than that for liquid hydrogen, so would be more demanding of the cryogenic equipment containing it. It has been demonstrated that it can be accomplished without great difficulty. The idea was tried in the late 1960's, and thought to be impractical. It has been revived, however, in connection with the "spaceplane" effort in the United States, and was considered promising. Unfortunately, this effort has now been tabled, and it remains to be seen whether work on slush hydrogen will be continued.

Metallic hydrogen is considered a possibility in some circles. Nardi (1984) suggests that this form of hydrogen would store 100 times the energy as compared with the same mass of liquid hydrogen. The great drawback is the high pressure required to maintain the solid state, and difficulties with its production. The question of stability at low pressure after formation remains unanswered.

With the possible exception of the use of glass microspheres, realistic considerations seem to dictate that vehicles must store hydrogen in one of the three modes used at present--as a compressed gas, as a liquid, or as a metal hydride.

4. Current Programs of Interest

Driven by increasing recognition of the need to reduce harmful emissions to the atmosphere, and legislative action toward this goal, hydrogen is attracting renewed interest today as a potential transportation fuel. Some current (post 1990) programs of particular interest will be reviewed here.

4.1 THE USE OF HYDROGEN AS A COMPLEMENT OF NATURAL GAS

There seems to be a consensus developing that the operation of engines with natural gas will be the transitional stage to operation with hydrogen. Swain (1993) used a mixture of 20% hydrogen (by volume) and natural gas (over 90% methane) and compared test results with those for natural gas alone on flame speed, brake thermal efficiency, nitrogen oxides, unburned hydrocarbons and carbon monoxide. Lynch has been experimentally studying the performance of mixtures of hydrogen and natural gas since 1988, and has coined the term "hythane"[2] to denote such mixtures. Fulton and Lynch (1993) have reported the results of such efforts, finding performance generally enhanced by the addition of hydrogen, particularly with leaner mixtures--for example, with 70% the fuel that could be burned with the same amount of oxygen employed. They refer to a *leverage* effect provided by the hydrogen. For a relatively minor reduction in torque, for example, significant reductions in total hydrocarbons and oxides of nitrogen were obtained. They conclude that hydrogen is a useful additive for natural gas.

4.2 THE NEED FOR A DEDICATED HYDROGEN ENGINE

It has been amply demonstrated that a wide variety of engines can be converted to run on hydrogen, and that, in general, performance has been enhanced thereby. Swain (1994) is pursuing the next step: the design and construction of a dedicated hydrogen fueled engine. The engine employs the unmodified cylinder block, crankshaft and oil pump from a Toyota 1.6 liter, four cylinder engine. The remaining components were modified or manufactured specifically for the use of hydrogen. A three valve cylinder head machined from an aluminum billet is used. The results of this work should go far toward answering the question as to whether the manufacture of an engine specifically designed to use hydrogen as a fuel is justified, or if merely modifying engines currently in production is adequate.

4.3 FURTHER CONSIDERATION OF SAFETY ISSUES

While it has been argued earlier that hydrogen, properly handled, is no more dangerous than gasoline or other high energy fuels, specific concerns do arise. Swain (1994), of the University of Miami, has initiated a program of such specific studies. A list of recent projects serves to define the nature and scope of the program:

- A safety analysis of hydrogen production techniques, with a view, in part, to the probability of accidental combustion. An analytical study of combustible gas cloud motion in various kitchen geometries.
- An experimental investigation of the permeability (to hydrogen) of non-metallic pipes.
- An experimental investigation of the flame visibility of gaseous fuels, with and without additives.
- An experimental investigation of the ignition of gaseous fuels on contact with hot metal surfaces.
- An experimental investigation of flame arrestors compatible with the use of hydrogen.
- An experimental and theoretical investigation of hydrogen compared to natural gas pipeline leakage.
- A theoretical study of the behavior of hydrogen leakage occurring in self ventilating buildings.

Without going in to detail at this point, it is safe to say that the results of these studies reinforce the earlier observation that hydrogen does not pose an undue danger in its usage. Swain (1994) concludes that hydrogen is much safer to use than propane, and about on a par with natural gas.

---

[2]Hythane is a registered trademark of Hydrogen Consultants, Inc. of Littleton, Colorado, USA.

4.4 HYBRID VEHICLES

Most hybrid vehicle considerations focus on a combination of a hydrogen fuel cell and batteries to power an electric vehicle. Nadal and Barbir (1994) describe such a concept. Fuel cell technology is presented in another chapter in this volume, so need not be treated here. Suffice it to say that the fuel cell, and therefore this vehicle, offers the ultimate zero pollutant emissions, as the low temperature hydrogen oxidation is accomplished without the generation of oxides of nitrogen. The authors do not discuss the onboard hydrogen storage problem, but it should be less than for the more conventional engine simply due to the lesser amount of hydrogen required as a result of the remarkably high efficiency obtainable with the fuel cell.

Fulton, et al (1993) suggest an interesting interim step toward this goal. They propose the use of a hydrogen engine as a "surrogate fuel cell". The engine would drive a motor/generator which would charge batteries for electrically augmenting vehicle operation. (Shaft power from the engine would be used directly for vehicle propulsion). Emissions would not be as benign as in the case of the fuel cell, but would easily meet the requirements for the "ultra low emissions vehicle" (ULEV) set by the California Air Resources Board. In fact, the only pollutant emitted in significant quantity from a lean burning hydrogen engine would be oxides of nitrogen. Thus, while not the "ultimate" system the fuel cell would be, the Surrogate system seems to have much to offer. It would suffer from the onboard storage problem; since the engine could run essentially at peak efficiency, however, the storage requirement might be less serious than with conventional engine operation.

4.5 SOLAR ENERGY CONVERSION

The problem of production of hydrogen for use in transportation is addressed elsewhere in this volume, so has not been treated here. It is desirable, however, to call attention to a very interesting project being conducted at the University of California, Riverside, by the Center for Environmental Research and Technology of the College of Engineering. The overall program integrates research in solar powered electrolysis for the production of hydrogen and hydrogen vehicle technologies; this provides a demonstration of a virtually zero emission, renewable energy source, and an ultra low emissions vehicle.

At the facility, sunlight is converted to electricity using a photovoltaic array (PVA). The current from the PVA travels to a 12 cell electrolysis unit where water is dissociated into hydrogen and oxygen. The hydrogen passes through a sensor to determine its purity; if the purity is acceptable, hydrogen is fed to a low pressure gas holder, if not it is vented to the atmosphere. When the gas holder is filled to capacity, a compressor switches on automatically, and hydrogen is compressed, and stored in a relatively high pressure storage station. During vehicle fueling, hydrogen is transferred from the high pressure station to the vehicle refueling station.

Two vehicles are being tested. One is a 1992 Ford Ranger with a modified 2.3 liter, 4-cylinder gasoline engine. The second is a hybrid electric vehicle with a hydrogen fuel cell. The modifications to the Ranger include rework of the cylinder heads and camshaft, adding a turbocharger and aftercooler, and adding a hydrogen injection system. These modifications should permit lean burning of the hydrogen, and a reduction in the oxides of nitrogen emissions to a near zero level. Onboard storage of gaseous hydrogen is at 3600 psi (245 atm), using tanks made from composite materials, as alluded to earlier.

The second vehicle is more of a test bed for experimentation with various types of fuel cells. Currently, a 1.5 kW phosphoric acid fuel cell is under study. Fuel cells in general promise approximately 2 to 2.5 times the efficiency of the internal combustion, and true, zero level of nitrogen oxides emissions. The electricity produced is charged to the batteries which operate the drive mechanism of the vehicle. Since the batteries can be recharged during the vehicles operation, the range is extended well over that of a vehicle operating on batteries only. Again, however, the limiting factor is the amount of hydrogen that can be stored onboard.

This project is scheduled to continue through 1994; a computer -based data acquisition system permits the collection and recording of extensive data which will allow the ultimate performance evaluation of each component in terms of efficiency, reliability and maintainability. This final evaluation should be an exceptional contribution to the knowledge base of hydrogen technology.

5. Some General Comments

The material presented thus far indicates that hydrogen's potential as an engine fuel has long been recognized. The failure to reach this potential is due to the abundance of gasoline, which must be recognized as an excellent transportation fuel in every way except for its contribution to environmental pollution. The severity of this problem, however, has now been recognized and it appears that severe steps must be taken to alleviate it. Not only will legislation be enacted to minimize harmful emissions, but mandates will be issued that will demand that a certain percentage of new transportation vehicles have emission levels unattainable with petroleum based fuels. In the State of California, for example, 2% of the new vehicles sold in 1998 must be have zero harmful emissions (ZEV); 2% must have "Ultra Low Emissions" (ULEV) and 48% must be "Low Emissions" (LEV), leaving 48% in the conventional category. The "figures-of-merit" for these categories are: 0.25 for the conventional, 0.075 for the LEV, 0.040 for the ULEV, and, of course, 0.00 for the ZEV. By the year 2003, only vehicles in the LEV, ULEV and ZEV categories may be sold in the state. The gasoline of today will not be able to meet these standards, and the "reformulated" gasolines now being developed will meet only the LEV standards. Thus, concern for the environment, implemented by legislative action will lead to a market for electric and hydrogen powered vehicles,- as other states adopt policies in line with those of California (at the present time, 12 other states have adopted the California standards). Coupled with this is the knowledge that petroleum is not a renewable source of energy; we no longer ask "whether-or-not" we shall run out of it, but "when" we shall run out of it. against this background, it is appropriate to speculate on hydrogen's role in transportation.

5.1 AIR TRANSPORTATION

Hydrogen is destined to play a dominant role in aviation in the future. Earlier experimental studies have demonstrated the suitability of hydrogen as an aircraft fuel; economic considerations are largely responsible for the lack of its utilization at present, or in the near future. Air travel at supersonic and hypersonic speeds, however, is another story. Winter (1990) states the case for liquid hydrogen in high speed air travel unequivocally: such flight is simply doomed without it. While the problems remain enormous, they are amenable to solution, and largely with present technology, aided and abetted by predictable advances in materials science. Hydrogen will fuel tomorrow's high speed air travel.

5.2 RAIL TRANSPORTATION

There seems to be a low probability that the propulsive systems of the railroads will change until such change is mandated by emission standards promulgated by law. As regulatory agencies are becoming continually more concerned with particulate emissions, however, this may occur sooner than presently anticipated. Both the coal - fired steam engine and the Diesel are prime offenders in this regard, and even the electric locomotive must be charged with the pollution generated along with the electricity to power them.

The hydrogen fuel cell would seem to be a favored candidate for this duty; no oxides of nitrogen would be produced, the high efficiency would minimize the fuel storage problem, and, in any case, weight would not be a limiting factor. The latter condition would also allow the use of the hydrogen fueled engine, but the elimination of the emission of nitrogen oxides favors the fuel cell wherever practical. Thus hydrogen has a future role in transportation by rail; the speculation is as to "when".

5.3 SURFACE VEHICLES

Automobiles, buses and trucks offer the most fertile field for speculation as to the role of hydrogen as a transportation fuel. There is no need, however, for speculation as to the engine operation, the environmental benefits, or the basic safety with hydrogen. As has been brought out in the earlier text, hydrogen is superior in the first two, and while its safety concerns may be different, they are no more severe than with conventional fuels. The chief concerns relative to large scale use of hydrogen are, rather, its production, its life-cycle cost, and its onboard storage. Comments on some of these considerations follow (production and costs are outside the scope of this paper, but are discussed elsewhere in the proceedings).

5.3.1 A Note on Engine Operation. As DeLuchi (1989) points out, one uncertainty relative to hydrogen fueled engine operation has to do with engine wear, and maintenance in general. There is just not the long term operational experience with hydrogen necessary to permit conclusions in this regard. Lubrication is a prime concern, and some question has been raised relative to the possibility that hydrogen is more explosive in combustion, thus increasing the stress on the engine. Reviewing the evidence and informed opinion available, DeLuchi suggests that hydrogen engines may have lower oil costs, and just slightly lower maintenance costs than gasoline engines.

5.3.2 Dual-Fuel Engine Operation. Hydrogen can be burned with gasoline and with diesel fuel with beneficial results; efficiency is increased in the neighborhood of 15%, and emissions are reduced. This would seem to be a possible transitional role for hydrogen, particularly for trucks and buses; it does not appear to be a long term solution to the need for a severe reduction of emissions.

5.3.3. Onboard Hydrogen Storage. While this aspect of hydrogen fueled vehicles has been discussed at length earlier in the text, it deserves further mention here to emphasize its severity; it is the most critical of the problems impeding the ultimate conversion to vehicular usage of hydrogen as a primary fuel, if one grants the continuing need for passenger vehicles with, approximately, a 300 mile (480 km) range. It is difficult to visualize other than liquid hydrogen for such duty, despite the cost penalty for the liquid state which does not appear susceptible to significant improvement. Notable improvement in the cryogenic technology necessary for the containment and delivery system of the liquid state must also be realized; for this, there is some confidence in the advancement of the technology, given the current rate of progress in materials science. For shorter ranges, less than 100 miles, both gaseous and hydride systems are viable.

This is perhaps the largest class of vehicle service, including as it does a high percentage of fleet vehicular operations.

5.3.4. The Fuel Cell. Emphasis in this paper has been on hydrogen as an internal combustion engine fuel and justifiable so, as the fuel cell is treated specifically elsewhere. As this paper is directed generally toward hydrogen as a transportation fuel, however, it is appropriate to emphasize the great potential for the fuel cell as a (perhaps "the") future engine of transportation. It promises not only zero toxic emissions, but a reduction of the storage problem of 40 or 50%, depending on the efficiency attainable. Surely there are problems to be solved, but the success of current operations with experimental buses offers good reason to be optimistic for the future.

6. Conclusions

The "off-again, on-again" research and development of hydrogen fueled internal combustion engines over the past 170 years, and more particularly the last 50, has resulted in a technology entirely capable of meeting the demands for low and ultra low emissions standards of the near future. Hydrogen fuel cells extend the capability to zero emissions vehicles.

Thus it seems clear that hydrogen has a viable role in transportation of the future, even though we can not specify at this time just what that role will be. It can be said with some certainty that hydrogen will propel the high speed aircraft (supersonic and hypersonic) of the future; indeed, the technological effort necessary to support such usage may form the basis for the development of the infrastructure necessary for large scale utilization of hydrogen for surface vehicles.

7. References

Alder, H.P. (Editor) (1987), "Hydrogen in Air Transportation. Feasibility Study for Zurich Airport, Switzerland". *Int. J. Hydrogen Energy*, **12**, 571-585.

Allen, Nelson (1982). Demonstration of a Dodge Pick Up Truck Operating on Hydrogen Using an SCI ALT 100 Cylinder as a Fuel Tank. 4th World Hydrogen Energy Conference, Pasadena.

Brewer, G.D. (1976), "Aviation Usage of Liquid Hydrogen Fuel - Prospects and Problems". *Int. J. Hydrogen Energy*, **1**, 65-88.

Brewer, G.D. (1978), "Hydrogen Usage in Air Transportation". *Int. J. Hydrogen Energy* **3**, 217-229.

Brewer, G.D. (1978), "Some Environmental and Safety Aspects of Using Hydrogen as a Fuel". *Int. J. Hydrogen Energy*, **3**, 461-474.

Brewer, G.D. (1979), "A Plan for Active Development of LH_2 for Use in Aircraft". *Int. J. Hydrogen Energy*, **4**, 169-177.

Das, L.M. (1990), "Hydrogen Engines: A View of the Past and a Look into the Future". *Int. J. Hydrogen Energy*, **15**, 425-443.

DeLuchi, M.A. (1989), "Hydrogen Vehicles: An Evaluation of Fuel Storage, Performance, Safety, Environmental Impacts, and Cost". *Int. J. Hydrogen Energy*, **14**, 81-130.

Escher, W.J.D. (1981), Private communication.

Feucht, K., W. Hurich, N. Komoschinski and R. Povel (1988), "Hydrogen Drive for Road Vehicles - Results From the Fleet Test Run in Berlin". *Int. J. Hydrogen Energy*, **13**, 243-250.

Fulton, J., F. Lynch and R. Marmaro (1993), "Hydrogen Engines as Surrogate Fuel Cells". *Proc. World Car Conference*, 2001, University of California, Riverside.

Gier, H.L. (1976), Hydrogen Fuel for the Railroads; A Proposal. *Proc. 1st. World Hydrogen Energy Conference, 7C-43*, Miami.

Gordon, R., (1984) , Composite Hydrogen Storage Vessels for Gaseous Powered Vehicles, *Hydrogen Energy Progress V*, **3**, 1225-1236, Toronto.

Hans, P.C. and G.H. Stine (1991), "Spaceplane Contributions to the Hydrogen Economy". *Int. J. Hydrogen Energy*, **16**, 427-431.

Hoffman, K.C., W.E. Winsche, R.H. Wiswall, J.J. Reilly, T.V. Sheehan and C.H. Waide (1969). *Society of Automotive Engineers*, paper no. 690232.

Hoffman, Peter (1981), "The Forever Fuel: The Story of Hydrogen". Westview Press, Boulder.

Ikegami, M., K. Miwa and M. Shioji (1980), "A Study of Hydrogen-Fueled, Compression-Ignition Engines", *Proc. 3rd World Hydrogen Energy Conference*, Tokyo, **2**.

Jones, L., C. Wuschke and T.Z. Fahidy (1983), "Model of a Cryogenic Liquid Hydrogen Pipeline for an Airport Ground Distribution System". *Int. J. Hydrogen Energy*, **8**, 623-630.

Jones, L.E., G.W. Hayward, K.M. Kalyanam, Y. Rotenberg, D.S. Scott and B.A. Steinberg, (1985), "Fuel Cell Alternative for Locomotive Propulsion". *Int. J. Hydrogen Energy*, **10**, 505-516.

Korycinnski, P.F. (1978), "Air Terminals and Liquid Hydrogen Commercial Air Transports". *Int. J. Hydrogen Energy*, **3**, 231-250.

MacCarley, C.A. (1976), "Development of a Sodium Borohydride Hydrogen Fuel Storage System for Vehicular Applications". *Monograph on Alternate Sources*. AIAA, Vandenberg Section, **20**, 315-321.

Mikolowsky, W.T., and L.W. Noggle (1978), "The Potential of Liquid Hydrogen as a Military Aircraft Fuel". *Int. J. Hydrogen Energy*, **3**, 449-460.

Momenthy, A. M. (1977), "Fuel Subsystems for LH_2 Aircraft: R & D Requirementa", *Int. J. Hydrogen Energy*, **2**, 155-162.

Narde, V. (1984), "On Metallic Hydrogen", *Int. J. Hydrogen Energy*, **9**, 543.

Price, R.O. (1991), "Liquid Hydrogen - an Alternative Aviation Fuel?". *Int. J. Hydrogen Energy*, **16**, 557-562.

Reilly, J.J. and R.H. Wiswall (1968), *Inorg. Chem.*, **7**, 2254.

Reilly, J.J. and G.S. Sandrock (1980), "Hydrogen Storage in Metal Hydrides", *Scientific American*, February, 1980.

Smith, E.M. (1989), "Slush Hydrogen for Aerospace Applications". *Int. J. Hydrogen Energy*, **14**, 201-213.

Steinberg, B.A. and D.S. Scott (1984), "Hydrogen vs. Diesel Fueled Locomotives: A Techno- Economic Appraisal". *Int. J. Hydrogen Energy*, **9**, 101-107.

Taube, M. et al. (1985), "A Prototype Truck Powered by Hydrogen from Organic Liquid Hydrides". *Int. J. Hydrogen Energy*, **10**, 595-599.

Tsujikawa, Y. and T. Sawada (1982) "Analysis of a Gas Turbine and Steam Turbine Combined Cycle with Liquified Hydrogen as Fuel". *Int. J. Hydrogen Energy*, **7**, 499-505.

Tsujikawa, Y. and T. Sawada (1984), "Off-Design Point Characteristics of the Precooled Gas Turbine Cycle with Liquified Hydrogen as Fuel". *Int. J. Hydrogen Energy*, **9**, 1011-1017.

Tsujikawa, Y. and M. Hirano (1988), "Effects of Precooling of Suction Air on the Performance of Liquid Hydrogen Fueled Supersonic Aircraft Engine". *Int. J. Hydrogen Energy*, **13**, 691-700.

Tupolev, A. (1988). Oral presentation at World Hydrogen Energy Conference No. 7, Moscow.

Turillion, P. (1982). "Design of Hydride Container for Hydrogen Storage". *Hydrogen Energy Progress*, **IV**, 1289-1308, Pasadena.

Ullman, A.Z. and W.D. Van Vorst (1976), "Methods of On-Board Generation for Vehicular Use". *Proc. 1st World Hydrogen Energy Conference*, **2**, 33-53, Miami.

Victor, D.G. (1990), "Liquid Hydrogen and the Greenhouse Effect". *Int. J. Hydrogen Energy*, **15**, 357-367.

Watson, H.C., E.E. Milkins, W.R.B. Martin and J. Edsell (1984), "An Australian Hydrogen Car". *Hydrogen Energy Progress*, **5**, 1549-1562, Toronto.

Wilkinson, K.G. (1983), "An Airline View of LH_2 as a Fuel for Commercial Aircraft". *Int. J. Hydrogen Energy*, **8**, 793-796.

Winter, C.J. (1990), "Hydrogen in High-Speed Air Transportation". *Int. J. Hydrogen Energy*, **15**, 579-586.

Wiswall, R.H. and J.J. Reilly (1972), *Inorg. Chem.*, **11**, 1691.

Yan, K.L. et al. (1985), "Storage of Hydrogen by High Pressure Microencapsulation in Glass". *Int. J. Hydrogen Energy*, **10**, 517-522.

FUEL CELL VEHICLE

F. BARBIR
Energy Partners, Inc.
1501 Northpoint Parkway, Suite 102
West Palm Beach, FL 33407, U.S.A.

Abstract. This paper describes the design and performance of a prototype zero emission electric vehicle, powered primarily by air breathing Proton Exchange Membrane (PEM) fuel cells using gaseous hydrogen as fuel. The fuel cell system is composed of the fuel cell stacks, hydrogen tank, air compressor, solenoid valves, pressure regulators, water pump, water tank, heat exchangers, sensors, programmable controller, and voltage regulator. The battery system provides power to the vehicle during periods of peak power demand such as vehicle acceleration or traveling at a high constant speed. The batteries also provide the power to initiate fuel cell startup. The vehicle has been designed, assembled, tested, and has traveled over 400 km solely on fuel cell power with satisfactorily performance. It has successfully proved the concept of a fuel cell powered zero emission vehicle. Further enhancements will improve the performance in terms of increased speed, acceleration, fuel efficiency, range and reliability.

1. Introduction

Automobiles are one of the major sources of air pollution in urban areas. As air pollution in heavily populated areas becomes unbearable, the search for cleaner alternatives emerges as an imperative.

The goal of this project was to design, develop and demonstrate a zero emission prototype vehicle powered by PEM fuel cells. PEM fuel cells are very efficient, compact and low weight, operate at almost ambient temperature, and use hydrogen - the cleanest fuel. As such the PEM fuel cells offer promise as the best future replacement for internal combustion engines in transportation applications.

Although the fuel cell was invented more than 150 years ago, it has not left the research laboratories yet, except for special applications such as the space program. Only recently, mainly because of environmental concerns, the fuel cell technology in general, and PEM fuel cells in particular, have come closer to commercialization. The technology exists, although it still needs some improvements, but what is really needed market and infrastructure development. The ultimate goal of this project is to bring fuel cell technology out of the laboratory into the marketplace for use in transportation as well as in distributed stationary power generation.

Y. Yürüm (ed.), Hydrogen Energy System, 241–251.

The project included development and manufacturing of the PEM fuel cell system and its integration into an existing lightweight vehicle, referred to as the Green Car, followed by testing of the vehicle's performance.

2. Vehicle Design

The Green Car was designed to incorporate a lightweight body with low aerodynamic drag and small frontal area (Figure 1). It is a hybrid electric vehicle powered primarily by PEM fuel cells. Figure 2 illustrates the functional block diagram of the hybrid propulsion system showing all of the major components, which include, fuel cell stacks, fuel cell support equipment, fuel cell voltage regulator, electric drive motor, battery bank, and microprocessor-based programmable controller (MPC). The motor is a highly efficient DC brushless motor capable of providing regenerative braking which can be used to charge the auxiliary battery bank. The vehicle operates primarily on fuel cell power and draws power from the batteries only for peak power requirements, such as acceleration. The fuel cell power system works in parallel with a battery bank. The fuel cells use hydrogen as the fuel and ambient air as the oxidant. The support components of the fuel cells consist of the high pressure gaseous hydrogen storage tank, air compressor, solenoid valves, pressure regulators, sensors, water pump, water tank, heat exchangers and piping system. The voltage regulator is required to condition the voltage output from the fuel cell stacks. It provides a constant voltage output independent of the large fuel cell voltage swing. Any excess power from the fuel cell stacks can be used to charge the batteries. Two DC/DC converters provide 12 VDC to power the vehicle accessories, instrumentation, and fuel cell support equipment. A microprocessor-based programmable controller starts, monitors, and shuts down the fuel cell system. It monitors all system parameters, which include temperatures, pressures, voltages, currents, and flow rates.

2.1 VEHICLE BODY

The vehicle body was obtained from Consulier Automotive, West Palm Beach, Florida. The high-strength, low weight, monocoque construction is made of foam core polycarbonate advanced composite material. The material is impervious to corrosion, does not rust, and will last indefinitely. Advanced composite materials are fully recyclable and can actually be ground up and reused. The weight of the body used for the Green Car is approximately 85 kg.

2.2 ELECTRIC MOTOR AND MOTOR CONTROLLER SYSTEM

The electric motor and controller that were selected for the Green Car were designed and built by Unique Mobility, Inc. The 26 kW motor is an air cooled, DC, brushless type motor and can deliver a constant torque of 74 Nm from 0 to 3,300 rpm. The maximum speed is 4,000 rpm, and maximum efficiency is 95%. This motor was selected because of its lightweight (approx. 20 kg) and compactness. The motor controller weighs another 20 kg, and is required for both commutation and control functions. The controller is capable of driving the motor both in the forward and reverse directions. Regenerative braking is also provided by the controller and the level of regenerative braking can be set by an external signal to the controller. The motor speed is controlled by the accelerator pedal which provides an analog voltage (-10 to +10 volts full reverse to full forward). Hall effect sensors in the motor are used for determining the timing information needed for motor commutation. These signals are processed by the controller and compared with the desired speed. In the event that the controller or motor overheats, the controller will automatically limit current by itself.

Figure 1. Energy Partners' Green Car

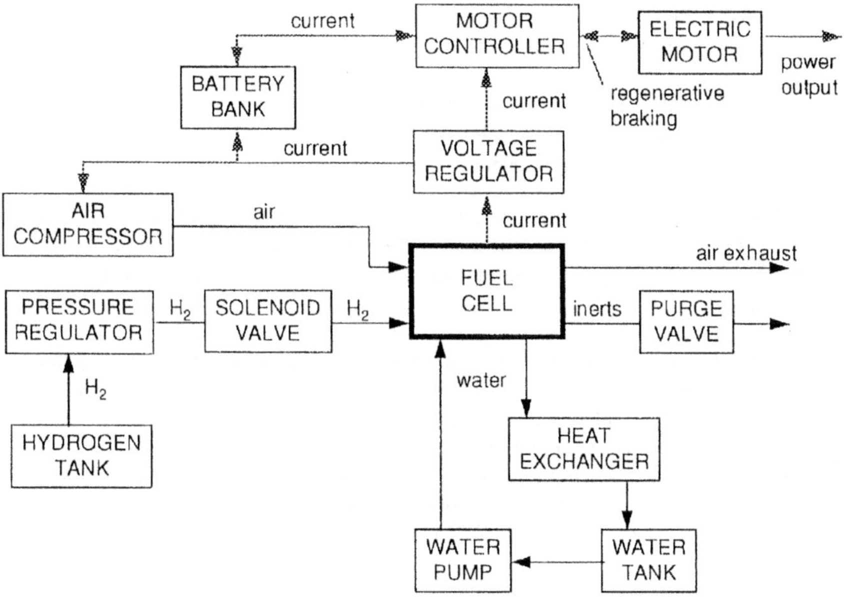

Figure 2 Functional block diagram of a hybrid fuel cell/battery electric vehicle (Energy Partners *Green Car*™)

2.3 TRANSMISSION SYSTEM

A variable speed transmission has been used in the design because of the motor's inability to operate at speeds in excess of 4,000 rpm without overheating. The transmission is a fully synchronized 5 speed manual transaxle with combined gear reduction, ratio selection and differential functions in one unit housed in a die-cast aluminum case. A reversing gear, although available, is not required since the electric drive motor is reversible. The motor is mounted directly on the transmission with a flexible coupling installed on the input shaft.

2.4 BATTERY SYSTEM

The battery system consists of ten 12 V batteries installed to provide both start-up power to the fuel cell system and auxiliary power for acceleration. The batteries are lead acid, sealed, maintenance-free, and gelled electrolyte type. Each battery weights 30 kg, and according to manufacturer specifications, can deliver 25 Amps for 200 minutes (1 kW-hr of energy) when totally charged.

2.5 FUEL CELL POWER SYSTEM

The Green Car fuel cell power system consists of the fuel cell stacks and their supporting subsystems, namely:

- Fuel storage and management subsystem,
- Oxidant management subsystem,
- Water management subsystem,
- Voltage regulation subsystem,
- Control & monitoring subsystem.

The fuel cell stacks are connected electrically in series and mechanically in parallel. Figure 2, the hybrid propulsion system functional block diagram, shows how the fuel cell power system is integrated into the electric vehicle.

2.5.1 Fuel Cell Stacks. The three proton exchange membrane (PEM) fuel cell stacks used for this project were designed and manufactured by Energy Partners. The PEM fuel cell uses a solid, fixed electrolyte and exhibits low temperature and pressure operation, rapid start-up, and rapid load response.

A fuel cell stack is composed of a series of cells separated by bi-polar current collector plates. The collector plates in the Green Car stacks have an active area of 780 cm^2 and are made of graphite with a polymer binder, which produces a corrosion resistant material. The collector plates contain internal manifolds and provide a proper flow field for fuel and oxidant, as well as providing pathways for water removal. Each cell consists of gas chambers and electrodes separated by Nafion™ 115, a proton conducting polymer membrane. Electrodes are made of water repellent carbon fiber paper with platinum based catalyst layer deposited on one side.

The fuel cell stack assembly hardware includes the fluid and compression endplates, which hold the stack together by sandwiching properly ordered cells and components with the use of tie-rods, nuts, and washers. The bus plates are made of copper and are designed so that wire cables can be attached easily.

Each of the stacks in the Green Car contains 60 cells for a total of 180 cells. Cooling of the stack is provided by water flowing through specially designed cells strategically distributed through the active section of the stack. The stack also contains an internal humidification section which provides the proper conditions for the electrochemical reaction and prevents dehydration of the membrane. Table 1 lists the Green Car fuel cell stack specifications.

2.5.2 Fuel Storage and Management Subsystem. The fuel cells require pure hydrogen (at least 99.9% purity). The hydrogen storage system chosen for the Green Car is compressed gaseous hydrogen at a storage pressure of 200 bar in a composite pressure vessel. This method was chosen because it is simple in design, the technology is well advanced, and when combined with a suitable pressure regulating device provides a passive, simple load following system. The compressed gas fuel storage system is designed to hold approximately 11,600 standard liters of pure hydrogen. The vessel is made of an inner 6061-T6 aluminum liner with hemispherical ends, wrapped with a composite spiral wound E-glass/epoxy resin. This design results in a lightweight structurally sound cylinder. The cylinder is designed to the latest applicable DOT specifications for compressed gas fuel storage for automotive applications.

The fuel management subsystem receives hydrogen from the pressure vessel and regulates the pressure down to system pressure of 240 kPa. The pre-regulated pressure is measured and supplied to the MPC to provide information on remaining fuel stores; the post regulated pressure signal is supplied to the MPC to ensure proper system pressure. Hydrogen is directly fed to the stacks in a dead-ended mode, with intermittent purging for removal of contaminants.

2.5.3 Oxidant Management Subsystem. In the Green Car air is used as the oxidant. The ambient air is drawn through a particulate filter by a compressor and raised in pressure to approximately 200 kPa. A rotary vane type compressor has been selected, rated at 800 slpm and 240 kPa. Power requirement at rated conditions is 3.5 kW. The air flow rate is kept constant at a level that supports the electrochemical reaction at rated power, a stoichiometric ratio of about 1.5. The stoichiometric ratio at lower power levels is therefore always higher than 1.5. The pressure and flow rate values for the air are supplied to the MPC by suitable sensors. The air is then directed through the heat exchanger and water/air separator and into the stacks. Oxygen in the air stream is used to support the fuel cell reaction while the rest of the stream carries the product water from the stack.

2.5.4 Water Management Subsystem. Water is used for cooling the fuel cell stacks and for humidification of the reactant gases within the stack. Water is drawn from a reservoir by a pump and directed through a particulate filter and into the fuel cell. The water moves through the dedicated cooling passages within the stack where it removes the waste heat. After flowing through the cooling passages, the cooling water moves through the humidification section of the stack where through hydrated membranes humidifies the reactant gas streams. The remainder of the cooling water exits the stack where its temperature and flow rate are measured. Depending on the temperature of the exiting water, a thermal control valve directs the water either through a suitable heat exchanger to release the heat or directly back to the water reservoir.

2.5.5 Fuel Cell Voltage Regulator System. The purpose of the fuel cell voltage regulator is to provide a constant voltage output to the drive train from the fuel cell stack by conditioning the large dynamic voltage range of the fuel cell. Under no-load condition, the fuel cell voltage can be as high as 180 volts, and under full load condition the voltage can drop below 120 volts. The regulator is adjustable in the range of 110 to 130 VDC and can be set at a point slightly above battery nominal voltage to maintain the batteries in a float charge condition without overcharging them. This satisfies the requirement of low power demand being supplied only by the fuel cell to the limit of the regulator output. Thus, the primary power of the vehicle is drawn from the fuel cell stack. Any excess current from the fuel cell is used to charge the battery bank. The capability of adjusting regulator output voltage is important because it limits the amount of maximum charge that the fuel cell stacks can deliver to the batteries. The efficiency of the voltage regulator is about 95%.

Since the regulator is connected between the fuel cell and battery bank, and the battery bank is directly connected to the motor drive system, the absence of fuel cell power will not interfere with the operation of the vehicle on battery power alone. Also, there is no possibility of reverse current flow from the battery bank or drive motor (in regenerative braking mode) to the fuel cell through the regulator. The regulator has internal circuitry to prevent operation at too low an input voltage and can be enabled or disabled by an external signal. In the event of a system shutdown by the controller, this last feature allows the load to be disconnected from the fuel cell.

2.5.6 Microprocessor-Based Programmable Controller. The microprocessor-based programmable controller (MPC) that is installed in the Green Car is capable of monitoring the sensor inputs for specific limit conditions (temperatures, pressures, flows and water level) and taking appropriate action in the event of a fuel cell malfunction. An additional dedicated analog multiplexer with 180 differential channels monitors the individual fuel cell voltages. The MPC also provides analog outputs for control of external devices, such as pumps and solenoids, and a display output capable of interfacing with a remote LCD display located on the vehicle dashboard. The MPC displays the fuel cell system status information on the instrument panel. In the event of a shutdown, the MPC stores the current system parameters and these can be read directly from the LCD display or downloaded to any IBM compatible computer through the RS232 communication port on the MPC.

The microprocessor is an Intel 8052H-BASIC chip operating at 12 MHz. The instruction set to implement the control logic is stored in an EPROM, where any subsequent changes to the control strategy can be made easily. A "watchdog" timer circuit is employed to prevent any uncontrolled operation in the event of a controller failure. All programming was done in BASIC and in assembly languages. However, only the fixed hardware dependent drivers were written in assembly language for optimum speed and minimum memory use.

The capability to monitor individual cell voltages is one of the most important functions of the MPC. The monitoring of total stack voltage alone does not provide sufficient information to detect a malfunction of an individual cell within the stack. A cell with a low voltage reading is an indication of a potential hazardous condition, such as a gas leak in that cell, a perforated membrane, or a reversal in cell potential. For air operation, the individual cell voltages are usually between 0.5 and 1.0 V. A shutdown is performed if any cell voltage is below 0.4 V or if the voltage difference between two adjacent cells is greater than 0.2 V. The differential voltage test is important because a defective or erratic cell can be detected before the fuel cell is subjected to a large load.

2.5.7 Instrumentation. The instrument panels on the vehicle dashboard and center console contains the digital meters displays, status and warning indicator lights, and switches for the driver to start and monitor the fuel cell. Meter displays include fuel cell amps, volts, and temperature; battery amps and volts; motor amps, temperature, and rpm; inlet air pressure; and hydrogen pressure. An LCD display with 2 rows of 20 characters located on the front dashboard provides the driver with status information on fuel cell performance. In the event of a fail-safe shutdown, the display will output the sensor reading that caused the shutdown. The center console contains status and warning lights for the stacks and fuel cell voltage regulator.

3. Vehicle Performance

After all the components were installed in the Green Car, a series of tests were conducted in order to determine the performance of the Green Car. First, the performance of each component and subsystem was tested independently. After necessary modifications, driving tests were conducted. The performance of the Green Car is described below, and summarized in Table 2.

3.1 SPEED AND ACCELERATION

Only limited driving tests have been performed due to unavailability of a suitable test track. The car has only been driven on the parking lot around the Energy Partners facilities. The Green Car has successfully completed over 400 km solely on fuel cell power. Average driving speed was about 50 km/h. Maximum speed, however, is determined by the maximum power of the main propulsion motor (26 kW), vehicle weight, drag and rolling resistance. A computer program has been developed to predict the Green Car performance based on these constraints. Figure 3 shows vehicle speed as a function of required power from the fuel cell system to the electric motor and auxiliary (hotel) loads. For 26 kW the maximum attainable speed is 95 km/h. During the test drive it was not possible to develop maximum speed, due to limitations of the test track. According to Figure 3, the fuel cells can provide power for constant speeds up to 70 km/h, which means that excess power has to be supplied by the battery bank. The battery bank starts to supply electric power as soon as the fuel cell voltage regulator output drops below battery nominal voltage (120 V).

Similarly, the acceleration performance is also determined by the main propulsion motor's capabilities. The Green Car needs approximately 10 seconds to accelerate from 0 to 50 mph. Acceleration needs more power than cruising at constant speed. Again, the fuel cell system is capable of providing up to 15 kW, and the remaining power required for acceleration has to be supplied by the battery bank.

3.2 FUEL CONSUMPTION AND RANGE

Fuel consumption in the fuel cell stacks (in standard liters per hour) is a function of current, according to the following equation, originally derived from the Faraday's Law:

$$FC = 0.45\ (I + I_{loss})\ N_{cell}$$

where:

I = current (Amps),
I_{loss} = current losses due to diffusion of hydrogen and oxygen through the membrane;
I_{loss} = 6.5 Amps (based on experimentally determined diffusion rates).
N_{cell} = number of cells; N_{cell} = 180.

Approximately 10 kW are needed for constant cruising speed of 50 mph only on fuel cell power, (Fig. 3) assuming that the batteries are fully charged, i.e., power from fuel cells is used only to run the main electric motor and hotel load. The fuel cell stacks generate 10 kW at 77 Amps and 130 V. Therefore, the calculated fuel consumption is 6,760 l/h. Specific fuel consumption per mile traveled at constant speed of 50 km/h is 135 l/km. Since the amount of hydrogen stored on-board is 11,600 standard liters, the maximum vehicle range at constant 50 km/h speed is 85 km if the vehicle runs solely on fuel cell power. In addition, the battery bank has a practical storage capacity of 35 Ah, which is enough for at least an additional 30 km, and makes the total vehicle range of 115 km for a constant speed of 50 km/h. For city driving, the predicted range obtained from the computer model is about 100 km.

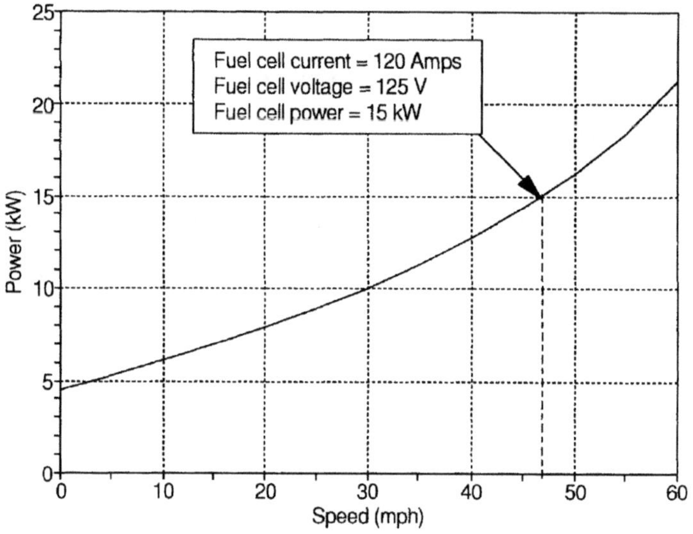

Figure 3. Fuel Cell Power Requirement for the Green Car Travelling at Constant Speed

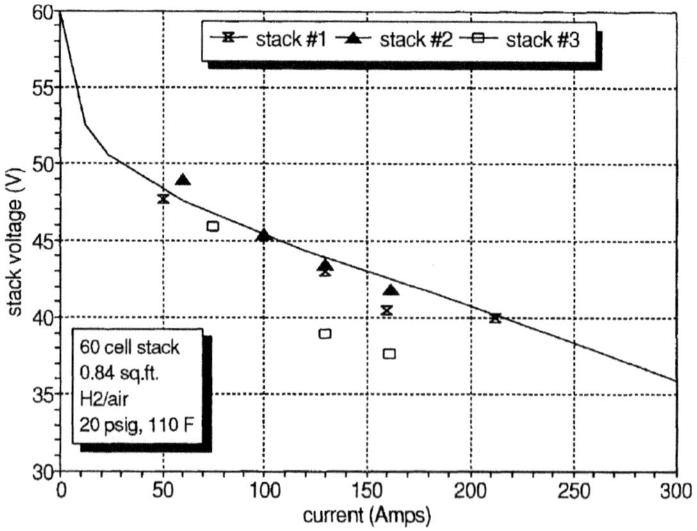

Figure 4. EP Green Car Fuel Cells Performance
(solid line = performance goal, markers = actual test data)

3.3 OTHER PERFORMANCE CHARACTERISTICS

Responsivity. PEM fuel cells have demonstrated very fast response to load changes, even at cold start-up. The start-up procedure takes about 15 seconds, measured from the moment the power switch is on until the fuel cell system is ready to supply power. This time is an arbitrary value programmed into the controller, that gives the controller time to check all the operating parameters. Regular shut-down takes less than one second, but the heat exchanger fans may continue to run until the cooling water temperature drops below a certain pre-set value (45 °C).

Emissions. The Green Car is truly a zero emission vehicle. The exhaust is actually depleted air from the fuel cell stacks, which also contains the product water from the electrochemical reaction. Water is released both in liquid form (about 0.1 lit/km) and as water vapor (100% saturated exhaust gas). There are no pollutants of any kind present, whatsoever.

Noise. Although an electric vehicle is supposed to be very quiet, the Green Car generates an unexpectedly loud "buzz." The moving parts in the vehicle include an electric motor with integral cooling, air compressor, compressor electric motor and cooling blower, water pump and heat exchanger fans. By far, the "noisiest" component is the rotary vane type air compressor, which generates a noise level of about 85 dB at 1 m distance.

Safety and reliability. Safety of hydrogen powered vehicles is usually a matter of concern. However, during the assembly and testing of the Green Car no accidents or any hazardous or potentially hazardous situations were experienced. Regular leak checks are required for all hydrogen piping and fittings, including the fuel cell stacks. The fuel cell is monitored and controlled by a programmable controller (MPC), which automatically performs a shut-down whenever it detects a parameter out of a prescribed range. During the test drives, intermittent shut-offs were experienced. In case of a shut-down, the vehicle continues to operate on battery power. The three main causes for these shut-downs were overheating of the compressor electric motor, fuel cell problems indicated by low cell voltage, and false-alarms caused by inaccurate voltage readings in the MPC. These causes were constantly corrected, and the time between two shut-downs has dramatically increased.

4. Factors Affecting the Performance

Overall performance was affected either by the component suppliers not being able to meet Energy Partners' specifications and/or by the fuel cell stacks not meeting the design goals when integrated into the vehicle. The major issues concerning performance are discussed below.

4.1 FUEL CELL STACKS

Manufacturing of the fuel cell stacks is still in an experimental stage, which explains the differences in performance between the three 60-cell stacks installed in the Green Car. Figure 4 shows the actual performance profiles of the three Green Car stacks under air operation obtained in the test stand. The solid line was generated from a computer model and shows the performance goal. One of the stacks was able to deliver 212 Amps at 40 Volts, but the maximum attainable current from each of the other two stacks was only 160 Amps. Since the stacks are connected in series, the stack with the lowest current rating set the system performance. Referring to Figure 4, at 160 Amps combined stack voltage is approximately 120 Volts, which implies a total power output of 19.2 kW. Although this is near the design goal of 20 kW, this performance was not achieved after installing the stacks in the Green Car, because the Green Car system lacks the capability of maintaining and controlling ideal operating conditions, such as temperature, pressure and flows, as compared with the test stand. The maximum power output of the Green Car fuel cell power system that has been achieved is approximately 15 kW (120 Amps at 125 V). The performance of the entire stack is usually held down by a few cells that experience water management problems. Since the stacks performance is monitored and controlled by an automatic programmable controller (MPC), low voltage in only one cell creates a shut-down of the entire fuel cell system, although the remaining 179 cells are capable of delivering much more power.

4.2 DRIVE MOTOR/CONTROLLER

The limitations of the motor/controller system are the low power level of 26 kW, and the inability to operate at speeds in excess of 4,000 rpm without overheating the motor. Because of this limitation, a variable speed transmission has been used. The car performance, in terms of maximum speed and acceleration would dramatically change with a more powerful motor. The goal to eliminate the variable speed transmission in favor of directly matching the motor to the differential is possible with the new generation electric motors (>60 kW and >7,500 rpm).

4.3 AIR COMPRESSOR

The compressor was selected to deliver 800 slpm at 240 kPa (which corresponds to a stoichiometric ratio of 1.5 at 20 kW design point). Power required for the compression is 3.5 kW (efficiency of the selected compressor is therefore less than 50%). However, during the tests, the air compressor motor was constantly overheating, and it was necessary to reduce power requirements below 3 kW. This was possible by lowering the air compressor motor's rpm setpoint, which in turn limited the output pressure to 180 - 200 kPa and air flow to 680 slpm. The performance of the fuel cell stacks was affected by this change.

5. Conclusions

The goal of this project was to power a lightweight passenger vehicle using a PEM hydrogen/air fuel cell system. The proof-of-concept vehicle, known as the Green Car™, was the result of substantially modifying an existing composite two-seater vehicle by installing three stacks producing 15 kW of pollution-free power, along with appropriate instrumentation, programmable controller, electric drive motor, auxiliary battery bank, and support hardware. With a curb weight of 1,360 kg, the demonstration vehicle has a top speed of 95 km/h, a range of 100 km in city driving, and can accelerate from 0 to 50 mph in 10 seconds with its 26 kW electric motor. The vehicle has been successfully driven more than 400 km on fuel cell power during the initial test period. The concept has been proved.

It is now clear that PEM fuel cells are serious candidates for the propulsion of zero emission vehicles. Although the performance of the Green Car is not spectacular, it exceeds the internal combustion engine powered vehicles in two important characteristics, namely fuel efficiency, and emissions. The Green Car has clearly demonstrated the feasibility of the fuel cell powered propulsion systems for transportation vehicles. The potential benefits are enormous, including reduced impacts on environmental quality, reduced dependency on petroleum imports, and the advancement of fuel cell technology into other application areas.

Phase I of the Green Car Project has been completed and Phase II is ready to begin. Its primary objective will be to improve the fuel cell stacks and all support systems. The final goal is to promote PEM fuel cell technology and accelerate its commercialization, making it accessible to all applications requiring a clean and efficient source of energy.

TABLE 1 FUEL CELL STACKS SPECIFICATIONS

Manufacturer	Energy Partners
Type	PEM
Membrane	Nafion™ 115
Collector Material	Graphite with binder
Reactants	Hydrogen/Air
Rated Power Output	20 kW
Voltage @ Rated Output	120 Volts
Current Density @ Rated Output	215 mA/cm^2
Cell Voltage @ Rated Output	0.7 Volts
Stack Efficiency @ Rated Output	54 % (LHV)
Average Operating Temperature	55 °C
Reactant Pressure	240 kPa
Cell active area	780 cm^2
Number of Cells	60 (per stack)
Number of Stacks	3
Stack Dimensions	35 x 35 x 50 cm
Stack Weight	105 kg each (315 kg total)

TABLE 2 GREEN CAR PERFORMANCE

PARAMETER	PERFORMANCE
Average cruising speed	50 km/h
Electric power required for cruising speed	10 kW
Vehicle maximum speed	95 km/h
Acceleration 0 to 50 km/h	10 sec
Fuel consumption at cruising speed	135 l/km
Vehicle range at constant speed	115 km
Vehicle range for city driving cycle	100 km
Refueling time	5-10 min
Start-up/shut-down time	15 sec / <1 sec
Emissions (gaseous)	11% O_2, 84% N_2, and 5% H_2O
Emissions (liquid)	H_2O: 0.1 lit/km
Noise	85 dB at 1 meter

APPLICATIONS OF HYDRIDES

GARY SANDROCK
SunaTech, Inc.
113 Kraft Place
Ringwood, NJ 07456
USA

ABSTRACT. This chapter surveys the applications for rechargeable metal hydrides, covering both those only demonstrated and those commercially realized. Applications include hydrogen storage (stationary and vehicular), hydrogen purification and separation, H-isotope separation, H_2 compression, various thermodynamic devices that handle heat (storage, pumping, refrigeration, mechanical conversion, temperature sensing) and hydride batteries. All of these applications have been demonstrated on a prototype level, but only temperature sensors and hydride batteries are really commercially successful to date. The reasons for the failure of some hydride applications will be discussed in relationship to shortcomings in available hydriding alloys and the R&D needed to overcome those shortcomings.

1. Introduction

The extraordinary properties of rechargeable metal hydrides have generated a large number of proposed applications. This review of metal hydride applications should be read in conjunction with my companion introductory chapter on HYDROGEN-METAL SYSTEMS presented elsewhere in this volume [1]. The reader should also be aware of a comprehensive review of mine published in 1992 [2]. This chapter will not attempt to duplicate that large review, but should be considered a condensation and updating of it.

Metal hydrides may be considered as reversible metal "sponges" for hydrogen gas, so the concept of hydrides as a safe and compact storage medium is obvious. Because hydride-forming metals are uniquely selective to hydrogen (and its isotopes) only, they can be used for the separation of H_2 from mixed gases and its purification. Hydrides involve true chemical reactions that are associated with significant amounts of heat, so there are a number of heat-associated thermodynamic hydride devices that have been developed from this principle. The electrochemical charging and discharging of hydride electrodes has resulted in the commercially successful rechargeable nickel-metal-hydride battery.

Unfortunately, many of the hydride applications to be discussed below have failed to reach true commercial success [3], in spite of many years extensive prototype work in national laboratories and industry around the world. One of the objects of this chapter will be to answer the question "why or why not?" as we discuss each application. The answers often involve the following hydride alloy property shortcomings:

Y. Yürüm (ed.), Hydrogen Energy System, 253–280.

1. Low hydrogen capacity
2. High alloy cost
3. High sensitivity to gaseous impurities
4. Disproportionation
5. Difficulty in initial activation or reactivation
6. Pyrophoricity when suddenly exposed to air

It is hoped that a frank admission that hydriding alloy shortcomings have limited the successes of some hydride applications will act as a stimulus to new generations of researchers in the field of metal hydrides.

2. Sources of Hydriding Alloys

It is fortunate for those interested in developing the applications for metal hydrides that the production metallurgy of hydriding alloys is now pretty well developed [4] and large scale production is under way [5]. Hydriding alloys are available from several companies around the world. Table 1 gives at least a partial list of the more established suppliers of MH alloy, in terms of country and types of alloy sold. The list shown in Table 1 should

Table 1
Hydriding Alloy Suppliers

Country	Company	Trade Name	AB_5	AB_2	AB	A_2B	Ref.
Austria	Treibacher	--	X				6
China	Beijing GRINM	--	X	X	X		7
China	Eastern Hydride	--	X	X			8
China	Nankai Univ.	--	X	X			9
China	Tianjin Xinbau	--	X				10
China	Zhongshan Tianjiao	--	X				11
Germany	GfE	Hydralloy	X	X	X	X	12
Germany	H.C. Starck	--	X				13
Japan	Chuo Denki	--	X		X		14
Japan	JMC New Mat'l	HYDMAC	X	X	X	X	15
Japan	Mitsui Mining	MMS	X				16
Japan	Santoku Metals	--	X				17
USA	Aldrich Chemical	HY-STOR	X		X	X	18
USA	Ergenics	HY-STOR	X	X	X	X	19
USA	Rhone-Poulenc	--	X				20
USA	Teledyne Wah Chang	--		X			21

not be considered 100% complete and accurate. There are companies entering and exiting the hydriding alloy business at times. The situation in China is especially active with the formation of many new companies and joint ventures. (I would appreciate any new information readers might offer to correct and expand this list).

Most of the companies listed in Table 1 offer a series of standard alloys. Sometimes the pressure-temperature-composition properties must be carefully matched to a given application. Many of the companies listed will help by attempting to prepare custom alloys designed to meet the special properties required.

3. Container Design Considerations

The design of hydride containers must take into account the factors of hydrogen embrittlement, heat transfer and expansion.

The basic container obviously should not be constructed of a material known to embrittle in the presence of hydrogen [1]. Generally speaking, copper, aluminum and stainless steel are preferred.

Charging and discharging rates of hydride reservoirs are dictated largely by the rates at which the reaction heat can be removed (during charging) or added (during discharging). This can be done with internal or external, liquid or air, heat exchangers. Because powder beds have poor inherent thermal conductivity, additional means of internal means of thermal conductivity enhancement must usually be included. Internal heat exchange fins, tubing and plates can be used. One of the most popular means for hydride containers is the use of aluminum foam [22]. Porous metallic-matrix hydrides (PMMH) can be made by the blending of hydride powder and high conductivity Al or Cu and pressing into desired shapes [23]. Hydriding alloy powder can also be coated with Cu to increase conductivity [24]. The container design shown in Figure 1 combines pressed aluminum

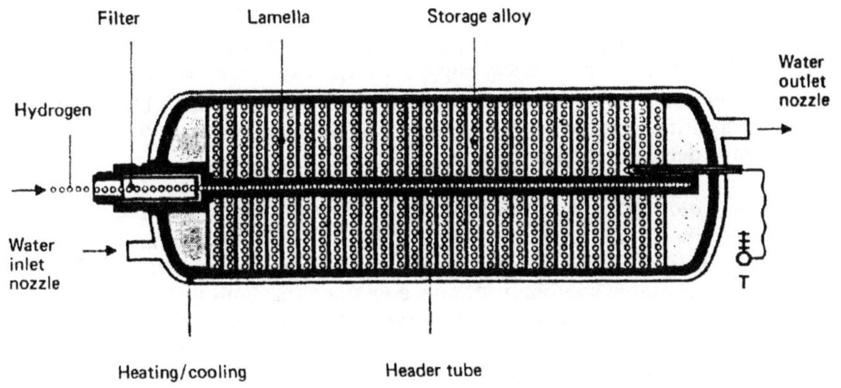

Figure 1 - Hydride Storage Container Design [25]

and hydriding alloy powder disks into an Al-finned tubing [25]. This German design (developed by HWT, now GfE [12]), although relatively expensive, combines both macroscopic heat exchange (fins) and microscopic heat exchange (fine Al network).

In the case of rapidly cycling hydride containers, it is often useful to store the heat generated during exothermic absorption for reuse during endothermic desorption. This can be accomplished by including a phase change material or thermal ballast within the hydride bed or in close thermal connection [26]. For near-ambient temperature hydride beds a low melting salt (such as Glauber's salt) or a simple thermal ballast such as Ni-powder [27] can be used. For high-temperature Mg-hydride beds, a mixed molten salt such as KBr-KCl-KNO_3 can be used [28].

Hydriding alloy particles expand up to 25% in volume during hydrogen absorption and contract the same amount during desorption. In the case of powder migration, packing can occur and lead to expansion stresses and strains on the container walls. This can lead to rupture of the container [29]. In addition, packed powder beds can have low gas permeability, contributing to poor charging rates. Therefore, container designs must take into account such expansion effects. The container can be lightly filled or designed for extensive elastic expansion [30]. The powder can be compartmentalized [31], lubricated to prevent locking [32,33] or pelletized using plastic, rubber or metal binders [2]. Elastic or collapsible internal structures can also be used to control expansion [34, 35].

4. Hydride Applications

The last 25 years (since the practical beginnings of intermetallic hydrides [1]) have seen numerous attempts to develop hydride applications. Prototype devices have been made in a number of areas: hydrogen storage, separation and compression; isotope separation and handling; various thermodynamic devices such as heat pumps, engines and temperature sensors; and hydride batteries. These areas will be reviewed, mainly in the form of a brief summary and extension of a recent review paper on the subject [2]. The new Savannah River Tritium Facility, which has components of hydrogen storage, compression and isotope separation will be discussed in Section 4.4 on Isotope Separation and Handling.

4.1 HYDROGEN STORAGE

The use of rechargeable metal hydrides for the storage of hydrogen gas offer a third (solid) alternative to cryogenic liquid and high pressure gaseous storage. The hydride option offers high volumetric density (approaching LH_2), low pressure (with inherent safety) and no need for refrigeration equipment and insulated tanks. The relative safety advantage of hydride storage seems to have driven most of the early stationary and vehicular storage demonstration programs.

4.1.1. *Stationary Storage* . The storage of H_2 in stationary hydride storage tanks can be considered a modest success because there are units available commercially today, ranging from small hand held cylinders to large custom-made tanks holding tons of alloy.

The first relatively large stationary storage container was built in 1974 by Brookhaven National Laboratory for an electric peak shaving demonstration by the Public Service Electric & Gas Company of New Jersey [36,37]. This unit contained internal water tubes for heat exchange and used TiFe hydride. A comparison of this unit with an equivalent

Table 2
Comparison of BNL PSE&G Hydride Tank with Compressed Gas Cylinders

Storage Method	Pressure, atm	H_2 Stored, kg	Total Weight, kg	Total Volume, m^3
Hydride Tank	37	6.4	570	0.144
10 T-Cylinders	164	6.5	658	0.49

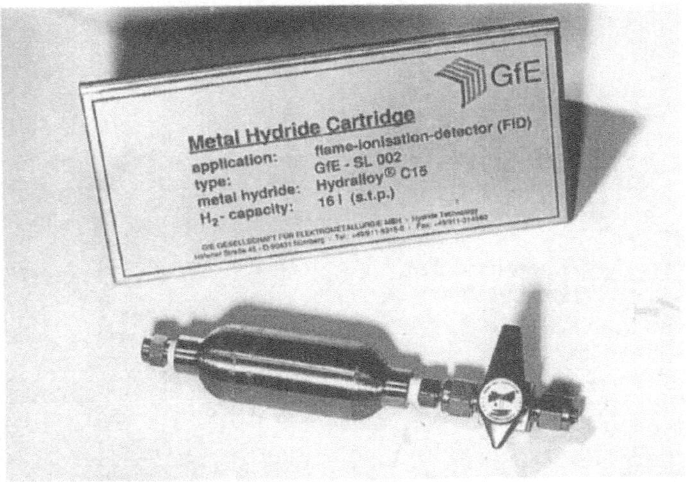

Figure 2a - **GfE 16L Hydride Tank** [12]

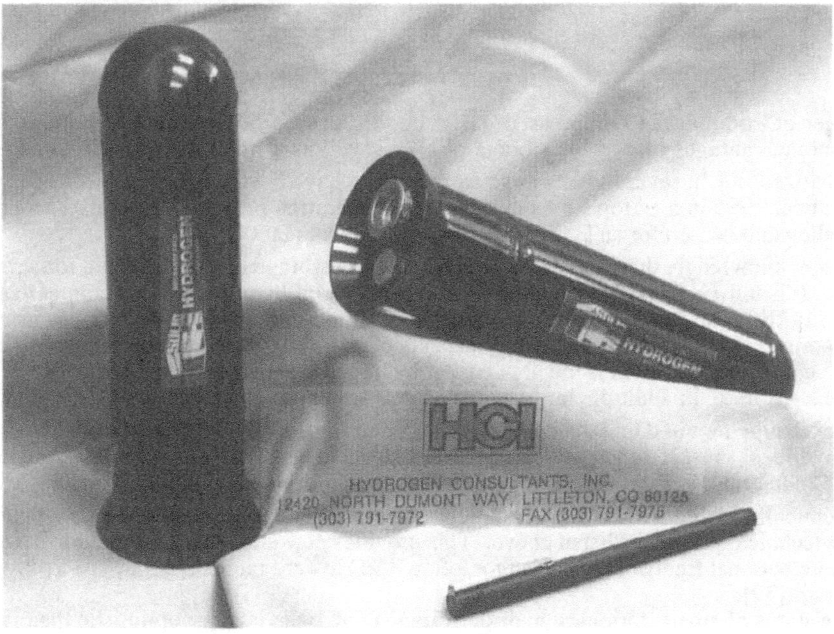

Figure 2b - **HCI 100L Hydride Tank** [40]

Figure 2c - Ergenics 1275L Hydride Tank [19]

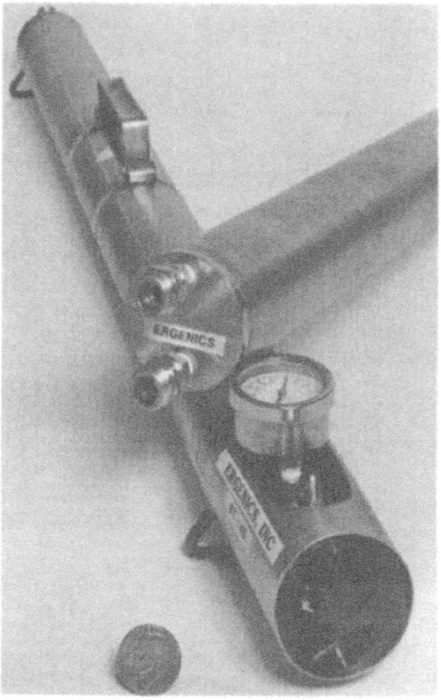

number of commercial compressed gas cylinders shows the pronounced volume and pressure advantages of hydride storage of H_2, as well as a small weight advantage. As summarized in [2], several other large storage units have been built. The last large unit, built about 1985 in a European Community demonstration program, contains 10 tons of AB_2 alloy and is used for an H_2 source in a metallurgical plant [25].

To my knowledge, there has been no large hydride storage unit constructed for several years. It is unlikely that bulk hydrogen storage in metal hydrides will be commercially viable in the near future, relative to other (gas, liquid and cryoadsorbants) because of economics [38]. Although the safety aspects of metal hydride storage are strong and positive, the main problem is the high cost of hydrogen storage alloys relative to the value of the H_2 stored. In addition, large hydride tanks are expensive because of the extensive heat exchange required and the overdesign required for the higher pressures of activation. The hydrides' sensitivity to gaseous impurities can also be a problem when relatively low purity industrial H_2 is stored. I do not believe that large scale H_2 / hydride storage will economically viable without new H-storage alloys, lower in cost and with improvements in the technical properties listed above. This provides new challenges for materials R&D. The International Energy Agency is promoting R&D toward radically improved hydriding materials [39].

The status of small stationary hydride storage containers is more optimistic than large units. The low volume and low pressure safety advantages of small hydride units, along with the ultra high purity of the H_2 discharged, have created at least a small successful commercial market. Standard hydride storage units ranging from a few liters to 2500 liters can be purchased from several suppliers: Ergenics (USA)[19], Hydrogen Consultants (USA)[40], GfE (Germany)[12], JMC New Material (Japan)[15], among others, probably. Some of the small available are shown in Figure 2. Most use ambient air heat exchange (Fig.2a & b), but there are also water heat exchange units available (Fig.2c). In most cases, the container manufacturers will consider custom units to best fist the users' needs. In any event, standard units can be loaded with selected alloys to meet the customers' PTC requirements.

4.1.2. *Vehicular Storage.* The high degree of safety associated with hydride storage of H_2 has resulted in a number of H_2-powered vehicle demonstration programs that used hydride fuel tanks. As shown in Figure 3, the hydride tank can be placed to receive the required heat of H_2 desorption from the waste heat of the engine (i.e., cooling water) [41]. A short early history of the many H_2 /hydride fueled vehicles is given in [2].

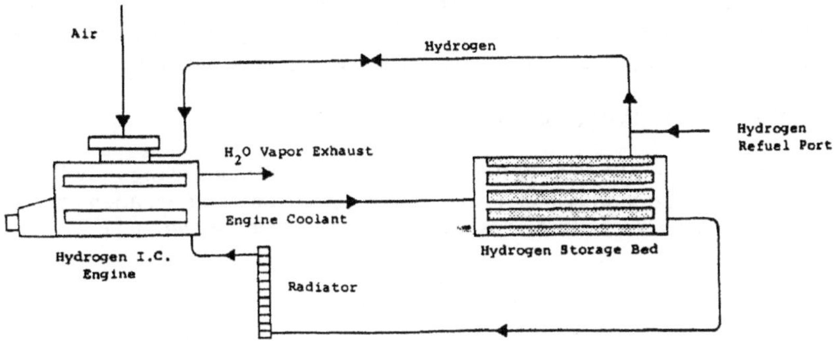

Figure 3 - Vehicular Hydride Tank [41]

Much of the active H_2-fueled vehicle work in the 1970s and 1980s was done by Daimler Benz in Germany, using hydride fuel storage. The most extensive project, running from 1984-1988 was the "Berlin Fleet" demonstration that involved ten vehicles: five dual gasoline-hydrogen passenger cars and five hydrogen vans [42]. In all cases the H_2 was carried in tanks as a hydride of the AB_2 alloy $Ti_{0.98}Zr_{0.02}V_{0.43}Fe_{0.09}Cr_{0.05}Mn_{1.5}$ [25]. The program was successfully completed. Except for one deactivation resulting from accidental charging with contaminated H_2, the hydride tanks worked well [42]. Since the end of the Berlin program, there have been other H_2-fueled vehicles, but few using hydride tanks. The only presently active vehicle program with hydride storage is the Mazda HR-X rotary-engine vehicle [43].

Recent H_2-fueled vehicles seem to be concentrating on LH_2 and very high pressure GH_2. As seen in Table 3, the reason seems to be the high weight penalty that must be paid with the use of a hydride tank. There are liquefaction energy penalties with LH_2 but the weight and volume advantages are obvious, if safety is ignored. High pressure gas relies on new, lightweight composite containers, e.g., carbon wrapped aluminum. There are significant compression energies involved to compress H_2 to 300 bars, but the weight

Table 3
Comparison of Automobile
Hydrogen Storage Tanks [44]

Storage Method	Weight, kg	Volume, L	Pressure, bar
Metal Hydride	320	170	50
Cryogenic Liquid	20	140	4
High-Pressure Gas	120	250	300

advantages over metal hydrides seem to be dominating recent vehicle programs.

If hydrides are to become a practical vehicle fuel carrier, we must develop new and improved lightweight hydrides. Improvements in activating Mg have helped (cf. [1], Sect. 3.3.5). Even though not enough thermal energy is available from a vehicle engine to completely dissociate MgH_2, some of the H_2 can be catalytically burned to generate the required heat [45]. More innovative approaches on hydrides and systems are needed.

4.2 HYDROGEN COMPRESSORS

It is possible to use a hydride bed as a hydrogen compressor using only heat. This is made possible by the sharply increasing hydride formation or dissociation pressure P with increasing absolute temperature T, as given by the van't Hoff equation [1]

$$\ln P = \frac{\Delta H}{RT} - \frac{\Delta S}{R}, \qquad \text{(Eq.1)}$$

where ΔH and ΔS are the enthalpy and entropy changes of the hydride formation. Thus, the basic principle is to absorb H_2 at low pressure when the bed temperature is low and then heat the hydride bed to a higher temperature for high pressure desorption of the H_2. A hydride/hydrogen compressor is a chemical compressor that has essentially no moving parts such as pistons or bellows and, strictly speaking, needs no mechanical energy input. In principal, H_2 compression can be done with any hydride bed with appropriate heat transfer connections to a low temperature heat sink and high temperature heat source. In practice, hydride compressors are specially constructed for high throughput (rapid heat transfer) and automated operation.

It is theoretically possible to achieve ultra high (GPa) pressures by heating a single hydride to very high temperatures. However, most practical hydriding alloys are intermetallic compounds and tend to disproportionate at high temperature and high H_2 pressure, resulting in loss of H-capacity [46,47]. Recent attempts toward an AB_5 40 GPa hydride compressor in Russia have run into this problem, along with materials and thermodynamic problems associated with the high pressure gas phase [48]. Even the use of disproportionation-free vanadium hydride for a hydride compressor has resulted in problems associated with a perplexing "plastic flow" phenomenon that results in the formation of dense metallic masses which expand and exert troublesome stresses on the compressor walls [47,49].

A more logical approach to generate high pressures with modest temperature is the use

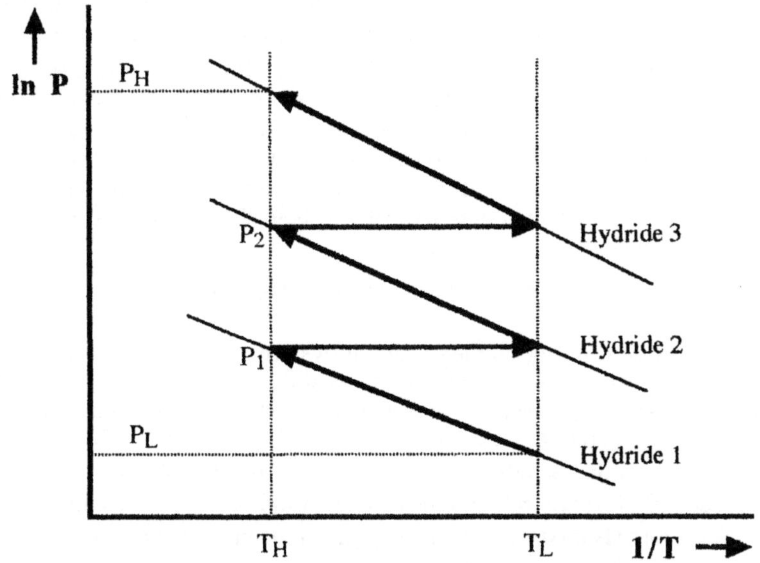

Figure 4 - **Ideal van't Hoff Diagram Representing a Three-Stage H_2 / Hydride Compressor**

of multi-staged hydride compressors, a concept developed to commercial practicality by Golben [50]. Instead of using a single hydriding alloy over a large temperature swing, a multi-staged compressor uses a series of two or more hydriding alloys over a smaller temperature swing, this minimizing the problems of disproportionation and inefficiencies associated with the filling of bed void volumes during single-stage compressor operation. The simple concept of a three-staged hydride compressor is shown by the idealized schematic van't Hoff diagram of Figure 4. Using only a low temperature heat sink at T_L and a modestly high temperature heat source at T_H (say 100°C), H_2 pressure is "stepped" through three different alloy beds, each containing a successively higher pressure hydride former and each undergoing more or less the same T_L <----> T_H temperature swing. The use of three hydride stages greatly increases the compression ratio P_H/P_L over that which can be achieved by a single stage alone, P_1/P_L, with the same ΔT. Recently, Ergenics made an H_2 / hydride compressor containing six AB_5 stages capable of compressing H_2 from 1 atm to 200 atm using a temperature swing of only 25°C to 85°C [19].

Laboratory H_2 / hydride compressors have been used for more than 20 years [51]. Both single-stage and multiple-stage compressors have been sold commercially by Ergenics [19] for more than 10 years. Some Ergenics units have seen millions of pumping cycles over several years' field operation. Hydride compressors are also now commercial in China [52]. In addition to the economic advantage of being operable with nearly "free", low-grade waste heat, the H_2 output of a hydride compressor is extremely pure [52]. However, the main disadvantage of hydride compressors is that they require reasonably pure input H_2 to avoid damage to the working alloys.

Although hydride compressors are a limited commercial success, they do not represent a large business and they have not penetrated much of the commercial market for H_2 compressors, mainly because of their susceptibility to impurities in the input H_2. R&D improvements to the impurity resistance of hydriding alloys are needed to drive this application further into the merchant H_2 industry. MH compressors should also have a role in the Hydrogen Energy scenarios of the future. For example, if the transportation industry concludes that high pressure GH_2 should be the automotive fuel medium, then hydride compressors should have a strong role in filling the high pressure tanks. Hydride compressors are a key component of other hydride applications and devices to be further discussed in this chapter.

4.3 H_2 SEPARATION, PURIFICATION AND GETTERING

The fact that hydride-forming elements and alloys are selective to the absorption of hydrogen (and its isotopes) allows them to be used for H_2 separation from mixed gas streams, purification and gettering. These three terms should be defined at the onset of this section. By H_2 separation, I mean the removal of H_2 from a gas mixture containing 1-99% non-H_2 gases, that separated H_2 to be used in moderately pure form form for various purposes. By H_2 purification, I mean the removal of less than 1% impurity gases from a 99+% starting H_2 mixture. By H_2 gettering, I mean the removal of less than 1% H_2 from a 99%+ non-H_2 gas.

4.3.1. *Separation.* H_2 separation must use a "flow-through" reactor because the the concentration of non-H_2 gasses during H_2-absorption in a "dead end" reactor results in the "inert gas blanketing" around hydride particles which greatly reduces local H_2 absorption dynamics (i.e., kinetics are dominated by the slow diffusion of H_2 through non-H_2 or H_2-depleted layers). Figure 5 schematically shows the basic two-step separation process using a flow-through reactor [53]. During the separation step the mixed gas ($X + H_2$) is passed through the cooled reactor and the H_2 is absorbed down to the alloy plateau pressure, leaving H_2-depleted X in the exhaust. The separation process proceeds by a absorption front which has been measured [54] and modeled [55]. For the discharge step, the flow-through valves are closed, the reactor heated and essentially pure H_2 discharged. Fig.5 shows a "temperature swing" cycle where the desired H_2 output pressure is greater than the input H_2 partial pressure. It is also possible to run an "adiabatic pressure swing" cycle using internal heat storage, but this results in an output H_2 pressure that is lower than the input partial pressure [54]. By performing a brief H_2 purge at the end of the separation step or discarding some of the initial H_2 discharge, the overall purity of the output H_2 can be increased. It should be noted that the cyclic hydride/dehydride separation process is quite distinct from the steady-state diffusion , hot Pd membrane process used widely today.

Although hydride separation of H_2 from mixed gas streams is not a widespread commercial process, the history of prototype demonstrations goes back nearly a century. Pd powder was used around the turn of the century to separate H_2 from N_2-H_2 mixtures [56]. The early 1970s saw the first trials with the newly discovered intermetallic

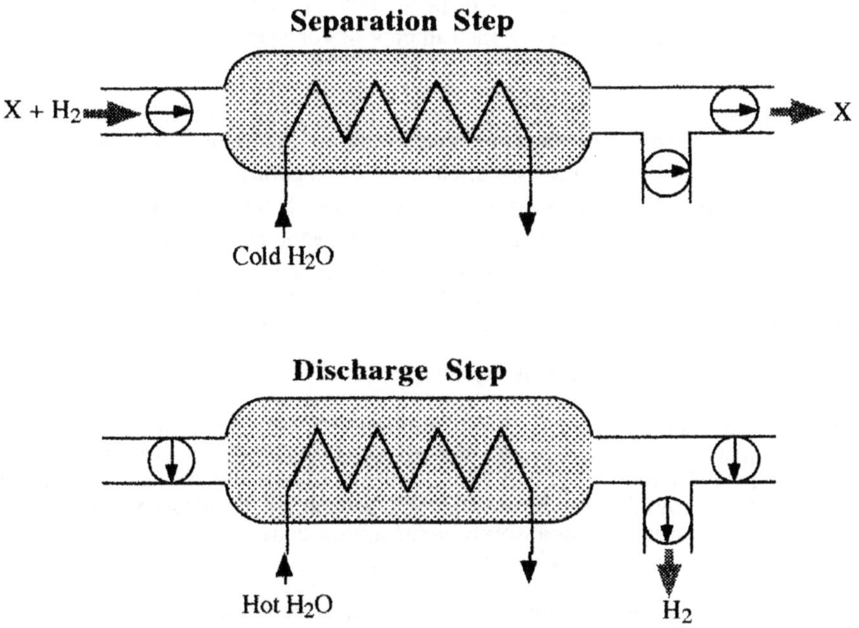

Figure 5 - **Separation of H_2 from Mixed Gas Using a Flow-Through Hydride Reactor** [53]

compounds [57]. Industrial scale demonstrations of H_2 separation from ammonia purge gas have been successfully demonstrated with intermetallic hydrides [54,58] and is now performed on a small commercial scale in China [59]. A few other experiments are listed in [2].

Hydride separation is only competitive with alternate processes (e.g., refrigeration and membrane diffusion) when the partial pressure of the input mixture is relatively low, typically < 50% [60]. One grave shortcoming of the hydride/dehydride process is the sensitivity of hydriding alloys to typical impurities in mixed gas streams, e.g., CO, H_2S, O_2, H_2O, etc. The growth of this application will require improvements in alloy surface engineering, as discussed in my other chapter in this volume [1]. The use of fluoride coatings offers promise [61].

4.3.2. *Purification.* H_2 purification relies on the fact that reactive impurities are strongly chemisorbed on the surfaces of the alloy particles and inert impurities that remain in the gas space can be purged away in the first few percent of the discharge. The result is exceedingly high purity H_2 outputs. Impurity level less than 100 ppb are easy to achieve [25,62,63]. Because the input impurity levels are typically well below 1%, there is no need to use the flow-through reactor necessary for separation processes. Simple dead-end reactors are used for purification.

Uranium hydride beds have been used in laboratories for nearly 50 years for the production of ultra high purity H_2. Many of the intermetallic hydride / H_2 storage units discussed in Section 4.1.1 are purchased specifically for their high purity H_2 output, e.g., for use as carrier gas for analytic instruments or for special materials synthesis or processing.

4.3.3. *Gettering*. Many sealed high vacuum or inert gas filled devices require getters to absorb the H_2 (as well as some other reactive gases) that tend to accumulate during service. In addition, flowing inert gas streams are often purified using getter purifiers. In order to reduce the H_2 partial pressure to very low levels (down to 10^{-6} mB, or lower), very stable Zr-based alloys are usually used [64]. H_2 getter technology is well-established commercially and dominated by the Italian company SAES Getters [65]. A number of good reviews on getters exist (e.g., [66]). Because most of the present getter applications are not directly related to Hydrogen Energy, the subject will not be further covered here.

4.4 ISOTOPE SEPARATION AND HANDLING

In addition to the normal hydrogen isotope protium (H), hydriding alloys will also readily absorb the heavy hydrogen isotopes deuterium (D) and tritium (T). However, the practical rates of absorption may vary from isotope to isotope and this fact makes possible using hydriding alloys for the chromatographic separation of hydrogen isotopes. For a given alloy, the differences in absorption/desorption behavior for different hydrogen isotopes originate from three fundamental factors [2]:

1. Different equilibrium pressures (hydride vs. deuteride vs. tritide)
2. Different diffusion rates in the metal or hydride phase
3. Different surface dissociation/recombination rates

Deuterides and tritides may show different plateau pressures from each other and from hydrides. The normal isotope effect says that the heavier the isotope the higher the plateau pressure. A classic example of the normal isotope effect is seen with Pd, as shown in Figure 6 [67,68]. But inverse isotope effects can be found where the deuteride or tritide plateau is lower than the hydride plateau, e.g., the dihydride / dideuteride phase of V [69]. Sometimes no significant isotope effect is observable at one temperature but then observable at other temperatures [70].

Rates of isotope diffusion in the metal or "hydride" phase vary with mass. Normally the heavier the mass (T>D>H), the slower the diffusion by a factor of the square root of the mass ratio. But there are many deviations from this simple rule.

The least understood isotope effect is the dissociation and recombination of such molecules as H_2, HD, D_2, DT, T_2, etc. These surface related processes can influence practical isotope separation.

In fact, all of the three factors listed above work together to influence kinetic separation in ways that are not fully understood. Isotopic exchange in a formed hydride is one way to study separation effects [71]. Another approach is to look at breakthrough fronts from chromatographic columns [72,73]. A brief history of isotope separation using metal hydrides is given in [2], concentrating on Pd, V and TiNi. Recent work in Japan with vanadium in a pressure swing absorption process has shown promise for the separation of H-D mixtures [74]. Modest separation of isotopes using Mg_2Ni was recently reported in Yugoslavia [75]. When radioactive tritium (a ß-emitter) is involved, special effects are involved such as autocatalytic surface effects [76] and isotherm distortions resulting from

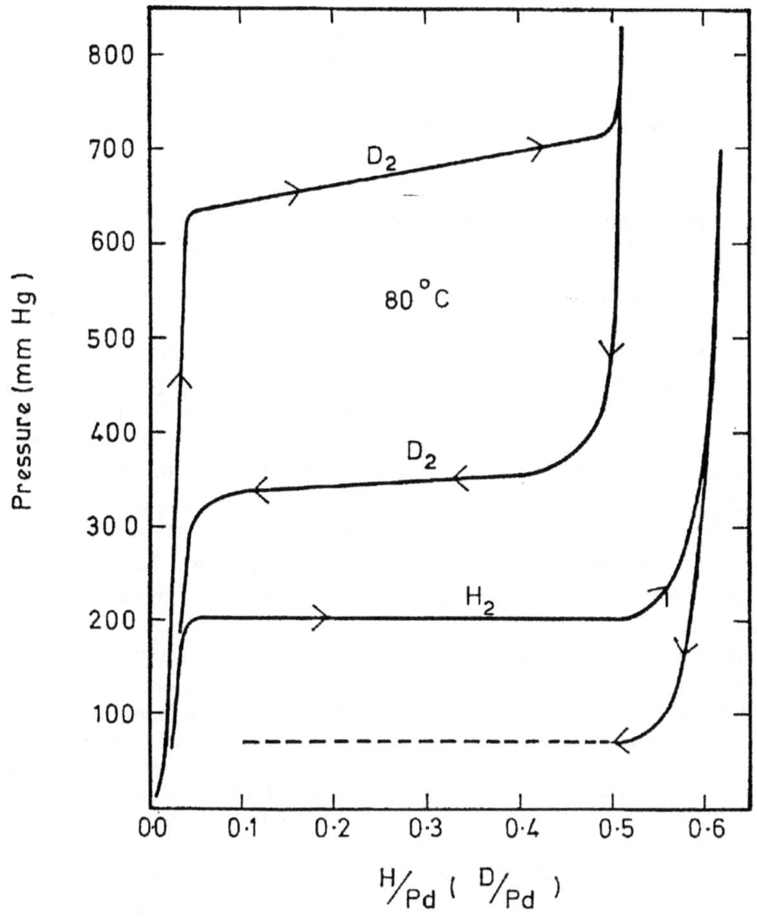

Figure 6 - Hysteresis Loops (80°C) for the Pd-H_2 and Pd-D_2 Systems [67,68]

internal ^{3}He generated in the lattice by T-decay [77].

Perhaps the single most impressive use of hydrogen isotope separation and handling via metal "hydrides" is the new tritium facility at the Savannah River Plant in the U.S. [77-83]. There hydride storage, compression, H-isotope separation from inert gas streams, isotope separation itself and vacuum pumping takes place in an industrial environment.

A schematic of the combined tritium storage and 2-stage compressor beds is shown in Figure 7. The main storage bed consists of $LaNi_{4.25}Al_{0.75}$. It offers two advantages over the old low pressure gaseous T_2 storage tanks: (1) Storage is very compact (300 times

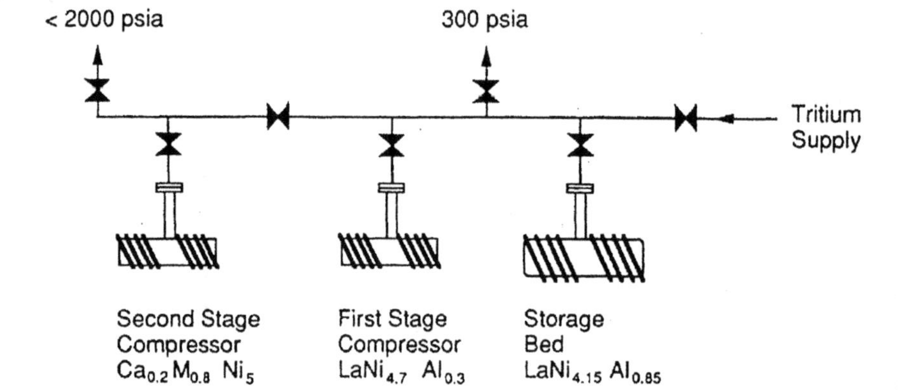

Figure 7 - Schematic Diagram of the Tritium Storage and Compression Beds at Savannah River [78]

smaller), and (2) the AB_5 lattice completely retains the ^3He decay product, thus eliminating the last minute Pd-Ag membrane purification required just before use in the old facility. The first stage of the compressor is $LaNi_{4.25}Al_{0.75}$.with an output pressure of 300 psia (20.4 atm abs). The second compressor compressor stage, $Ca_{0.2}Mm_{0.8}Ni_5$, discharges T_2 at pressures up to 2000 psia (136 atm). Mm signifies the rare-earth mixture mischmetal [1].

Hydrogen isotopes are separated from inert gas (He or N_2) mixtures using column of Pd supported on high porosity Kieselgur (a diatomaceous earth). Similar chromatographic columns are used for the separation of the isotopes themselves, by processes described earlier by Lewis [68]. Because of the normal isotope pressure effect of Pd (Fig.6) and the faster lattice diffusion of the lighter isotopes, H and D are preferentially absorbed by the Pd leaving T enriched in the exiting gas.

A final hydride device at Savannah River is a continuous 2-bed vacuum pump. A cooled bed maintains a <20 torr vacuum on the downstream side of Pd-Ag membranes that are sill used for H-isotope separation from inert gas (e.g., N_2 gloveboxes). When that bed is filled another takes over and the first is heated for T_2 recycling.

Although the Savannah River Tritium Plant is a highly specialized, high technology operation, its successes with hydride technology clearly show the potential for other industrial applications as well as viability in Hydrogen Energy scenarios.

4.5 CLOSED THERMODYNAMIC SYSTEMS

The fact that hydrides exhibit strong interrelations among temperature, H_2 pressure and reaction heat (Eq.1) has resulted in a family of hydride devices that constitute closed thermodynamic systems, as shown schematically in Figure 8. The center element of each device consists of an H_2 / hydride compressor which converts heat to pressurized gaseous H_2, as described in Sect. 4.2. This "pneumatic energy" can be converted to mechanical energy, using an expansion engine, to produce heat engines, actuators or temperature sensors. Conversely, the H_2 gas can be reacted in another hydride bed to produce heat,

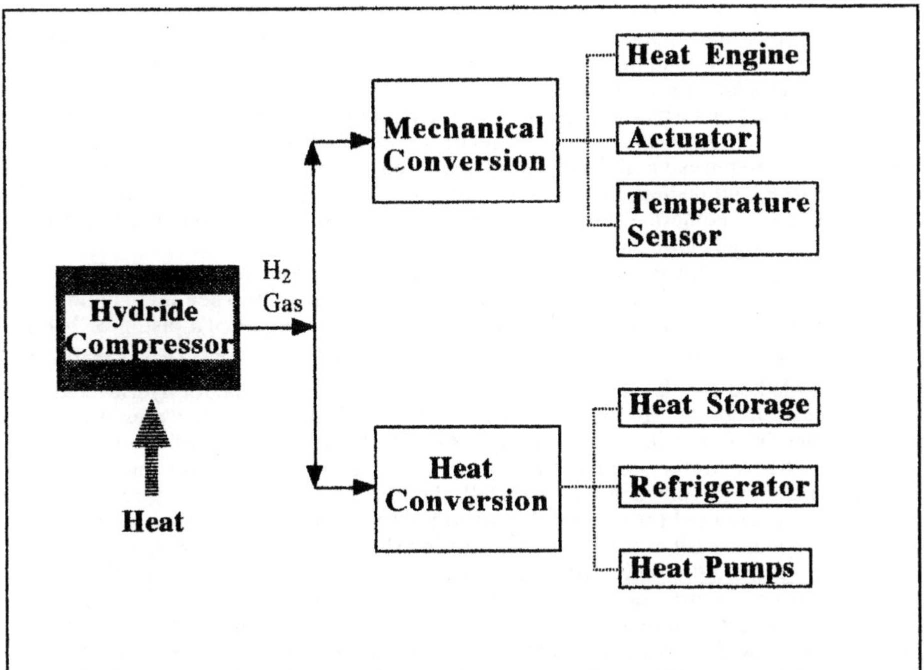

Figure 8 - The Family of Closed Thermodynamic Hydride Devices

leading to heat storage, refrigeration and heat pumping devices. In all cases the H_2 inventory is contained permanently in the system as a sort of working fluid. Sometimes this general class of closed thermodynamic hydride-based systems are called "heat transformers" [84].

The literature on this subject is immense and this brief review cannot hope to give it justice. Ref.[2] reviews much of the early work up to the mid-1980s. More recent work is summarized by Lynch [85].

4.5.1. *Heat Engines, Actuators and Temperature Sensors*. All three devices convert the hydride-compressed high pressure H_2 to mechanical energy. The difference is the degree of work done by the three title devices. Heat engines are meant to provide a large and continuous output of mechanical energy, actuators an occasional burst of mechanical energy and temperature indicators just enough to provide a signal of some sort.

Hydride heat engines are useful in the conversion of low-grade heat, such as incident solar energy, to mechanical (or electrical) energy. The modest temperatures involved (<100°C) can result in significant H_2 pressures from a hydride compressor. Using a piston, bellows, turbine, Sterling or other expansion engine, the pneumatic (PV) gas energy can be converted to mechanical energy which, in turn can be converted to shaft motion (linear or rotational) and/or electrical energy via a generator. For continuous output, the heat engine must have at least two hydride beds. At a given time, one bed is heated to provide the high pressure H_2 for the expansion engine and the other bed is cooled to receive the low pressure, work-depleted H_2 from the expansion engine.

Periodically, the roles of the two hydride beds are reversed.

The main practical application of hydride heat engines has been in the area of remote, solar powered water pumping [86-88]. As far as I know, these activities have represented only prototype demonstrations and have not reached the commercial stage. The problem is the complexity and expense of hydride heat engines relative to photovoltaic technology (e.g., photovoltaic + electric motor). It seems to me that the steadily decreasing costs of photovoltaics requires breakthroughs in improved hydriding alloys and simple device designs before hydride heat engines will compete in the solar field. There may be special situations (e.g., geothermal or industrial) where heat alone is available and therefore hydride engines may have a chance.

Hydride actuators are much more simple devices than hydride heat engines. They need to contain only one hydride bed and operate only occasionally when that single bed is heated. On each "actuation" the H_2 gas pressure typically moves a piston or bellows intended to open a valve, window or other device. When cooled the H_2 gas is reabsorbed in the hydride bed and the action is reversed. The greatest potential for hydride actuators lies in the field of fire protection. The first hydride actuators for pressurized fire extinguishers were patented in the early 1980s [89]. Recently, a fast-acting, hydride-based, resettable water sprinkler has been developed for industrial and home fire protection use by Ergenics in the U.S. [19]. Unlike the fusible link (non-resetting) fire sprinklers that are used throughout the world today, the hydride sprinkler automatically shuts off the water when the fire has been extinguished, thus minimizing water damage.

Hydride temperature indicators need to do only minimal work, e.g., simply close an electrical switch or provide a analog voltage output. In general, hydride temperature indicators cannot compete with cheap and accurate "point" indicators such as thermocouples and resistance thermometers. However, they have served as a valuable "linear" temperature sensor in the form of an aircraft fire detector. This device, patented 30 years ago by Lindberg [90], is shown schematically in Figure 9 [91]. A TiH_2 wire is

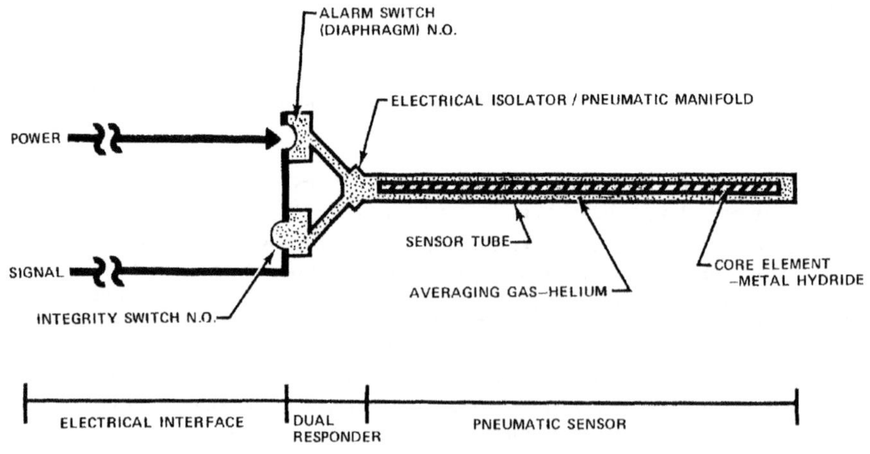

Figure 9 - Schematic of Hydride-Based Fire Detector [91]

contained in a several meters long sealed tube which is hermetically attached to a pressure switch. The tube can be wound in and around those parts of the aircraft (especially the engine) that are susceptible to in-flight fires. During a fire, a discrete length of the TiH_2 wire is heated which quickly outgases H_2 and increases the internal pressure of the sensor. This results in the closing of an electrical switch which activates a warning light in the cockpit. The entire sensor is filled with He "averaging gas" which serves as an indicator of possible leaks and helps speed up response during a fire. This device has been manufactured and placed in most military and commercial aircraft for many years [92]. Still made by two companies [93,94], the aircraft fire detector probably represents the single most successful commercial hydride device.

4.5.2. *Heat Storage.* First proposed by Libowitz [95], the high reaction enthalpies associated with hydriding and dehydriding suggests hydrides might be useful as heat storage media. As shown by the simple solar-home schematics of Figure 10, the concept

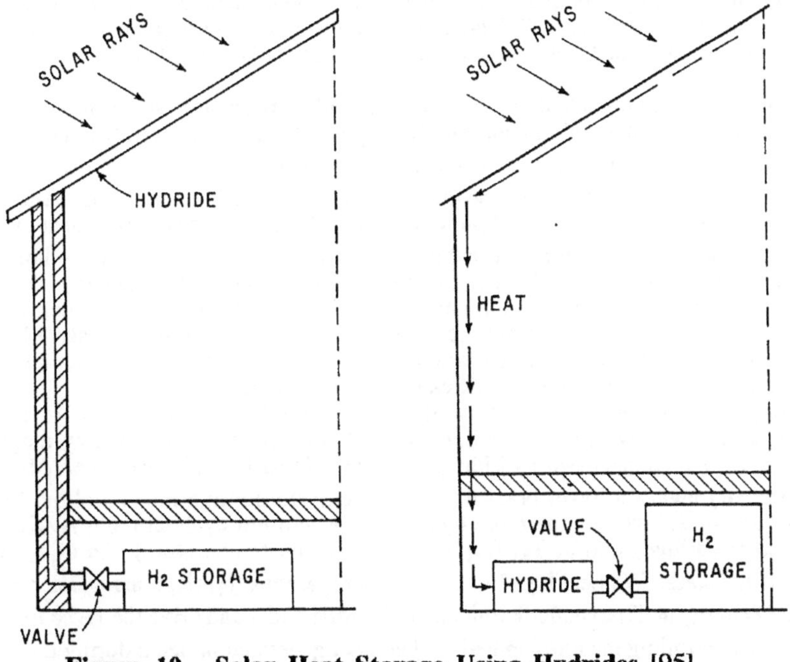

Figure 10 - **Solar Heat Storage Using Hydrides** [95]

is conceptually quite simple. Heat (in this case solar) is used to endothermically dissociate a hydride to form H_2 gas. The hydride can be located either in the solar panels (Fig.10, left) or in a separate container to which the solar heat is piped (Fig.10, right). The heat is then recovered later by exothermically reforming the hydride. Solar energy storage via surface activated MgH_2 has been demonstrated in Germany [84] and waste heat storage and reuse in automobiles has also been demonstrated in Germany [96]. Table 4 shows some of the gravimetric and volumetric advantages of heat storage via hydrides (both low

Table 4
Hydride Reaction Heat vs. Latent Heat Storage [85]

Medium	Gravimetric Density, MJ/kg	Volumetric Density, GJ/m3
Low Temperature		
Neopentyl glycol	0.131	0.139
Glauber's salt	0.251	0.366
Vanadium dihydride	0.356	1.608
High Temperature		
Lithium fluoride	1.037	1.875
Li-Ca fluoride eutectic	0.791	1.659
Calcium dihydride	4.371	9.305

and high temperature), compared to some corresponding latent heat storage materials. The major disadvantage of hydride heat storage is the relatively high cost of the hydrogen storage alloys. This fact has limited the application to only a few demonstration projects.

4.5.3. *Refrigerators and Heat Pumps.* Both hydride refrigerators and heat pumps operate on the same principal involving an H_2-interconnected pair of hydride forming alloys. The H_2 gas forms a working fluid, so to speak, that is moved back and forth between the two beds in a precise thermodynamic sequence to result in heating or cooling. Although the H_2 can be moved by a mechanical pump [2], most designs are driven by a heat introduced into a thermal hydride compressor that effectively constitutes one of the paired beds. Thus most hydride refrigerators or heat pumps can operate on solar or waste heat. The hydride beds are fixed (unlike the common pumped liquid <----> vapor systems) and require the heat to be transferred to and from the beds via heat transfer fluids or passive heat pipes [84]. Hydride refrigerators are environmentally friendly in that they contain no ozone-damaging CFCs and require no significant electric input, if any.

It might be instructive to follow the classic thermodynamic cycle of a hydride refrigerator, as shown in the idealized van't Hoff construction and flow diagrams of Figure 11. We connect two hydriding alloy containers by an H_2 gas line. Alloy 1 (the compressor) makes a lower pressure, higher temperature hydride and alloy 2 makes a higher pressure, lower temperature hydride. We start our cycle at point A with alloy 1 charged (hydrided) and at T_M (typically near room temperature) and alloy 2 in the unhydrided state. We heat bed 1 to a higher temperature T_H (e.g., using solar heat) and input Q energy at T_H to endothermically dehydride bed 1 and drive the H_2 so liberated to bed 2 thus hydriding the alloy in bed 2. The heat generated in bed 2 during this <u>recharge step</u> is rejected to the ambient heat sink T_M. Bed 1 is then cooled to T_M and the <u>cooling step</u> begins by allowing the H_2 to flow from bed 2 back to bed 1. The desorption of H_2 from bed 2 results in its cooling to subambient T_L and the absorption of Q at T_L, our desired cooling effect. During this step, bed 1 heat is rejected to ambient T_M. The cycle is now complete and can be sequenced again. As can be seen, this cycle is a batch process with cooling only on about half the cycle. If continuous cooling is required, two pairs of hydride beds must be operated in parallel and out of phase.

As can be seen from Fig.11, 1Q of heat is added at T_H, 2Q rejected at T_M, resulting in

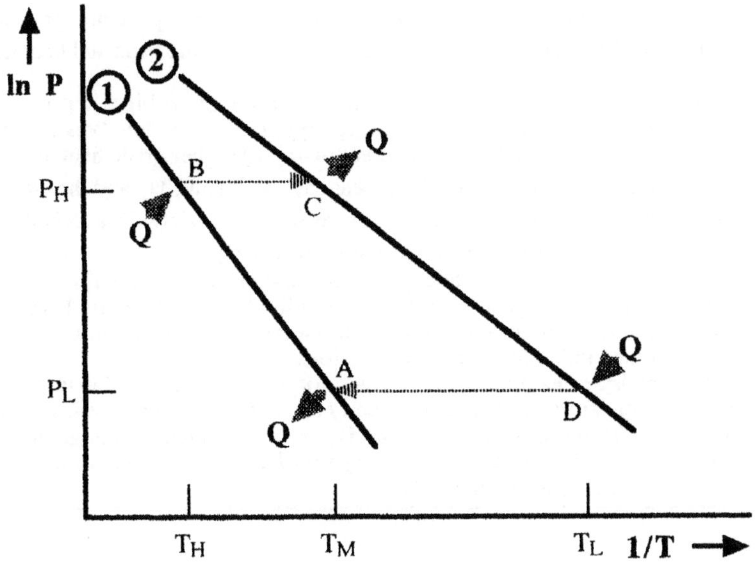

(a) van't Hoff Diagram

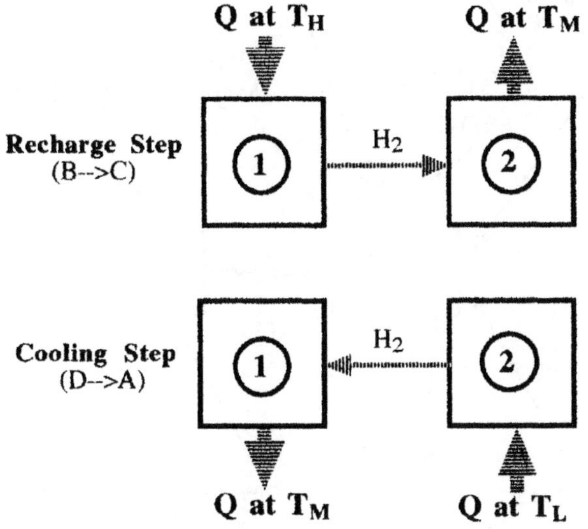

(b) Gas and Heat Flow Diagram

Figure 11 - Thermodynamic Cycle of a Hydride Refrigerator

1Q of refrigeration at T_L. The idealized cycle shown is not quite realistic because of various inefficiencies such as hydride pressure hysteresis, H_2 pressure gradients in powder beds and flow lines, sensible heat losses in heating hydride beds and heat transfer fluids, etc.

Using differently paired hydrides and similar constructions to Fig.11, it is possible design heat pumping cycles [2]. The temperature upgrading cycle involves the input of 2Q heat at T_M, rejection of 1Q at T_L and the creation of 1Q higher grade heat at T_H. The heat upgrading cycle involves the inputs of 1Q heat each at T_H and T_L and the creation of 2Q heat at T_M. In all of these refrigerators and heat pumps, it is possible to build multi-staged units using not two but three or more hydride beds [97].

Numerous hydride refrigerator and heat pump prototypes have been built over the last 15 years [2]. It is difficult to do justice to the many individuals involved in this extensive work, but I will mention a few of the principal pioneers in this field, with apologies in advance to many omitted individuals and organizations.

The most active hydride heat pump and refrigerator work has been in Japan, with Suda leading most of the academic, as well as indirectly the industrial work. Many of his key papers are listed in [97] and also reviewed briefly in [85]. Much of this great work in Japan has now diminished. Some work at Sanyo on hydrogen refrigeration continues toward, hopefully, commercial products [98].

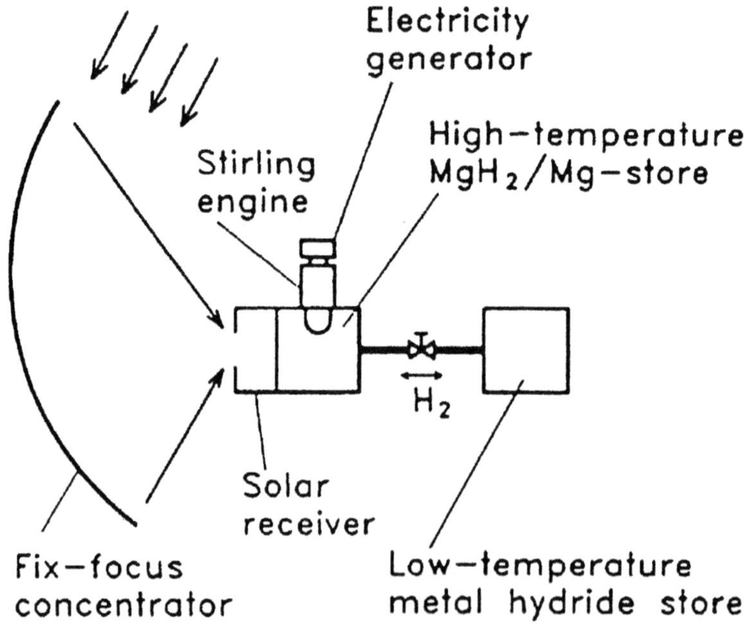

Figure 12 - Solar Hydride Heat Engine and Storage [84]

A number of activities have developed in Europe. In Germany, Groll has lead much prototype work directed toward environmental friendly devices such as solar energy storage, heat pumping and refrigeration [84]. An example of a solar hydride sterling engine / heat storage system is shown in Figure 12. In France, Dantzer has worked extensively on modeling the thermodynamics of heat pumps and developing experimental and analytic understanding understanding the engineering problems of mass and heat transfer in powder beds [99, 100]. In Israel, Ron has developed improved heat transfer and championed the cause of hydride air conditioners for vehicles [23, 101]. In Sweden, Lewis was one of the earliest pioneers in the area [102].

U.S. activities started in the middle 1970s with Gruen's HYCSOS project at the Argonne National Laboratory [22]. HYCSOS demonstrated refrigeration, heat pumping and heat engines. About the same time, the individual inventor Terry [103] began work in close association with Ergenics, Inc. [19]. Several prototype air conditioners have been constructed at Ergenics, recently including one designed for cooling military shelters [104] and one for cooling automobiles that uses waste engine heat [105]. Lynch (HCI) has recently built a -20°C hydride refrigerator designed for use in the future space station, as well as a hydride "backpack" cooler for astronauts working in space or lunar environments [85].

In spite of many prototype units built over nearly 20 years, hydride refrigerators and heat pumps have not reached commercial production. They cannot yet compete with with the well entrenched and relatively simple liquid compressor / evaporator systems. To be fair, hydride devices certainly have not been properly engineered for low cost mass production. Unfortunately they are more complicated than existing systems. They require "batch" operation and the careful sequential movement of H_2 and heat transfer fluids. It has been argued that hydriding alloy shortcomings (including alloy cost and H-capacity) are hindering application [3], and this is certainly partly true. But it is still difficult to displace a well entrenched system. The advancement of the hydride refrigerator and heat pump will have to center around the dramatic improvement of economics and associated energy and environmental advantages.

4.6 LIQUID H_2 APPLICATIONS

All Hydrogen Energy scenarios will certainly involve the widespread production, storage and use of LH_2. Metal hydride technology can have an important role in all three of these areas.

Production of LH_2 involves the Joule-Thompson expansion of high pressure GH_2. The use of metal hydride compressors for the necessary compression of the GH_2 was demonstrated by 1973 [106]. This concept, shown in Figure 13, has potential not only for large scale LH_2 production but also small liquefiers and cryocoolers for space vehicles and infrared satellite cameras [107,108].

The slow boiloff of LH_2 during storage can be captured by hydride absorption beds [109]. By absorbing the boiloff (i.e., not allowing pressure buildup in the container), the temperature of the LH_2 can be kept constant at 20K and the H_2 captured by the hydride can be recovered for use or later reliquefied by applying heat to the hydride bed. During LH_2 use large surges of boiloff occur during the cooling of transfer lines. It has been experimentally demonstrated on an engineering scale that such surges of boiloff can be captured by hydriding alloy beds [110].

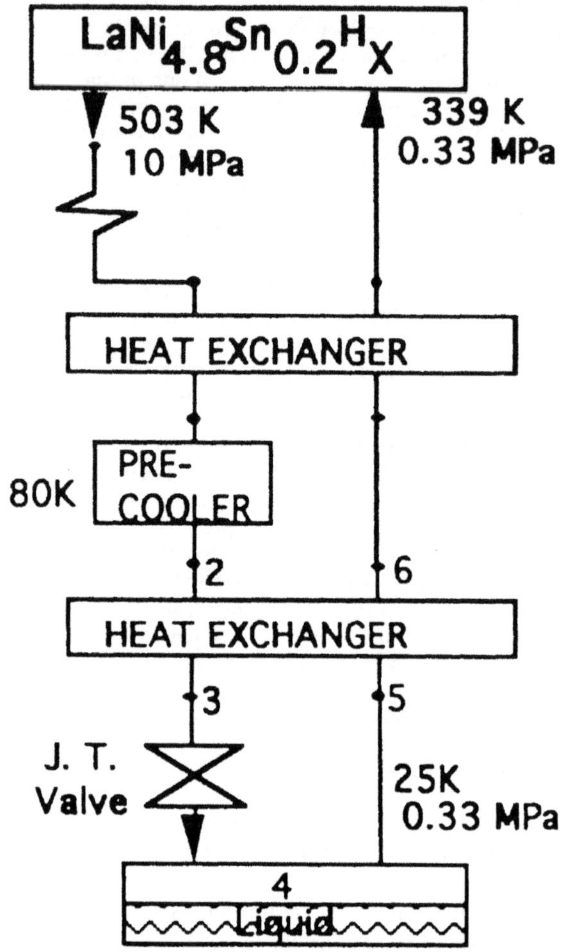

Figure 13 - AB_5 Hydride-Compressor-Based Cryocooler Capable of H_2 Liquefaction [107]

4.7 RECHARGEABLE BATTERIES

The most rapidly growing application for metal hydrides, by far, is in the area of nickel metal hydride (NiMH) rechargeable batteries. Although, this application does not directly relate to Hydrogen Energy, the subject of this conference, it should be briefly mentioned for completeness. More complete reviews are available, particularly by Sakai [111,112].

The NiMH battery is electrochemically similar to the well-known nickel cadmium (NiCad) battery. In both batteries, the positive (Ni) electrode undergoes the following reversible electrochemical reaction (charging to the right in the following equations):

$$Ni(OH)_2 + OH^- \longleftrightarrow NiOOH + H_2O + e^- \quad \text{(Eq.2)}$$

The complementary reaction for the negative MH electrode is as follows:

$$M + xH_2O + xe^- \longleftrightarrow MH_x + xOH^- , \quad \text{(Eq.3)}$$

where M is a hydriding alloy and MH_x is its hydride. For reference, the corresponding negative Cd electrode reaction is as follows:

$$Cd(OH)_2 + 2e^- \longleftrightarrow Cd + 2OH^- . \quad \text{(Eq.4)}$$

Thus, combining Eq.2 and Eq.3 in a typical aqueous KOH electrolyte, we have an NiMH battery. Combining Eq.2 and Eq.4 in the same electrolyte, we have a NiCad battery.

The construction and operating voltage of NiMH and NiCad batteries are almost identical and the cyclic lives of the two are (after long NiMH R&D) now similar (typically 500-1000 cycles). The NiMH batteries have certain advantages over NiCad batteries: (1) somewhat higher gravimetric energy density (perhaps 20-50%); (2) higher charging and discharging rate capabilities, especially at low temperature; (3) freedom from the memory effect (NiCad loses effective capacity if only partially cycled several times); and no toxic materials (Cd is a highly regulated toxic metal). On the negative side NiMH batteries are presently more expensive than NiCad.

The NiMH battery is rapidly displacing a significant fraction of the NiCad production, especially in crowded Japan where the environmental advantage of Cd-free batteries are desired. More than 100 million NiMH batteries will be made in Japan in 1994 (perhaps 15% compared to NiCad production). The NiMH battery production figures are also growing throughout the rest of the world with strong interest in the U.S., Europe and China. Most of the Japanese NiMH batteries are small (AA to C size) and are made using AB_5 negative electrodes, e.g., alloys on the order of $MmNi_{3.5}Co_{0.8}Mn_{0.4}Al_{0.3}$, where Mm is the rare earth mixture mischmetal [111]. Similar cells are being made at Energizer in the U.S. Larger NiMH batteries are also being made in the U.S. for testing in electric vehicles and typically contain multi-phased, disordered alloys of elements such as Ti, Zr, V, Ni, Cr, Co and Fe [113].

A predecessor of the NiMH battery is the Ni-H_2 battery. Instead of the hydrogen being stored in an MH electrode, it is simply stored as high pressure H_2 gas in the void space of the cell. These cells are very expensive but extremely reliable, so they are widely used today in solar-powered communication satellites. A low-pressure version of the Ni-H_2 battery was proposed in 1974 where the hydrogen was stored as a non-electrode metal hydride [114]. However, this design was not practical because H_2O vapor within the cell damaged the MH. Recently, Ergenics has developed a modification of the Ni-H_2 battery called the "Segmented Metal Hydride" battery where the hydriding alloy is physically separated from the cell via an H_2 conduit, valve and H_2-purifier [115]. By physically separating the hydrogen from the positive Ni electrode, a greater measure of safety is achieved. In addition, lightweight hydrides (such as Mg or Mg_2Ni) that are completely incompatible with KOH electrolytes can be used for increased energy density.

4.8 OTHER APPLICATIONS

In addition to the several hydride applications discussed above, there are a number of others: (1) neutron moderators and shielding (e.g., YH_2), (2) atomic H-donators for specialty chemical reactions (e.g., $NaBH_4$), (3) water reactive H_2 sources (e.g., CaH_2), (4) electrolyzer and fuel cell catalysts, (5) hydride / dehydride processing for the production of specialty metal powders (e.g., Nd-Fe-B magnet powders), (6) wind energy storage and (7) many others clever inventors will discover in the future. Most of these specific applications are not directly related to Hydrogen Energy and need not be discussed further here. For those interested, further reading on these applications can be found in Refs.[1,2} and the citations therein.

5. Conclusions

An impressive array of real and potential hydride applications have been developed over the last 25 years. Many have an obvious potential for direct and indirect application and use in Hydrogen Energy scenarios of the future. Some applications are clear commercial successes already, for example hydride-based aircraft fire detectors and nickel metal hydride batteries. There are also hydride applications that are commercial but represent rather small volumes: small stationary H_2 storage containers and purifiers, H_2 / hydride compressors and H-isotope separation units. Finally there are hydride applications that have not proceeded to commercial viability in spite of extensive R&D and prototype work: large stationary H_2 storage tanks, vehicular H_2 storage tanks, H_2 separation systems, heat engines, refrigerators and heat pumps.

Why have the latter applications not succeeded so far? I believe the answer lies, at least in part, in the fact there are still critical shortcomings in hydride properties we have not yet solved. To improve the chances for wider application of hydrides in the existing H_2 industry, as well as for the coming Hydrogen Energy era, we need to seriously address the following property areas, in order of priority:

1. Develop hydrides with much improved weight percent capacity at ambient temperature.
2. Develop much lower cost hydriding alloys ($ per unit H_2 stored).
3. Improve resistance to surface damage by impurities, perhaps by surface engineering.
4. Learn to reduce or prevent intermetallic hydride disproportionation.
5. Improve the ease of alloy activation.
6. Develop measures to eliminate the pyrophoricity of fine hydride powders.

These are not easy tasks and will require much R&D diligence and perseverance. I hope all present and future "hydriders" will keep these targets in mind. If we succeed, we can contribute importantly and effectively to the coming of the Hydrogen Energy era, as well as address some of the needs within the present industrial H_2 industry.

6. Acknowledgements

I would like to thank the many colleagues who have contributed directly and indirectly to this chapter by offering education and advice over the years. They are too numerous to mention, but they all know who they are. I regret that I could not cite all of you. Because of space limitations, many deserving papers are omitted.

I wish to give special thanks to a few individuals that provided specific material for this

Chapter and Ref. [1]: H. Buchner (Daimler-Benz, Germany), D.H. DaCosta & P.M. Golben (Ergenics, USA), B. Friedrich (GfE, Germany), E.L. Huston (Energizer Power Systems, USA), G.G. Libowitz (USA), F.E. Lynch (HCI, USA), D. Noréus (U. of Stockholm, Sweden) and S. Suda (Kogakuin U., Japan).

7. References

1. G. Sandrock, "Hydrogen-Metal Systems", this volume
2. G. Sandrock, S. Suda, L. Schlapbach: "Applications", in Hydrogen in Intermetallic Compounds II, Topics in Appl. Phys., **67** (1992) 197
3. S. Suda, G. Sandrock: Z. Phys. Chem., **183** (1994) 149
4. A. Percheron-Guegan, J-M. Welter: "Preparation of Intermetallics and Hydrides", in Hydrogen in Intermetallic Compounds I, Topics in Appl. Phys., **63** (1988) 11
5. B. Friedrich: "Large Scale Production and Quality Assurance of Hydrogen Storage (Battery) Alloys, J. Mat. Eng. and Performance, Feb., 1994
6. Treibacher Chemische Werke AG, A-9330 Treibach, Austria (Contact: A. Bouvier)
7. Beijing General Research Institute for Non-Ferrous Metals, No. 2 Xinjiekouwai St., Beijing 100088, China (Contact: Z. Lu)
8. Eastern Hydride Technology Co., Zhejiang University, Hangzhou,China (Contact: Q-D. Wang)
9. Nankai University, 94 Weijin Road, Tianjin 300071 China (Contact: Y. Zhang)
10. Tianjin Xinbao Special Metals Co., Tianjin, China (Contact: Mr. Ma; U.S. Contact: J. Wolfe, MPV Lanthanides, 401 E. Pratt St., Suite 2512, Baltimore, MD 21202)
11. Zhongshan Tianjiao Rare Earth Minerals Co., Zhongshan, China (Contact: P. Li)
12. GfE Gesellschaft für Electrometallurgie M.B.H., Höfner Strasse 45, D-8500 Nürnberg 80, Germany (Contact: B. Friedrich); (U.S. Contact: E.F.M. Williams, Shieldalloy Metallurgical Corp., West Blvd., Newfield, NJ 08344, USA)
13. H.C. Starck GmbH, Im Schieeke 78-91, D-3380 Goslar, Germany (Contact: H. Sickenberg)
14. Chuo Denki Kogyo Co., No. 1-1-12 Toranomon, Minato-ku, Tokyo, Japan
15. JMC New Material, Inc., JMC Annex Bldg., No. 8-4 Kaomi-cho, Nihonbashi, Chuo-ku, Tokyo 103, Japan (Contact: Y. Matsubara)
16. Mitsui Mining and Smelting Co., Battery Materials Division, 1436-3 Takehara-cho, Takehara-shi, Hiroshima 725, Japan (Contact: M. Sasaki; U.S.Contact: Y. Takayanagi, Mitsui Mining and Smelting Co. (USA), 461 Fifth Ave., New York, NY 10017-6256)
17. Santoku Metal Industry Co., No. 14-34, 4-chome, Fukae-Kitamachi, Higashinada-ku, Kobe, Japan (Contact: Y. Inoue)
18. Aldrich Chemical Co., 1001 W. Saint Paul Ave., Milwaukee, WI 53233, USA
19. Ergenics, Inc., 247 Margaret King Ave., Ringwood, NJ 07456, USA (Contact: D. DaCosta)
20. Rhone-Poulenc Basic Chemicals, Inc., Prospect Plains Road, Cranbury, NJ 08512, USA (Contact: C. Bounds)
21. Teledyne Wah Chang Albany, P.O. Box 460, Albany, OR 97321, USA (Contact: R. Peterson)
22. I. Sheft, D.M. Gruen, G. Lamich: "HYCSOS, a Chemical Heat Pump and Energy Conversion System based on Metal Hydrides", Rept. ANL-79-8 (Argonne Nat. Lab., Argonne, IL, 1979)
23. M. Ron, D. Gruen, M. Mendelsohn, I. Sheft: J. Less-Common Met., **74** (1980) 445
24. H. Ishikawa, K. Oguro, A. Kato, H. Suzuki, E. Ishii: J. Less-Common Met., **120** (1986) 123

25. O. Bernauer: Int. J. Hydrogen Energy, **13** (1988) 181
26. G.D. Sandrock, E. Snape: "Reaction Heat Storage Method for Hydride Tanks", U.S. Pat. 4,566,281 (Jan. 28, 1986)
27. P.S. Rudman, G.D. Sandrock, P.D. Goodell: J. Less-Common Met., **89** (1983) 437
28. S. Hynek, W. Fuller: "Hydrogen Storage Within the Infrastructure", Hydrogen Energy Progress X, (Int. Assoc. Hydrogen Energy, 1994) p.1341
29. F.E. Lynch: J. Less-Common Met., **74** (1980) 411
30. F.R. Block, A. Dey, H. Kappes, K. Reith: J. Less-Common Met., **131** (1987) 329
31. P.P. Turillon, G.D. Sandrock: "Hydrogen Container", U.S. Pat. 4,133,426 (Jan. 9, 1979
32. N.R. Baker, F.E. Lynch: "Hydrgen Sorbant Flow Aid Composition and Containment Thereof", U.S. Pat. 4,600,525 (July 15, 1986)
33. Q-D. Wang, J. Wu, C-P. Chen, Z-P. Li: J. Less-Common Met., **131** (1987) 399
34. P.P. Turillon, G.D. Sandrock: "Hydride Storage Containment", U.S. Patents 4,134,490 and 4,134,491 (Jan. 16, 1979)
35. P.M. Golben, W.F. Storms: Flexable Means for Storing and Recovering Hydrogen", U.S. Pat. 4,396,114 (Aug. 2, 1983)
36. J.M. Burger, P.A. Lewis, R.J. Isler, F.J. Salzano, J.M. King, Jr.: Proc. 9th IECEC, (ASME, New York, 1974), p.428
37. G. Strickland, J.J. Reilly, R.H. Wiswall, Jr.: "An Engineering Scale Storage Reservoir of Iron Titanium Hydride", Proc. Hydrogen Economy Miami Energy Conf., (Int. Assoc. Hydrogen Energy, 1974), p.S9
38. C. Carpetis: Int. J. Hydrogen Energy, **7** (1982) 191
39. H.-J. Pfriem: Int. J. Hydrogen Energy, **11** (1991) 329
40. Hydrogen Consultants, Inc., 12420 North Dumont Way, Littleton, CO, 80125, USA (Contact: F. Lynch)
41. K.C. Hoffman, J.J.. Reilly, F.J. Salzano, C.H. Waide, R.N. Wiswall, W.E. Winsche: Int. J. Hydrogen Energy, **1** (1976) 133
42. K. Feucht, W. Hurich, N. Komoschinski, R. Povel: Int. J. Hydrogen Energy, **13** (1988) 243
43. Anon., "Mazda Environmental Presentation - Hydrogen Cars", Mazda Motor Corp., Japan, 1993
44. Anon., "Hydrogen: an Alternative Fuel", Daimler-Benz AG, Öffentlichkeitarbeit, Postfach 80 02 30, D-7000 Stuttgart 80, Germany
45. L. Marinescu-Pasoi, U. Behrens, G. Langer, W. Gramatte, A.K. Rastogi, R.E. Schmitt: Int. J. Hydrogen Energy, **16** (1991) 407
46. G.D. Sandrock, P.D. Goodell, E.L. Huston, P.M. Golben: Z. Phys. Chem. NF, **164** (1989) 1285
47. R.W. Marmaro, F.E. Lynch: "Investigation of Long Term Stability in Metal Hydrides", Rept. NAS9-18175 (Hydrogen Consultants, Inc., 1991)
48. Yu.K. Baichtok, N.N. Korostyshevsky, M.H. Sosna: "Chemical Compression of Hydrogen up to 40 GPa: Problems of Materials and Design", Proc. 10th World Hydrogen Energy Conf., (Int. Assoc. Hydrogen Energy, 1994), p.1029
49. R.C. Bowman, B.D. Freeman, E.L. Ryba, R.E. Spjut, E.A. Liu, J.M. Penso, F.E. Lynch: Z. Phys. Chem., **183** (1994) 245
50. P.M. Golben: "Multi-Stage Hydride-Hydrogen Compressor", Proc. 18 IECEC, (A.I.Ch.E., New York, 1983), p.1746
51. J.J. Reilly, A. Holst, R.H. Wiswall, Jr.: Rev. Sci. Inst., **42** (1971) 1485
52. C-P. Chen, Z. Ye, J. Wu, Q-D. Wang: Z. Phys. Chem., **183** (1994) 251
53. G.D. Sandrock, E.L. Huston: Chemtech, **11** (1981) 754
54. J.J. Sheridan III, F.G. Eisenberg, E.J. Greskovich, G.D. Sandrock, E.L. Huston:

J. Less-Common Met., **89** (1983) 447
55. A.V. Kruglov, A.N. Perevezentsev, B.M. Andreev: Int. J. Hydrogen Energy, **19** (1994) 363
56. E.W. Morley: Smithsonian Contributions to Science, **29** (1892-1903) 56
57. J.J. Reilly, R.H. Wiswall, Jr.: "Alloys for the Isolation of Hydrogen", U.S. Pat. 3,825,415 (July 23, 1974); "Separation of Hydrogen from Other Gases", U.S. Pat. 3,793,435 (Feb. 19, 1974)
58. Q.-D. Wang, J. Wu,C.-P. Chen, Z. Ye: J. Less-Common Met., **131** (1987) 321
59. Q.-D. Wang: Zhejiang University, Hangzhou, China, private communication
60. E.L. Huston, J.J. Sheridan III: "Applications for Rechargeable Metal Hydrides", Industrial Applications of Rare Earth Elements, ACS Symposium Series 164 (ACS, Washington, 1981), p.223
61. G. Sandrock: "Hydrogen Separation and Purification Using Metal Hydrides", Res. Repts. Kogakuin Univ., **26** (Apr.1994) 29
62. H. Wenzl, K.H. Klatt: "The Use of FeTi-Hydride for Production and Storage of Supra Pure Hydrogen", Hydrides for Energy Storage, (Pergamon, 1978), p.323
63. T. Gamo, Y. Moriwaki, N. Yanigihara: J. Less-Common Met., **89** (1983) 495
64. C. Boffito, B. Ferrario, D. Martelli: J. Vac. Sci. Technol., **A1** (1983) 1279
65. SAES Getters, S.p.A., via Gallarate 215, 20151 Milan, Italy
66. T.A. Giorgi: Japan J. Appl. Phys., Suppl. **2**, Pt. 1 (1974) 53; T.A. Giorgi, B. Ferrario, B. Storey: J. Vac. Sci. Technol., **A3** (1985) 417
67. A. Sieverts, W. Danz: Z. Phys. Chem., **38B** (1937) 46,61 (cited in Ref.68)
68. F.A. Lewis: The Palladium Hydrogen System, (Academic Press, 1967), p.118
69. R.H. Wiswall, Jr., J.J. Reilly: Inorg. Chem., **11** (1972) 1691
70. J, Tanaka, R.H. Wiswall, J.J. Reilly: Inorg. Chem., **17** (1978) 498
71. D.H.W. Carstens, P.D. Encinias: J. Less-Common Metals, **172-174** (1991) 1331
72. L.K. Heung, R.S. Tran, K.J. Stoner: J. Less-Common Metals, **172-174** (1991) 1313
73. G.S. Nichols: J. Less-Common Metals, **172-174** (1991) 1338
74. S. Fukada, N. Mitsuishi: Int. J. Hydrogen Energy, **18** (1993) 227
75. R. Nikolic, K. Zmbov, M. Veljkovic: Int. J. Hydrogen Energy, **18** (1993) 743
76. T. Schober, N. Iniotakis: J. Less-Common Metals, **172-174** (1991) 1345
77. A. Nobile, Jr., R.T. Walters, W.C. Mosley: J. Less-Common Metals, **172-174** (1991) 1352
78. M.S. Ortman, L.K. Heung, A. Nobile, R.L. Rabun: J. Vac. Sci. Technol., **A8** (1990) 2881
79. A. Nobile: Fusion Technol., **20** (Sept.1991) 186
80. T. Motyka: "The Replacement Tritium Facility", J. Fusion Technol. (Suppl.) (1992)
81. A. Horen, K. Mallory: Metal Hydride Based Separation - -Large Scale Operations, Ibid.
82. J. Klein, K. Mallory, A. Nobile: Tritium Measurement Technique Using 'In-Bed' Accountability", Ibid.
83. K. Hueng, H. Owen, R. Hsu: Tritium Confinement in a New Tritium Processing Facility at the Savannah River Site", Ibid.
84. M. Groll, A. Isselhorst, M. Wierse: Int. J. Hydrogen Energy, **19** (1994) 507
85. F.E. Lynch: J. Less-Common Metals, **172-174** (1991) 943
86. J.J. Reilly, G.D. Sandrock: Scientific Amer., **242** (Feb.1980) 118
87. K. Nomura, E. Akiba, S. Ono: J. Less-Common Metals, **89** (1983) 551
88. Anon: "Hydride Pump Uses Low Grade Heat", World Water, June, 1986, p.47
89. P.M. Golben: Thermally Activated Metal Hydride Sensor/Actuator", U.S. Pat. 4,377,209 (Mar. 22, 1983

90. J.E. Lindberg, Jr.: "Method of Making Heat Sensors", U.S. Pat. 3,153,847 (Oct. 27, 1964)
91. Anon.: "Systron-Donner's Reliable Fire Detection System for DC-10 and A300 Aircraft", Systron-Donner Corp., cf. Ref.93
92. D.E. Warren, K.A. Faughnan, R.A. Fellows, J.W. Godden, B.M. Seck: J. Less-Common Metals, **104** (1984) 375
93. Systron-Donner Corp., 2731 Systron Dr., Concord, CA, 94518, USA (Contact: Robert Fellows)
94. Aerosafe International, 3033 Richmond Parkway, Richmond, CA, 94806, USA (Contact: Don Warren)
95. G.G. Libowitz: "Metal Hydrides for Thermal Energy Storage", Proc. 9th IECEC (ASME, New York, 1974), p.322
96. H. Buchner, H. Säufferer: "Parking Heater and Method using Hydrides in Motor Vehicles Powered by Hydrogen", U.S. Pat. 4,214,699 (July 29, 1980)
97. S. Suda, Y. Komazaki, H. Narasaki, M. Uchida: J. Less-Common Metals, **172-174** (1991) 1092
98. I. Yonezu, F. Fujitani, T. Yonesaki, T. Imoto, N. Hiro, K. Nasako, T. Saito: "Development of an F-Class Refrigeration System Using Hydrogen-Absorbing Alloys", Hydrogen Energy Progress X (Int. Assoc. Hydrogen Energy, 1994), p.1547
99. P. Dantzer, E. Orgaz: Int. J. Hydrogen Energy, **12** (1986) 797
100. M. Pons, P. Dantzer: J. Less-Common Metals, **172-174** (1991) 1147
101. M. Ron: J. Less-Common Metals, **104** (1984) 259
102. G. Anevi, L.Jansson, D. Lewis: J. Less-Common Metals, **104** (1984) 341
103. L.E. Terry: "Hydrogen-Hydride Absorption Systems and Methods for Refrigeration and Heat Pump Cycles", U.S. Pat. 4,055,962 (Nov. 1, 1977)
104. P.M. Golben, E.L. Huston: "Design and Fabricate a Metal Hydride Heat Pump with a Cooling Capacity of 9000 BTU/H", Ergenics report to U.S. Army, AD-A213 756 (NTIS, Springfield, VA, 1989)
105. D.H. DaCosta: "Metal Hydride Heat Pumps", Proc. Refrigeration and Air Conditioning Workshop (Oak Ridge National Laboratory, 1993)
106. H.H. van Mal: Chemie-Ing.-Tech., **45** (1973) 80
107. B. Freeman, E. Ryba, R. Bowman, J. Phillips: Progress Toward the Development Hydrogen Sorption Cryocoolers for Space Applications", Hydrogen Energy Progress X (Int. Assoc. Hydrogen Energy, 1994), p. 2041
108. A. Johnson, J.A. Jones: "Fast, Low-Duty-Cycle Sorption Refrigerators", NASA Tech Briefs (Jan.1994), p.58
109. H.H. van Mal: "Stability of Ternary Hydrides and Some Applications", Ph.D. Thesis, Technische Hogeschool Delft, 1976, p.74
110. M.J. Rosso, M. Golben: J. Less-Common Metals, **131** (1987) 283
111. T. Sakai, K. Muta, H. Miyamura, N. Kuriyama, H. Ishikawa: "Nickel-Metal Hydride Batteries Using Rare-Earth Based Hydrogen Storage Alloys", Hydrogen Storage Materials, Batteries, and Electrochemistry, Proc. Vol. **92-5** (The Electrochemical Society, 1992), p.59
112. T. Sakai, M. Matsuoka, C. Iwakura: "Rare Earth Intermetallics for Metal-Hydrogen Batteries", Handbook of the Physics and Chemistry of Rare Earths, **21** (in press)
113. S.R. Ovshinsky, M.A. Fetcenko, J. Ross: Science, **260** (9 April 1993) 176
114. J.D. Dunlop, M.W. Earl, G. Van Ommering: "Low Pressure Nickel Hydrogen Cell", U.S. Pat. 3,850,694 (Nov. 26, 1974)
115. P.M. Golben, T.D. Meyer: "Extended Cycle-Life Metal Hydride Battery for Electric Vehicles", U.S. Pat. 5,250,368 (Oct.5, 1993)

HYDROPYROLYSIS AND CATALYTIC HYDROPYROLYSIS OF OIL SHALES

E. EKİNCİ*, M. ÇITIROĞLU*, A. AKAR*, E. PÜTÜN**, C.E. SNAPE***

* İTÜ Kimya-Metalurji Fak. 80626 Maslak, İstanbul, Türkiye.
** Anadolu Üniversitesi, Kimya Müh. Bölümü, Eskişehir, Türkiye.
***University of Strathclyde, Dep. of Pure and Applied Chem. G1 1XL, Scotland, UK.

ABSTRACT. Oil shales have been pyrolysed under a number of regimes and the resultant oils characterised by chromatographic fractions, GC, GC-MS, NMR analyses. In addition, to confirm the classification of oil shales investigated as Type I kerogen, biomarker compounds in the bitumen were identified and the oil shales characterised by single pulse excitation solid state ^{13}C NMR.

The use of a sweep gas in fixed bed pyrolysis (steam and nitrogen used) increased tar yields considerably compared to those obtained in a static or self generated atmosphere. However, as well as overcoming mass transfer limitations, steam promoted cracking reactions indicated by the higher concentrations of n-alkane and increased aromatic character of the non-paraffinic material in the pyrolysis oil.

The use of high pressure hydrogen (hydropyrolysis) gave rise to further increases in tar yield but conditions were chosen to minimize gas yield and gave a high selectivity to liquid products as in the atmospheric pressure pyrolysis experiments. In hydropyrolysis, a dispersed sulphide Mo catalyst gave rise to further increase in tar yield for the samples investigated. The hydropyrolysis tars contain similar concentration of paraffins, aromatics and polars as the lower yields tars but are slightly more aromatic character. The chemical reaction occurring in hydropyrolysis can be accounted for a mechanism involving hydrogen radical mediated bond cleavage.

1.0 Introduction

It is well known that the quality of fuels are generally accepted to increase in the order of solid, gas and liquid states. One of the reasons why petroleum crisis in 1973 and 1979 felt in such intensity was this fact. Otherwise, solid and gaseous fuel reserves have much greater available reserves than the liquid counterpart, namely crude oil.

There are well proven technologies to substitute crude oil, based on various solid inputs such as coal, oil shale, asphaltite and plants. The ease of synfuel production is partly dependent on the H/C ratios of the solid inputs. The total hydrogen consumption involved in the synthetic fuel production is also one of the important cost items. For these reasons hydrogen supply and price will be one of the determining factors in the synthetic crude production (1).

Y. Yürüm (ed.), Hydrogen Energy System, 281–296.

1.1 PYROLYSIS

A much simpler and therefore, in principle, economically more attractive method of extracting tar/oil from oil shale or coal is (solvent free) pyrolysis [2].

Pyrolysis of coal and oil shale are complex functions of the experimental conditions, under which the pyrolysis process proceeds. The most important factors, which affect the yield and composition of the volatile fraction liberated, are: coal rank, kerogen type, maceral composition, particle size, temperature (i.e. temperature-time history), heating rate, atmosphere, pressure and reactor configuration [3, 4, 5] .

Coal and oil shale are a heterogeneous, mainly organic, material which decomposes upon exposure to high temperatures. This process is called pyrolysis. Pyrolysis of coal and oil shale are divided into a hydrogen-rich volatile fraction, consisting of gases, vapours, and tar-components, and a carbon-rich solid residue [3] .

The importance of kerogene structure on the conversion behavior of oil shales becomes more apparent when the conversion behavior of oil shales from different deposits, geologic ages and depositional environment.

According the Gavin (1924), "Oil shale is a compact, laminated rock of sedimentary origin, yielding over 33% of ash and containing organic matter that yields oil when distilled, but not appreciably when extracted with the ordinary solvents for petroleum". The term "oil shale" will be used to denote an organic-rich rock that contains little or no free oil [6, 7]. Thus, shale oil is defined as the oil produced from an oil shale on heating. Therefore, shale oil and pyrolysis are defining partners.

For purpose of simplicity, oil shales are classified into three categories of type I, II and III which may be classified according to van Krevelen plot using H/C and O/C ratios. Moreover, type I kerogens has a tendency to yield more liquid products than type II and III. The mechanism of pyrolysis reactions are proposed to take place in a manner given in Figure 1[8].

The reactions, which occur upon pyrolysis of oil shale, are illustrated in Figure 2[9] The coal and oil shale structure will decompose by the braking of the least stable bonds within the structure. The least stable bonds are the methylene, oxygen and sulphur-bridges between the aromatic building blocks. The result of this bond breaking is the formation of a large number of radical components, of which part is volatile under the experimental conditions. These radicals are very reactive and can undergo secondary reactions like cracking and carbon deposition, both inside and outside the particle. Stabilisation of a radical, primarily via hydrogen addition, leads to a volatile component. The evolution of tar is controlled by mass transport in which the tar molecules evaporate into the light gas species and are carried out of the particle at rates proportional to their vapour pressure and the volume of light gas. High pressure reduces the volume of light gases and hence reduces the yield of heavy molecules with low vapour pressure [10]. Polymerisation and condensation reactions, occurring via recombination of both volatile and non-volatile radical components, result in the formation of the solid char particle [3, 11]. High pyrolysis heating rates produced chars with large macroporosites, more open pore structures, and larger macropore surface areas [12].

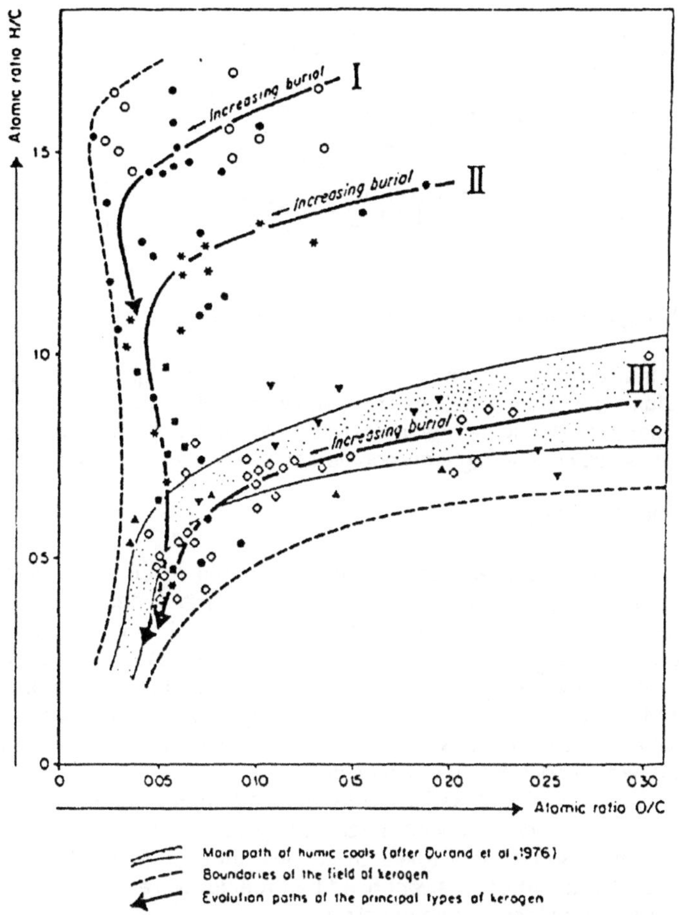

Type	Age and/or formation	Basin, country	
I	Green River Shales	Utah, USA	•
	Algal Kerogens		O
II	Lower Toarcian shales	France, Germany	●
	Silurian shales	Algeria, Libya	■
	Various shales		●
III	Upper Creloceus	Cameroon	✧
	Lower Mannville shales	Alberta, Canada	Δ
	Lower Manville shales	Alberta, Canada	∇

Figure 1. The van Krevelen diagram traces the evaluation paths of type I, II, and III kerogen as a function of increasing burial (increased time-temperature exposure)

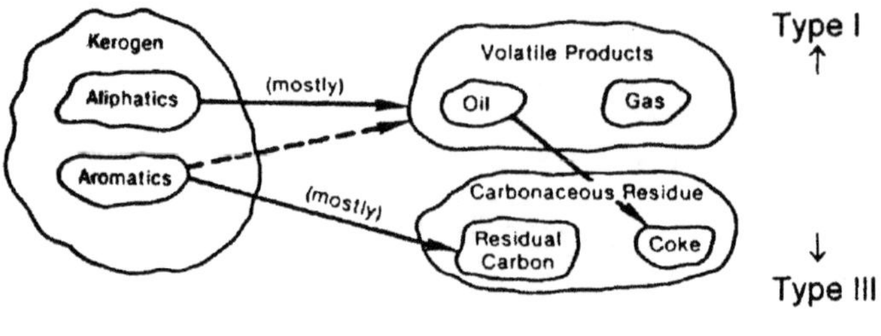

Figure 2. A proposed reaction model for oil shale pyrolysis

Due to the oil shale or coal quickly reaching a high temperature, chemical transformation and their kinetics are somewhat different from those leading to petroleum formation during catagenesis over prolonged periods of geological time. For example, shale oils contain higher proportion of alkenes are not found in petroleum [8].

1.2. HYDROPYROLYSIS

Coal being a more hydrogen deficient fuel compared to oil shale was subjected to intense hydropyrolysis and other hydrogenation research. It has been realised as early as 1896 that coal may be converted to an oil-like product by chemical reduction using hydrogenation. The first patent on this field was taken on the conversion of brown coal to heavy crude oil by Bergius et. al. [11] in 1914. Since then numerous patents were received and research has been conducted on hydrogenation of coals. Even though there are some differences between coal and oil shale a lot of relevant features exit between the hydropyrolysis of the two fuels.

Early results showed that whilst the tar yield from oil shale in a hydrogen swept fixed bed reactor increased with hydrogen pressure the principal product was hydrocarbon gas. Subsequent research showed that the light oil produced from the pyrolysis of coal in a hydrogen swept fixed bed reactor were rich BTX compounds and would be suitable for blending into gasoline.

Benzene yield from North Dakota lignite was dependent on the type and pressure of the sweep gas. When the coal pyrolysed at 1023 K under 10 MPa helium with a tar vapour residence time of 0.6 s, benzene yield was 0.97%, the corresponding methane and total hydrocarbon yields were 4.28% and 6.6% respectively. Substituting hydrogen from helium in the reactor at the same pressure and tar vapour residence time, increased benzene, methane and total hydrocarbon yields to 7.3, 24.0 and 40.4% respectively. The authors concluded that pyrolysis under hydrogen pressure (or Hydropyrolysis) increases hydrocarbon yield due to the reaction of hydrogen with the hydrocarbon free radicals evolved from the coal thus preventing their polymerisation. The hydropyrolysis tars were found to be lighter than tars evolved under an inert atmosphere due to radical stabilisation by the hydrogen[13, 14].

In their study on Colorado oil shale Hershkowitz et.al. [9] elucidated a mechanical discussion on inert and hydrogen atmosphere pyrolysis. They have found that a total of 87% of the organic kerogen in Colorado oil shale has been recovered as oil in hydrogen atmosphere compared to 67% in an inert atmosphere. This was explained in terms of hydrogen inhibiting the conversion of aliphatic carbon to residual carbon.

One of the important hydropyrolysis work on oil shales showed that the kerogen was converted predominantly the temperature range of 698 to 798 K temperature range at pressures up to 3.4 MPa. The organic carbon conversions, exceeded 90% at the temperature between 823-873 K, which is of course not a measure of conversion to liquid fuel[9].

Eastern oil shale of US is one of the type II kerogen, therefore they are hydrogen deficient. For this reason, a few studies have been done. The results show that the higher oil yields from oil shale were obtained in the presence of hydrogen at elevated pressure. Both oil and gas yields increased with hydrogen pressure for Eastern oil shales. For increasing temperature carbon conversion to gas was found to increase but pressure was not influential. The highest oil yield was reported to be at the highest pressure tested of 7 MPa . The level of oil yield was improved to levels in excess of 200% Fischer assay (FA) [15, 16, 17].

1.3. CATALYTIC HYDROPYROLYSIS

Coal hydropyrolysis is a more attractive route than direct liquefaction for the production of liquid fuels and chemical feedstocks . The disadvantages are that liquid (tar) yields are generally much lower, making char the principal product for which some application, apart from providing process heat and hydrogen, must be found [18]. Unless catalysts are employed to modify the reaction chemistry, tar yields are usually limited to 30 - 40 wt % for high volatile bituminous coals. Moreover, large quantities of methane are often produced in hydropyrolysis making hydrogen consumption high [19].

In the early 1960s, in a study of dry (solvent free) hydrogenation of coal in an entrained flow reactor, it was demonstrated that coal coated with a 0.5-1.0% loading of ammonium molybdate could be hydrocracked under hydrogen pressures greater than 10 MPa at temperatures of 723-823 K resulting in an overall conversion of 72% (daf basis) and a tar yield of 46%. The other product was a 32% gas yield which, other than hydrocarbons, would also include other products such as CO_2, CO, H_2S and water[20] .

Subsequent research by British Coal into the use of dispersed catalysts in hydropyrolysis showed that it was possible to lower the temperature of maximum BTX formation away from the gasification zone, thus increasing liquid yield at the expense of methane formation[14, 21, 22] .It was found that molybdenum disulphide was a suitable catalyst for the process . The catalyst was loaded from an aqueous solution of ammonium dithiodioxymolbdate which decomposes at 523 K under hydrogen to form MoS_2 in-situ[23, 24] .

Hydrogen and oil shale under pressure depresses char forming reaction and favors gas and liquid formation. To increase in the pyrolysis products to tar, use of catalyst is necessary. For the case of coal processing, ammonium molybdate and molybdenum disulphide have been reported to be used to increase the liquid yield.

In some cases, two stage catalytic hydropyrolysis system are employed in which catalytic pyrolysis is followed by cracking for some target chemicals.

The extent of hydrogenation reactions are determined mainly by factors such as temperature, hydrogen pressure, contact time and most of all, by the type of catalyst. The choice of catalyst not only affects the total conversion but selectively determines the end product since different catalysts may promote different sequences of reactions.

2.0 Experimental

In this study, we are going to illustrate the effect hydropyrolysis and catalytic pyrolysis on product yields and compare the results with other pyrolysis methods. Proximate and elemental analysis of oil shales are show in Table 1.

Table 1. Proximate and Ultimate Analysis of Oil Shales

wt %	Seyitömer	Göynük A	Oil Shale B	Cleveland
Moisture	3.5	2.6	5.7	2.18
Ash	69.0	17.5	24.4	73.84
VM	20.1	69.4	56.3	15.09
FC	7.4	10.5	13.6	8.89
C	10.9	59.07	49.92	16.0
H	1.13	7.63	6.42	1.9
S	0.1	3.15	3.32	1.9
N	0.3	0.63	1.07	0.8
H/C[a]	1.62	-	1.52	1.15

H/C[a] Bitumen

The pyrolysis reactors were fixed-bed tubular and (modified Heinze retorts) in which 4 K/min heating was employed and the sample was taken from room temperature to 823 K. Self generated atmosphere, nitrogen and steam sweep (up to 22 cm/sec) were used.

Fluidised bed experiments were carried out 30 mm id 300 mm long semi continuos bed. Sample was heated at 5 K/min. from room temperature to 823 K at twice the minimum fluidisation velocity (6 cm/sec).

Hydropyrolysis unit was conducted on the lines of the reactor used extensively by British Coal on hydrogenation studies of coals. The reactor was made from 80 cm long 1.4 cm o.d., 0.8 cm i.d. incoloy tubing. It was resistively heated using a 2.3 kVa transformer. The pressure of hydrogen gas was increased using a Haskel model 75 GT gas booster. 15 MPa pressure, 10 l/min. gas flow rate (at atmosphere pressure) and, 5 K/sec heating rate ambient to 793 K were the main experimental conditions.

Detailed information on experimental setups, procedures and runs can be obtained for fixed bed in [25] for fluidised bed in [26] and for hydropyrolysis in [27]. For each experimental run, liquid, gaseous and char yields are obtained. Characterization studies mainly included proximate and ultimate analysis, solid state single pulse excitation ^{13}C NMR, solution state NMR and chromatographic investigation.

3.0 Results and Discussion

In order to set comparative basis results of self generated (Heinze and Incoloy retorts) and their N_2 and steam sweep experiments for Göynük and Seyitömer pyrolysis results are completed in Table 2 .

From these results, it is evident that reactor geometry has a distinct effect on pyrolysis results which may originate from sample size variation and uniformity of temperature profile. It is also clear that sweeping the retort environment with N_2 or steam increase the oil yield nearly up to the volatile matter content. The effect of steam being greater than N_2. The differential is more pronounced at lower sweep velocities. The reasons of the improvement in due to sweep gases positivly effect on the mass transport limitation. Sweeping the environment shortens the residence times of volatiles to reduces their chances to get involved with char forming reactions catalysed by inorganic matrix.

The extra effect with steam is attributed to its involvement in the prevention of retrogressive char forming reactions of alkane via dehydrogenation to alkenes and subsequent cyclization.

However, some researchers propagate that steam effect over nitrogen as sweep environment is due to trapping of light oil which captures lower molecular mass products than improve the yield by chemical means.

The compositional variation of the tars obtained from static, nitrogen and steam sweep conditions are shown in Figure 3. It is clear that compared to self generated atmosphere the yield of aliphatics and aromatics increased markedly and polars decreased slightly for both nitrogen and steam.

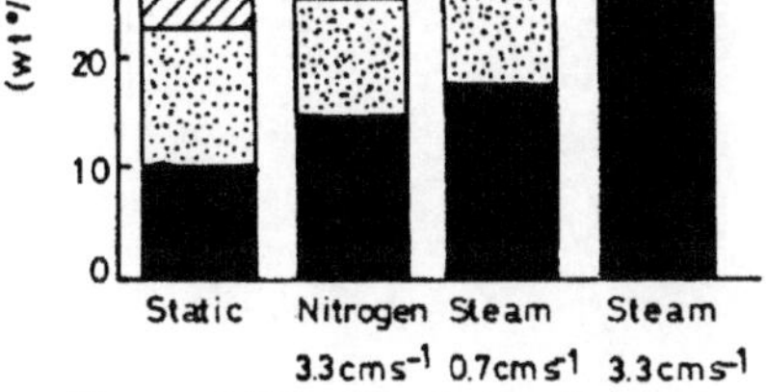

Figure 3. Compositions of Göynük shale oils prepared in Heinze retort

Table 2 Product Distrubutions From Pyrolysis Experiments

Oil Shale	Type of reactor	Sweep gas	Gas Flow cm/s	Pressure MPa	% wt daf/dmmf basis Char	Oil	Total Gas*
Goynuk-A	Heinze	Static	-	-	55	33	12
	Incoloy	Static	-	-	48	38	14
	Heinze	N_2	3.3	0.1	47	39	14
	Incoloy	N_2	22	0.4	39	51	10
	Heinze	steam	0.7	0.1	46	40	14
	Heinze	steam	1.3	0.1	40	44	16
	Heinze	steam	3.3	0.1	25	55	20
Seyitomer	Heinze	Static	-	-	59	30	10
	Incoloy	Static	-	-	50	33	12
	Incoly	N_2	22	0.4	42	42	10

* by difference

3.1 HYDROPYROLYSIS OPERATING CONDITIONS

Previous research on the hydropyrolysis of coals had determined that the optimum conditions for maximizing oil yield whilst minimizing methane production were at a hydrogen pressure of 15 MPa and a pyrolysis temperature close to 773 K[14, 28]. Tables 3 and 4 show the pyrolysis and hydropyrolysis yields obtained from Cleveland and Göynük-A oil shales under a range of process conditions.

Table 3 Product Yields and Ratios from Single Stage Runs in Nitrogen

Sample	Pressure MPa	Gas vel. m/s	Char	Oil	Methane	C_2-C_4
			wt% daf			
Cleveland	0.1	2.0	53	41	0.8	1.2
	0.53	0.2	49	32	2.9	3.6
Göynük-B	0.1	1.0	23	62	1.2	1.9

Table 4 Product Yields and Ratios from Single Stage Hydropyrolysis Runs

Sample	Pressure MPa	Gas vel. m/s	Char	Oil	Methane	C_2-C_4
			wt% daf			
Cleveland	0.51	0.2	27	68	1.8	2.3
	0.15	0.07	24	61	1.2	2.2
Göynük-B	0.50	0.2	13	68	1.3	1.9
	0.15	0.3	8	81	1.5	2.6

The product yields and ratios from hydropyrolysis were determined 5 to 15 MPa. The standard temperature program used for the hydropyrolysis experiments was; ambient to 793 K at 5 K /s held at maximum temperature for 10 min. The pattern of product yields and ratios are simple and confirms the earlier studies, char yield steadily decreasing as the oil yield increased and hydrocarbon gases slowly increasing with increasing hydrogen pressure. The hydrogen pressure protecting the oil yield from polymerisation and condensation reactions so that the oil yield increased from 68 to 81% for Göynük-A oil shale. This is greater oil yield than all fixed and fluidised bed pyrolysis of oil shale with and without sweep indicating the favorable effect of hydrogen as stabilizing agent for the volatile radicals.

Further more, the use of low pressure hydrogen resulted in lower overall conversion as well as lower oil and gas yields. On the other hand use of hydrogen at lower temperature resulted in higher oil yield than use of nitrogen.

Table 5 list the yields from the hydropyrolysis experiments. At 823 K, the temperature used in hydropyrolysis to maximize oil yields, the hydrocarbon gas yield is appreciable. To increase the selectivity to oil (% oil/% hydrocarbon gases), the temperature was reduced to 893 K, the same as that used in recent studies on coals[19,23]. Table 5 indicates that the same oil yield was obtained at both temperatures but, as anticipated, the hydrocarbon gas yield was lower at 793 K. As discussed previously [24], the beneficial effect of high pressure hydrogen in terms of mediating bond cleavage reactions is offset by the increased mass transfer to volatiles released within the oil shale bed.

The use of the dispersed Mo catalyst slightly increased the total conversion to ~ 100%, the oil yield increased to 82% daf shale and hydrocarbon gas yield was reduced by 4% (Table 5). In fact, the gas yield was reduced probably as a consequence of volatile release being shifted to slightly lower temperatures. Despite the total conversion being in excess of 90% daf shale without catalyst, the increase in oil yield is much higher than that typically achieved for low-rank coals [19]. One reason why catalyst has got a favorable effect on liquid product distribution is the low ash nature of Göynük oil shale which is found to have a higher beneficial effect compared to Seyitömer oil shale.

Table 5 Product yields and ratios from hydropyrolysis runs Goynuk-A oil shale

	Temperature		wt% daf	
Catayst	K	Conver.	Oil	C_1 -C_4
No	793	95	73	8
No	823	94	72	9
Yes	793	100	82	4

A number of major compositional differences have been found between the hydropyrolysis and atmospheric pressure pyrolysis oils. For Göynük oil shale, the hydropyrolysis oils contain fewer asphaltenes and more polars and have lower oxygen and sulphur content (Table 6) due to mass transfer and hydrogenolysis effects, respectively. In addition, the hydropyrolysis oils contain fewer alkene groups and IR has indicated that carboxyl concentrations are lower. Due to the combined effects of aromaticity increasing with conversion and hydrogenolysis reactions [29].

Table 6. Compositional Parameters for Some of the Oils Produced from Göynük-A Oil Shale

						% wt/wt		
	C	H	N	S	AR	Alkanes	Aromatics	Polars
Göynük-A Oil Shale								
-static	82.5	11.6	1.1	1.9	55	30	18	38
-hydropyrolysis	83.9	10.0	1.7	1.5	34	30	22	38
-cat. hydropyrolysis	85.0	11.3	1.7	0.5	25	25	28	35

AR: Product Aromaticity

Overall aromaticity balances give useful insights into the conversion chemistry, especially for highly aliphatic immature Type I kerogens. Static pyrolysis gave rise to a large net aromaticity increase (Table 6) consistent with earlier studies [30]. Both the use of effective sweep gasses and hydrogen at pressure nullify this change and, with the catalyst in hydropyrolysis, an overall decrease occurs indicating that hydrogenation has occurred [29].

The alkanes obtained via hydropyrolysis contain a greater proportion of low molecular mass compounds than their pyrolysis counterparts. This is probably due to a combination of hydrocracking of higher molecular mass alkanes and the additional alkanes released as a consequence of covalent bond cleavage during hydropyrolysis having lower molecular mass distribution

3.2 USE OF CATALYST IN HYDROPYROLYSIS

Table 6 list the char, oil and hydrocarbon gas yields from the hydropyrolysis experiments on Göynük oil shale. For the shale, normal hydropyrolysis gives roughly 10% higher oil yield and total conversion than the best achieved in atmospheric pressure pyrolysis. As discussed previously [18, 24], the beneficial effect of high hydrogen pressure in terms of mediating hydrogenolysis reactions is partly offset by the increased mass transfer within the coal bed to volatiles release.

The use of the sulphided Mo catalyst increased the already high total conversion for Göynük oil shale to virtually 100% (Table 5). Moreover, the oil yield increased by 10% to over 80% daf shale due to a significant reduction in gas yield probably resulting from a shift in the release of volatiles to significantly lower temperatures. For bituminous coals [30], much grater increases in oil yield have been achieved with catalyst due to a combination of the lower conversions without catalyst and probably the pyrobitumen generated during hydropyrolysis being more effective at transferring hydrogen atoms from the catalyst sites than that from lower rank coals.

3.3. REACTION MECHANISMS OPERATIVE IN HYDROPYROLYSIS

The higher tar yields obtained in hydropyrolysis compared to atmospheric pressure pyrolysis could be due to the resultant hydroaromatic groups in mobile tar stabilizing free radicals formed at higher temperatures via classical donor-solvent mechanisms similar to those for direct coal liquefaction [31, 32]. Alternatively, the tar formed could be involved in radical hydrogen transfer promoting the scission of strong bonds in the macromolecular structure of coals and oil shales to form moieties of even lower molecular weight reaction I (Figure 4).

In mechanistic work directed at understanding the models of bond scission available for relevant coal structures. Malhotra and McMillen[33, 34] observed experimentally that the selectivity and efficiency with which donatable hydrogen is used to cleave aryl-aryl linkages in coal surrogates is highly dependent on the nature of the donor. The observed variations in efficiency were well reproduced by a mechanistic numerical model incorporating the various competing H-transfer process. The basic trend elaborated by the model is that an increase in the level of polycyclic aromatic hydrocarbon in the solvent can increase the efficiency of H-utilisation markedly. Specifically, the addition of H-atoms to positions on aromatic clusters (reaction I) is favoured over ring hydrogenation which can result in inefficient hydrogen utilisation if the hydroaromatic rings are cracked to yield alkyl groups (reaction II).

Reaction I

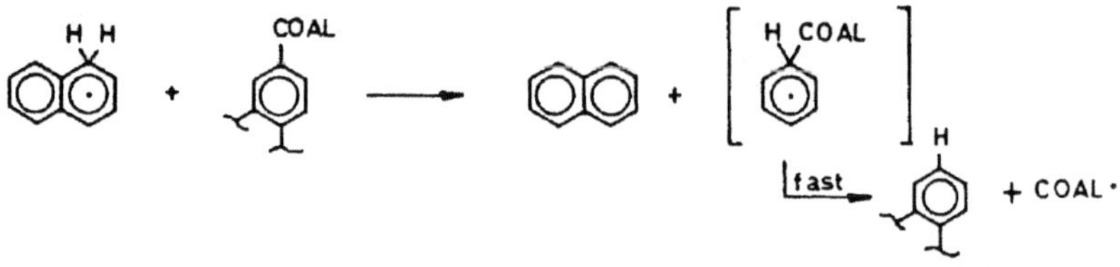

Reaction II

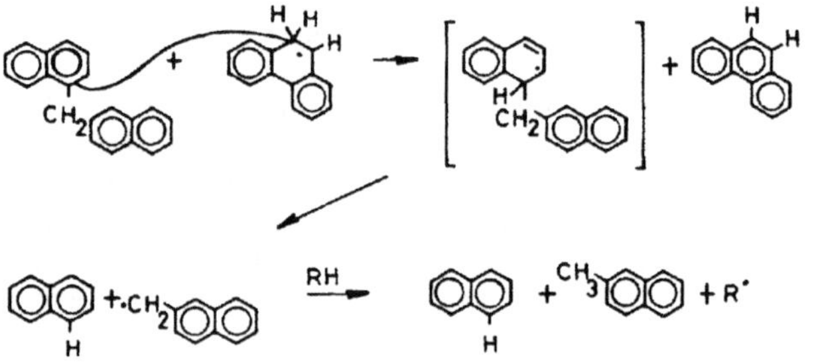

In hydropyrolysis, a much higher concentration of hydrogen radicals are generated than in normal pyrolysis and these radicals can therefore promote bond cleavage by the radical H-transfer process discussed above. When a catalyst is employed, it is likely that the active molybdenum sulphide increases the level of dissociation of hydrogen molecules and is responsible for hydrogenation of aromatic rings. The latter will then lead to elimination of more oxygen and sulphur because labile aliphatic C - heteroatom bonds are being formed.

3.4. HYDROGEN CONSUMPTION

From the yields and compositions of the products obtained from Göynük oil shale at 773 K and 15 MPa pressure the expression given below was used to estimate that the hydrogen consumption with dispersed MoS2 was approximately 3.9% (daf) compared to 2.3% without catalyst.

Most of the increased hydrogen consumption for the catalysed experiments is due to the greater tar production through a combination of heteroatom removal and C-C bond cleavage, the increase in hydrocarbon gas yield only accounted for 0.2% of the extra 2.2% hydrogen consumed.

The results of aromaticity measurements on each of the products from the different pyrolysis regimes yield important information regarding the reaction mechanisms of the conversion chemistry being investigated.

Pyrolysis using self-generated atmosphere or low sweep gas velocities resulted in a net dehydrogenation of the oil shale as anticipated from earlier studies. These observations are in accordance with results published by other workers on similar materials [35, 36].

The use of high velocity inert sweep gas or high pressure hydrogen without a sulphided catalyst did not result in any loss of aromatic carbon confirming that retrogressive reactions had been avoided has been avoided. However when the oil shale was pyrolysed in the presence of a sulphided molybdenum catalyst and high pressure hydrogen the net aromaticity of the system decreased, indicating that hydrogenation of oil shale had occurred to a significant extent.

The highly aliphatic nature of the atmospheric pressure pyrolysis oils agrees well with the hypothesis proposed by Miknis and coworkers [35, 36] that it is principally the aliphatic carbon in the oil shale that contributes to pyrolysis oils. Thus, a major advantage of hydropyrolysis is that it enables most of the aromatic carbon to be released from oil shale. The mechanisms by which the release occurs is a topic for further investigation but it will inevitably involve hydrogenation and subsequent cleavage of resultant weak C-C and C-heteroatomic bonds.

REFERENCES

1. E. EKİNCİ, Economic Considerations of the Oil Shale Related Conversion Processes in Composition, Geochemistry and Conversion of Oil Shales, Ed. Snape, C.E., Kluwer Academic Pub. (in press) 1994.

2. C.E. SNAPE, Similarities and Differences of Coal Reactivity in Liquefaction and Pyrolysis, Fuel, Vol. 70, pp. 285-288, (1991).

3. P.J.J. TROMP, Coal Pyrolysis, Ph.D. Thesis, University of Amsterdam, Netherlands, (1987).

4. C.S. SCOUTEN, Fuel Science and Technology Handbook, (J.G. SPEIGHT ed.), chapter 29, Marcel Dekker, (1990).

5. A. EKSTROM, C.J.R. FOOKES, H.J. LOEH, C.H. RANDALL, C. ROVERE, J. ELLIS, P.T. CRISP, Chemical and Pyrolysis Characteristics of Two Types of Oil Shale From the Condor Deposit in Queensland Australia, Fuel, Vol. 66, pp. 1133-1138, (1987).

6. C.S. SCOUTEN, Fuel Science and Technology Handbook, (J.G. SPEIGHT ed.), chapter 25, Marcel Dekker, (1990).

7 J.G. SPEIGHT, The Chemistry and Technology of Petroleum, Marcel Decker, New York, (1980).

8. B.P. TISSOT, D.H. WELTE, Petroleum Formation and Occurrence, 2nd ed., Springer-Verlag, Berlin, (1984).

9. F. HERSHKOWITZ, W.N. OLMSTEAD, R.P. RHODES, K.D. ROSE, Molecular Mechanism of Oil Shale Pyrolysis in Nitrogen and Hydrogen Atmospheres, American Chemical Society, pp. 301-316, (1983).

10. P.R. SOLOMON, D.G. HAMBIEN, M.A. SERIO, Z.Z. YU, S. CHARPENAY, A Characterization Method and Model for Predicting Coal Conversion Behavior, Fuel, Vol. 72, pp. 469-488, (1993).

11. Liquid Fuels From Coal, National Coal Board Report, Planning, Assessment and Development Branch, (1978).

12. ZYGOURAKIS, K., Effect of Pyrolysis Conditions on the Macropore Structure of Coal-Derived Chars, Energy and Fuel, Vol. 7, pp. 33-41, (1993).

13. R.K. HESSLEY, J.W. REASONER,J.T. RILEY, Coal Science; An introduction to Chemistry, Technology, and Utilization, John Wiley and Sons, (1986).

14. C.J. LAFFERTY, Catalytic Hydropyrolysis of Coal, Ph.D. Thesis, University of Strathclyde, Glasgow, UK. (1991).

15. M.J. ROBERTS, D.M. RUE, F.S. LAU, Pressurised Fluidised-Bed Hydroretorting of Six Eastern Shales in Batch and Continuous Laboratory-Scale Reactors, Fuel, Vol. 71, pp. 335-341, (1992).

16. M.J. ROBERTS, D.M. RUE, F.S. LAU, Pressurised Fluidised-Bed Hydroretorting of Raw and Beneficiated Eastern Oil Shales, Fuel, Vol. 71, pp. 1433-1439, (1992).

17. M.C. MENSINGER, D.M. RUE, M.J. ROBERTS, Environmental Data from Bench-Scale Pressurised Fluidised Hydroretorting of Eastern Oil Shales, Fuel, Vol. 71, pp.1377-1383, (1992).

18. BOLTON, C. RIEMER, C.E. SNAPE, F.J. DERBYSHIRE, M.T. TERRER, Effect of Low Temperature Catalytic Hydrogenation on Pyrolysis and Hydropyrolysis of a Bituminous Coal, Fuel, Vol.67, pp. 901-904, (1988).

19. C.E. SNAPE,C.J. LAFFERTY,G. EGLIGTON,N. ROBINSON,H.P. STEPHENS, R.G. DOSCH, Novel Dispersed Catalyst for Hydropyrolysis and Low Temperature Hydrogenation of Coals, Inter. Conf. on Coal Science, Vol.2, pp. 755-758, Tokyo, (1989).

20. Schroeder, U. S. Patent No. 3 926 775, (1975).

21. British Coal Corporation, Direct Conversion of Coal to Liquid Products, Final Report to European Coal and Steel Commission roject No. 7220 - EC / 27, (1988).

22. B.O. LI, C.B. DANHEUX,R. CYPRES, Catalytic Hydropyrolysis by Impregnated Sulphided Mo Catalyst, Fuel, Vol. 70, pp. 254-257, (1991).

23. C.E. SNAPE, C. BOLTON,R.G. DOSCH,H.P. STEPHENS, High Liquid Yields from Bituminous Coal via Hydropyrolysis with Dispersed Catalyst, Energy and uels, Vol.3, pp. 421-425, (1989).

24. C.BOLTON, C.E. SNAPE, R.J.O. BREIN, R. KANDIYOTI, Influence of Carrier Gas Flow and Heating Rates in Fixed Bed Hydropyrolysis of Coal, Fuel, Vol.66, pp. 1413-1417, (1987).

25. M. CITIROGLU, C.E. SNAPE,C.J. LAFFERTY,E. EKINCI, E., BARTLE, K. D., Pyrolysis of Turkish Goynuk Oil Shale Under Different Regimes, Erdol und Kohle, Vol. 43, pp.442-444, (1990).

26. E. EKINCI,M. CITIROGLU,E. PUTUN,C.J. LAFFERTY, C.E. SNAPE, Effect of Steam and Hydrogen Pressure on Fixed- Bed Pyrolysis of Lignites and Oil Shales, ICCS, pp. 520-523, (1991).

27. M. CITIROGLU, Pyrolysis Processes and Characterisation Studies on Three Solid Fuels, Ph.D. Thesis, Istanbul Technical University, 1993.

28. G. FYNES,J.O.H. NEWMAN, Optimisation of 300 bar Hydropyrolysis Yields, Coal Research Establishment, National Coal Board Report, (1982).

29. E. EKINCI,M. CITIROGLU, E. PUTUN,C.J. LAFFERTY,C.E. SNAPE, Effect of Steam and Hydrogen Pressure on Fixed- Bed Pyrolysis of Lignites and Oil Shales, ICCS, pp. 520-523, (1991).

30. C.E.SNAPE,C.J. LAFFERTY, Effect of Coal Rank on Low- Severity Hydropyrolysis, Prepr. Am. Chem. Soc. Div. Fuel Chem. Vol. 35(1), pp. 1-8, (1990).

31. E. GORRIN, Chemistry and Coal Utilisation (M. A. ELLITON ed.), John Wiley and Sons, (1981).

32. D.D WHITEHURST,T.O. MITCHELL,M. FARCASIU, M., Coal Liquefaction, Academic Press, New York, (1980).

33. D.F. MCMILLEN,R. MALHOTRA, S.J. CHANG,S.E. NIGENDA, Hydrogenolysis in Coal Liquefaction and Pyrolysis: The Relative Importance of Solvent Radicals and Free Hydrogen Atoms, Acs, Div. of Fuel Chem. Vol. 30, pp.297-307, (1985).

34. D.F. MCMILLEN, R. MALHOTRAS.J. CHANG,W.C. OGIER, S.E. NIGENDA, R.H. FLEMING, Mechanisms of Hydrogen Transfer and Bond Scission of Strongly Bonded Coal Structures in Donor-Solvent Systems, Fuel, Vol. 66, pp.1611-1620, (1987).

35. F.P. MIKNIS,P.J. CONN, A Common Relation for Correlating Pyrolysis Yields of Coals and Oil Shales, Fuel, Vol. 65, pp. 248-250, (1986).

36. F.P. MIKNIS, Combined n.m.r. and Fischer Assay Study of Oil ShaleConversion, Fuel, Vol. 71, pp. 731-738, (1992).

THE EURO-QUEBEC HYDRO-HYDROGEN PILOT PROJECT (EQHHPP)

Joachim GRETZ
Head of Energy Environment Interaction Unit. CEC Coordinator of the EQHHPP.
European Commission, Joint Research Centre, Ispra (VA) Italy.

ABSTRACT. The concept of a hydrogen-based, clean, renewable energy system, conceived by the Joint Research Centre Ispra of the Commission of the European Communities, is currently being investigated by European and Canadian Industries and coordinated by the JRC-Ispra of the Commission of the European Communities and the Government of Québec.

The 100 MW pilot project is to demonstrate the provision of clean and renewable primary energy in the form of already available hydroelectricity from Québec converted via electrolysis into hydrogen and shipped to Europe, where it is stored and used in different ways: vehicle, aviation and maritime propulsion, steel fabrication, advanced techniques of hydrogen storage and electricity/heat cogeneration.

Phase I, the first investigations of the hydro-hydrogen concept, was an assessment undertaken in 1986 by german industrial partners under the leadership of the Commission of the European Communities. The study indicated that such a concept could be built with today's technology and showed a cost of ~10 cents $_{ECU}$/kWh of liquid hydro-hydrogen produced in Canada and delivered in Germany.

Phase II, the detailed system definition indicated costs of the electrolytic hydrogen, produced with hydropower which would be available at 2 cents$_{ECU}$/kWh, shipped to and stored in a European port, of 14,8 cents$_{ECU}$/kWh in the form of liquid hydrogen.

The present Phase III,0 is a demonstration programme on the utilization of hydrogen in the above mentioned fields: a MAN city bus with converted natural gas engine (Germany); a VAN HOOL city bus with converted diesel engine (Belgium); a MACCHI-ANSALDO city bus with fuel cell (Italy); a city bus with STIRLING engine (Ireland); three city buses operated with HYTHANE (Quebec); a passenger boat with fuel cell (Italy); a hydrogen iron ore/scrap plasma ore melter-reducer plant (Ireland); a fuel cell cogeneration plant (Germany) and an internal combustion engine cogeneration plant (Belgium). In addition detailed studies on safety measures and codes, socio-economic studies on the comparison of hydrogen with conventional fuels are undertaken.

1. Rationale

Given the physical and technical characteristics of hydrogen and renewable energies the hydro-hydrogen energy system is basically syllogistic.

Hydrogen

- is a carbonfree fuel which oxidizes to water as combustion product. The generated water becomes, together with renewable primary energy for splitting it a source of clean and abundant energy in a carbon-free, natural cycle;
- can be transported and stored in gaseous and liquid form and used as fuel in industry,

Y. Yürüm (ed.), Hydrogen Energy System, 297–316.

households, power stations, motor car and aviation;
- being a secondary energy - an energy vector - it has to be generated with primary energy i.e. direct or indirect solar energy such as hydropower and wind energy if the overall system is to be clean and renewable.

As it stands today, the most convenient process for the generation of hydrogen is electrolysis which requires electricity as primary energy.

Natural hydropower and hydropower of already existing dams are amongst the cleanest and most promising renewable electricity sources. 23% of the solar energy falling on the earth operates the hydrological cycle, i.e. evaporation of water, precipitation and storage in water and ice.

Some characteristics of hydropower are given hereafter:
- 23% of the solar energy falling on the earth, i.e. ~ 350 10^6 TWh/y, operates the hydrological cycle, i.e. evaporation of water, precipitation and storage in water and ice;
- the World's topological and technically exploitable estimated hydro-energy potential is ~ 20 10^3 TWh/y, i.e. 0.0057% of the hydrological cycle energy;
- the yearly average utilization factor of hydroenergy is ~ 65% (in 1986, worldwide installed 380 GW generated 2200 TWh electricity);
- today's hydropower electricity generation of about 2200 TWh/y represents ~ 21% of the World's electricity generation;
- the table gives the World's estimated hydropower potential and its approximate distribution:
  . Asia (without UDSSR) : $5.3 \cdot 10^3$ TWh/y;
  . South America : $3.8 \cdot 10^3$ TWh/y;
  . Africa : $3.1 \cdot 10^3$ TWh/y;
  . North America : $3.1 \cdot 10^3$ TWh/y;
  . Ex UDSSR : $2.2 \cdot 10^3$ TWh/y;
  . Greenland : $1.5 \cdot 10^3$ TWh/y;
  . Europe : $1.4 \cdot 10^3$ TWh/y;

  (for comparison, Europe's (EC) electricity consumption (1984) was 1.3 103 TWh/y).

The worldwide water evaporation from the oceans and rivers is ~ 5 1014 m^3 per year. If mankind's present total energy consumption of sustained 11 TW would be effectuated by hydrogen, the ensuing yearly water evaporation would be some 2.5 10^{10} m^3/year i.e. about 1/20000 of the natural evaporation. Massive use of hydrogen may put local constraints, however.

Given the long lead/times and the dynamics of the introduction of energy technologies, time is likely to become a precious raw material in view of the menacing deterioration of the environment. In the light of the existing state of the art of the technology - all components and systems exist - the time if ripe to undertake, in parallel with R & D for components and systems, pilot and demonstration projects.

2. The Euro-Quebec Hydro-Hydrogen Pilot Project

The concept of a hydrogen-based, clean, renewable energy system, was conceived by the Joint Research Centre Ispra of the Commission of the European Communities in 1986. It is currently being investigated by European and Canadian Industries and coordinated by the JRC-Ispra of the Commission of the European Communities and the Government of Québec.

2.1. THE CONCEPT

The 100 MW (el) pilot project is to demonstrate the provision of clean and renewable primary energy in the form of Canadian hydropower converted via electrolysis into hydrogen and shipped to Europe, where it is to be stored and used in different ways: electricity/heat cogeneration, vehicle and aviation propulsion, steel fabrication and hydrogen enrichment of natural gas for use in industry/households (Figure 1).

For reasons of thermodynamic properties, availability of technology and end use, two different modes of vectorization have been investigated namely liquid hydrogen (LH_2) and methylcyclohexane (MCH) in order to have hydrogen in both forms, liquid and gaseous, available for the end use requirements.

2.2. MILESTONES

The project was foreseen to be carried out in 4 Phases:
- Phase I : Assessment; completed by March 1987.
- Phase II : Detailed system definition, 01.01.1989 - 31.03.1991.
- Phase III,0 : Hydrogen application demonstration programme; 1991-1993.
- Phase III : Detailed engineering & specifications; planned to last 1-2 years.
- Phase IV : Construction; planned to last 4-5 years.

2.3. PHASE I

First investigations on the hydro-hydrogen concept have been undertaken in 1986 by some twenty german interested companies under the leadership of the Commission of the European Communities and the overall management of the DECHEMA, the Deutsche Gesellschaft für Chemisches Apparatewesen, Chemische Technik und Biotechnologie e.V.

The scope of the investigation was to establish the main characteristics of system and components and to give first indications of costs of a complete system: provision of clean and renewable primary energy in the form of Canadian hydropower converted via electrolysis into hydrogen and shipped to Europe, where it is stored and used in different ways: electricity/heat cogeneration, vehicle and aviation propulsion and hydrogen enrichment of natural gas for use in industry and households.

Three transport/storage modes have been investigated: methylcyclohexane as energy carrier shipped in normal oil product carriers, ammonia shipped in special tankers and liquid hydrogen shipped in special LH2 ships and in aeroplanes.

Cost evaluations indicated costs of the electrolytic hydrogen, produced with hydropower at 2 can. Cents/kWh, stored in an European port, of 69 Dpfg/Nm^3 for the methylcyclohexane mode, 57 Dpfg/Nm^3 for the ammonia mode, 80 Dpfg/Nm^3 for the LH_2 mode by ship transportation and 77 Dpfg/Nm^3 by air transportation. These averaged equivalent costs of 20 Dpfg/kWh of clean, renewable and stored energy suggested competitivity within the not too distant future (1 ECU = 200 Dpfg, 1993).

The results of the study have been considered encouraging and it was recommended to pursue the investigations.

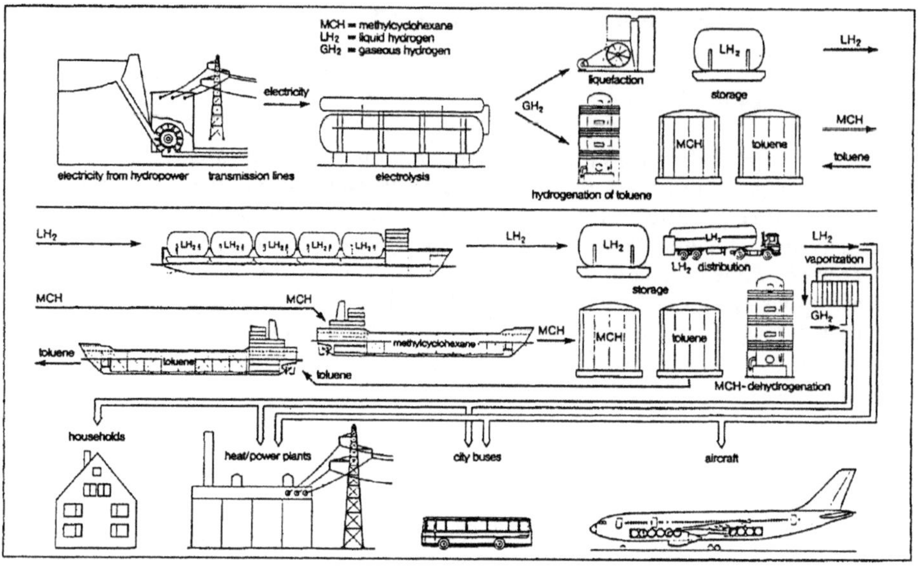

Figure 1. The concept.

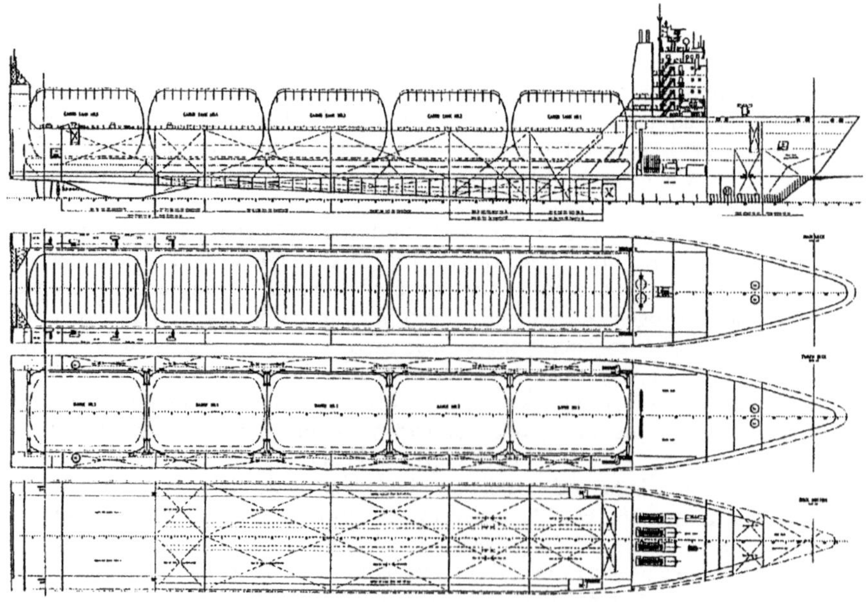

Figure 2. The barge carrier.

2.4. PHASE II. DETAILED SYSTEM DEFINITION

In Phase II of the project the Phase I results have been updated, cost calculations refined, environmental analyses undertaken and questions of safety and regulations investigated.

2.4.1. *The Partners*. The industrial partners involved in the project have been (alphabetically):

Air Liquide Canada (CND), AEG AG (D), Ansaldo Ricerche (I), Autobus MCI (CND), Blohm & Voss AG (D), BMW AG (D), CONOC Continental Contractors (D), Daimler-Benz AG (D), DECHEMA (D), Ecole Polytechnique (CND), Electrolyser Inc. (CND), FEDNAV Ltd. (CND), Fenco Lavalin Inc. (CND), Fraunhofer Institut für Systemtechnik und Innovationsforschung (D), Gaz Métropolitain (CND), GERAD (CND), Germanischer Lloyd AG (D), Hamburger Hochbahn AG (D), Hamburgische Elektrizitäts - Werke AG (D), Hamburger Gaswerke GmH (D), Holinger GmbH (D), Hamburgische Gesellschaft für Wirtschaftsförderung GmbH (D), Hydrogen Industry Council (CND), Hydrogen Systems NV (B), Institut Français du Pétrole (F), Joint Research Centre Ispra of the Commission of the European Communities (I), L'Air Liquide SA (F), Linde AG (D), Messerschmitt-Bölkow-Blohm GmbH (D), Messer-Griesheim GmbH (D), Paul Scherrer Institut (CH), Pratt & Withney (CND), Reederei August Bolten (D), SNC/FW Ltd. (CND), Staatliche Materialprüfungsanstalt der Universität Stuttgart (D), STCUM (CND), Technische Hochschule Darmstadt (D), Technische Universität Hamburg-Harburg (D), The LGL Group Ltd. (CND), Thyssen-Nordseewerke GmbH (D), Uhde GmbH (D), Union Eléctra Fenosa SA (E), Universidad da Las Palmas de Gran Canaria (E), VTG-Paktank GmbH (D), Université Concordia (CND), Université Laval (CND).

2.4.2. *Project Management*. Phase II of the project was managed by a Joint Management Group (JMG) consisting of the Ludwig-Bölkow-Stiftung (LBS), Ottobrunn, Germany, and Hydro-Québec (HQ), Montreal, Québec.

2.4.3. *Reference Case (LH2 Vector)*. For the reference case, i.e. the LH2 vector:

- hydropower : 100 MW;
- electrolysis (net) : 74 %;
- annuity (8% interest, 15 yr pay back) : 11,7 %;
- load factor : 95 %.

The main characteristics are:

- hydrogen delivered in Hamburg : 74 MW = 614 GWh/y;
- hydrogen transmission efficiency : 74 %;
- plant investment costs : 415 Mio ECU;
- specific hydrogen energy costs : 14,8 centsECU/kWh.

2.4.4. *Barge Carrier for Liquid Hydrogen (LH2) Transport*. The new component in Phase II is the LH_2 ship. The LH_2 is transported in 5 vessels, containing 3000 m^3 LH2 each, fixed on barges which are loaded on a barge carrier, thereby transporting altogether 15.000 m^3 LH_2, see Figure 2. The containers are super-insulated with no boil-off for 50 days. The floating barges permit individual towing on the waterways from the arrival sites to the users.

2.4.5. *Safety.* On 18.12.1990 the BAM (German Federal Institut for Material Research) declared that liquid hydrogen would be not more dangerous than LNG and LPG and that it has no objection to the transport of liquid hydrogen in 2 G type ships (IMO rules).

2.4.6. *Costs.* The cost of liquid hydrogen, produced with hydropower at 2 $cents_{ECU}$/kWh, shipped to Europe and stored in the port of Hamburg is 14,8 $cents_{ECU}$/kWh. Sensitivity analysis indicates a 19% impact on the product cost by doubling electricity cost.

The specific product costs and their breakdown for the LH2 vector (reference case) are given in Figures 3 and 4.

The specific product costs of gaseous hydrogen (GH_2) from the MCH vector would be 12 $cents_{ECU}$/kWh. If dehydrogenated with hydrogen, the specific product costs would be 15,3 $cents_{ECU}$/kWh. Therewith, the product cost of the MCH vector would be higher than those of the LH_2 vector if it is dehydrogenated with clean hydrogen which has to be the case if the product is to be clean in all its steps from production to utilization. Furthermore, the user's profile requires 62 MW LH_2 and 12 MW GH_2. To satisfy this, 62 MW of the GH_2 in case of the MCH vector would have to be liquefied by part of the GH_2 delivered at Hamburg, a procedure which would bring the product cost of the LH2 from the MCH vector up to 22 $cents_{ECU}$/kWh.

A standardized calculation for depreciation was used with an interest rate of 8% and a capital payback period of 15 years resulting in a constant annuity of 11,7% (1990 money). The operating costs include a.o. energy consumption and consumables.

2.4.7. *Vector Selection.* Weighing the pros and cons of the LH_2 vector vs. the MCH vector, i.e. considering the advantages of MCH:
- unlimited storage period;
- transport and storage in existing normal oil product ships and containers;

against its disadvantages:
- the energy intensive dehydrogenization and the liquefaction done at the user's site whereas the energy intensive liquefaction of LH_2 is done with abundant hydropower in Québec;
- MCH as well as toluene are petrochemical products and therewith environmentally less avantageous than LH_2;
- its product (GH_2) is not adapted to the user's profile, about 80% of hydrogen use being in form of LH_2;

the decision was taken to retain LH_2 as vector for the realization of the project.

2.4.8. *Economics.* The hydrogen costs of 14,8-15 $cents_{ECU}$/kWh are pictured in Figure 5 together with the gasoline prices in Europe (average of the 12 EC countries, August 1990) of 8,5 $cents_{ECU}$/kWh which are made up of 3 $cents_{ECU}$/kWh for the crude oil itself, its transportation, refinement, handling and distribution and of 5,5 $cents_{ECU}$/kWh for taxes.

With the above values and including the "submerged" costs of protection/repair of the damaged environment resulting from the use of fossil fuel (the value of 2,5 $cents_{ECU}$/kWh is the average from various literature sources), the "Fuel cost iceberg"

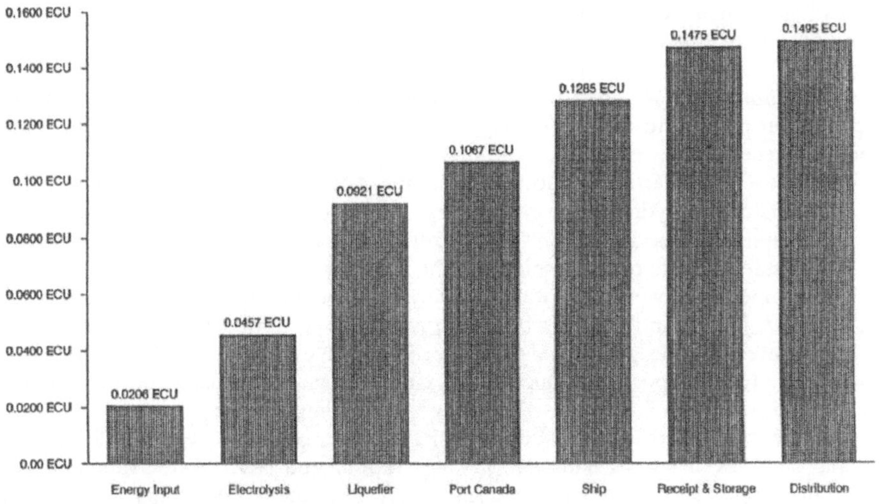

Figure 3. Specific costs ($¢_{ECU}$/kWh).

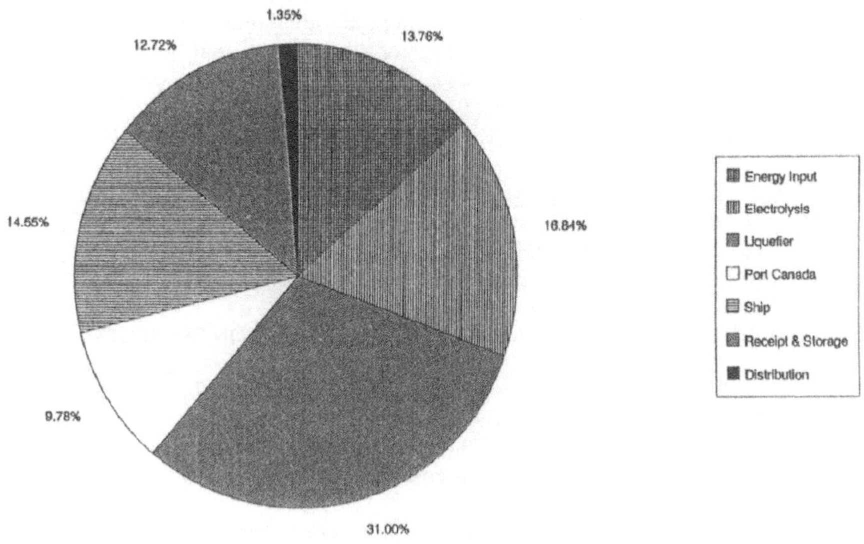

Figure 4. Specific cost distribution.

would look as shown in Figure 5, indicating the merits of clean technologies.

With the internalization of external costs, hydrogen energy would be about (1,45) 2,9 times higher than hydrocarbon energy costs, not taking into account the imponderable costs of climate effects.

2.4.9. *Environmental Aspects*. As it is the case for any energy system, the construction of hydropower plants and the use/management of hydrogen has an impact on the local and the global ecosystem:

- hydropower: the usually quoted worldwide potential of 15-20 10^3 TWh/y is only 0,0057% of the hydrological cycle energy and is based on topological, technical and environmental concerns. In any case, careful site selection and adquate measures have to be undertaken in order to minimize the impacts;
- water vapour: the worldwide water evaporation from the oceans and rivers is 5 10^{14} m^3 per year. If mankind's today's total energy consumption of sustained 11 TW would be effectuated by hydrogen, its yearly evaporation would be 2,5 10^{10} m^3 i.e. about 1/20000 of the natural evaporation. Once hydrogen will be massively used, local considerations are obligatory like it is the case for today's wet cooling towers;
- end use: whereas the combustion of hydrogen does not produce CO_2, CO, SO_2, VOC and particles, it entails emission of water vapour and NO_x. Water vapour emissions from airoplanes may be harmful since they generate - depending on the cruising altitude and latitude - ice clouds with ensuing greenhouse effects. The problem is of utmost importance and actually under investigation; cruising altitude management is one answer.

The formation of NO_x from the atmospheric components N_2 and O_2 is a function of flame temperature and its duration. Considering the wide flammability range of hydrogen its combustion can be influenced by the design of the engine so that the NO_x emission can be reduced.

2.5. PHASES III AND IV. DETAILED ENGINEERING AND CONSTRUCTION

Since the end of Phase I of the project financial engineering is under way to gather funds for Phases III and IV, the costs of which would be in the order of 500 MECU. Furthermore, an investigation is under way on the management and organization structure for the realization of the project.

2.6. PHASE III,0. HYDROGEN APPLICATION DEMONSTRATION PROJECTS

In parallel to Phases III and IV, demonstration projects have been launched on the utilization of hydrogen in five fields where the use of hydrogen exhibits its attractiveness. In Europe, these projects are executed by industry on the basis of contracts following a tender action by the Commission with a cost sharing of at least 50% by the industrial partners.

Furthermore, studies on safety measures and codes, socio-economics and on the comparison of hydrogen with conventional fuels are undertaken.

The project management is undertaken jointly by the ARGE EQHHPP LBST&CONOC (Europe) and HYDRO-QUEBEC (Quebec).

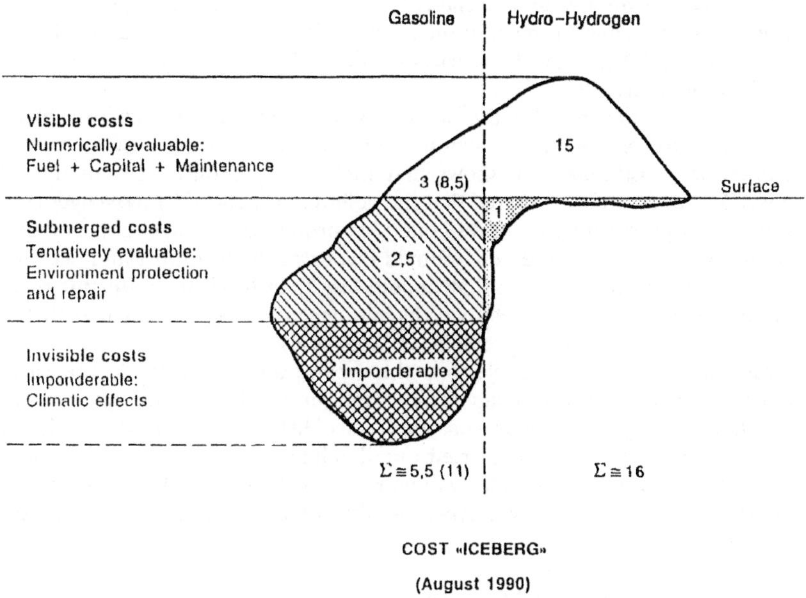

Figure 5. Fuel cost comparison.

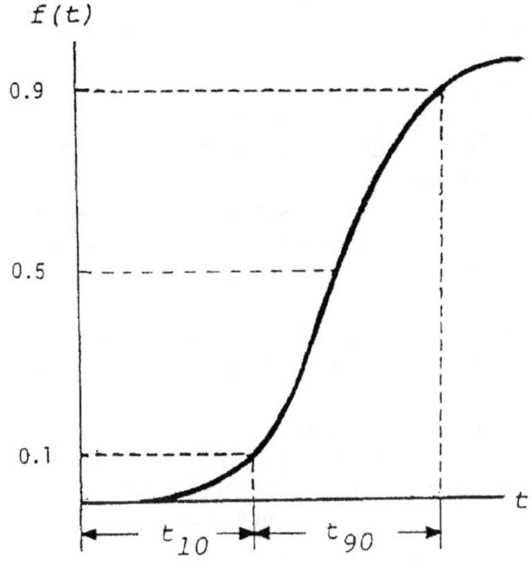

Figure 6. Market penetration of technologies.

2.6.1. *Vehicle Propulsion for Urban Transport.* It is estimated that in large urban areas motor vehicles are responsible for most of the air pollution emissions accounting for 50% of the nitrous oxides (NO_x), 70% of the carbon oxides (CO and CO_2) and 50% of the volatile organic compounds (VOC).

Public transport is predestined for the introduction of clean and therewith expensive fuel since fuel costs are only a small fraction of the overall operating costs of buses. A typical bus operation costbreakdown shows that fuel costs are only 7,3% of the total operation costs. Moreover, it is estimated that the pollution produced by passenger cars is five times greater per passenger transported than that produced by conventional buses.

Consequently, authorities in large cities should favour public transport and encourage their transit corporations to use fuels that pollute less than fossil fuels. In this regard, hydrogen is an energy vector especially worthwhile because it is a clean fuel that can provide solutions to the air pollution problems in large cities.

On this basis, six demonstration projects in the urban transport sector will be carried out in the countries of the European Union and in Québec. They cover various technologies (liquid or gaseous hydrogen, fuel cells, hythane, electricity) which are complementary. The projects will be carried out simultaneously in several cities where different variants of hydrogen technology will be developed and tested. The projects are all aimed at the development of prototypes capable of functioning in the context of commercial operations.

2.6.1.1. *City Bus with a Converted Natural Gas Internal Combustion Engine; Liquid Hydrogen Storage (Germany).* A full size regular floor city bus of the MAN SL 202 series was converted to liquid hydrogen operation by MAN Nutzfahrzeuge, Nürnberg, Germany. The six cylinder, 12 liter swept volume natural gas serial engine MAN E2866UH was converted to quantitative (air/fuel ratio of lambda = 1) operation. The engine is suited for gasified liquid hydrogen and unleaded gasoline and will provide a power output of approx. 140 kW at 2,200 rpm when operated with hydrogen. The engine can operate alternately with both fuels.

Tests of a one cylinder 2 liter engine were successfully carried out in June and July of 1993 on the BMW liquid hydrogen engine test bed in Munich. The results obtained from these tests were transferred to the full size engine. This engine together with three super-insulated LH_2 storage tanks of 200 l geometrical volume each (total of 570 l of LH_2 volume), with the LH_2 fuel supply and with appropriate safety equipment were integrated into the bus body. The operating range with LH_2 fuel will be at least 200 km. The operation mode with unleaded gasoline will allow to transfer the bus from one demonstration site to another.

In-company testing of this LH_2 city bus making use of an LH_2 refuelling station for very low transfer times of around 10-15 minutes specially developed by Linde AG will start early in 1994.

This testing and optimization period will extend throughout all of 1994, providing the necessary experience for an improvement of the bus system and the refuelling station infrastructure. Demonstration of the bus system in regular public transport will start in January 1994 for 8 months in the Bavarian city of Erlangen and continue for another two 8-month periods in Munich, thus terminating by the end of 1996.

When operating with gasoline the engine uses lambda = 1 regulation and, with the help of a three-way catalytic converter, falls inside the EURO II exhaust gas limits. In hydrogen mode the engine is operated at just under lambda = 1 and, by using the three-way catalytic converter as a reduction catalytic-converter, very low NO_x emissions can be achieved. This has been seen from measurements taken on the hydrogen engine testbed at BMW with the single-cylinder engine. In this way the engine can be kept within the proposed EURO III exhaust limits.

[Partners: Autous Oberbayern GmbH (AO); Bayerische Motoren Werke AG (BMW); Erlanger Stadtwerke AG (ESTW); Linde AG; MAN Nutzfahrzeuge AG; Stadtwerke München (SWM); Technischer Überwachungsverein Bayern-Sachsen (TÜV)].

2.6.1.2. *City Bus with a Converted Diesel Internal Combustion Engine; Liquid Hydrogen Storage (Belgium).* A full size regular floor city bus of the Van Hool A 120 type is being converted to LH_2 storage, fuel supply and engine operation. The engine is a six cylinder 12 liter swept volume MAN D 2566 Diesel adapted to qualitative operating mode (excess air/fuel ratio with a lambda value of about 2).

The tasks carried out by Hydrogen Systems N.V. involve replacing the injection nozzle with spark plugs, adding an ignition distributor and installing new pistons. Tests have shown an effective power output of 85 kW at 2200 rpm. The fuel supply and storage system is designed, fabricated and integrated by Messer-Griesheim GmbH and comprises a packaged liquid hydrogen (LH_2) storage system of 125 liter, an electric LH_2 evaporation system as well as all necessary connecting supply infrastructure (e.g. piping system) and relevant control and safety components (e.g. valves, sensors, protective devices such as blowers and control panel).

For the engine optimization tests, partly carried out at the test site of Oostmalle (Belgium), the bus was first equipped with a metal hydride storage. For demonstration to interested parties at Oostmalle the bus was then equipped with a 125 liter LH_2 storage tank, mounted inside the bus. Later on, for public demonstration the 125 liter LH_2 tank will be replaced by two 200 liter LH_2 storage vessels rooftop mounted in order to provide a better operating range.

The public demonstration of this ultra low emission bus concept will be carried out in Belgium, in the cities of Brussels and Ghent, in 1994 and 1995.

[Partners: Hydrogen System NV (HS); Messer-Griesheim GmbH (MG)].

2.6.1.3. *City Bus with a Fuel Cell; Liquid Hydrogen Storage (Italy).* A full size regular Macchi-Ansaldo floor city bus presently equipped with a diesel-electric hybrid traction system has been converted into a fuel cell hybrid bus. A PEM fuel cell fed with gasified liquid hydrogen will provide a constant power of about 35 kW to the electric traction system which includes lead acid buffer batteries. While the fuel cell provides the average power consumed by the bus, peak power for acceleration and hill climbing is provided by the buffer batteries which are then charged when the bus is idling or during braking.

The system integration work, both for the entire bus and for the fuel cell system which incorporates 9 fuel cell stacks of 5 kW each manufactured by De Nora Permelec is done by Ansaldo Ricerche s.r.l. The liquid hydrogen storage and mobile fuel supply system is designed, fabricated and integrated by Messer Griesheim GmbH. The super insulated vessels contain a total volume of 600 liter of LH_2, providing a range of approximately 300 km.

An approximately 20 months public demonstration period is planned in the north-Italian city of Brescia, starting in 1994. This demonstration, performed by the local transport authority (ASM Brescia), will provide experience in the operation of the fuel cell system in combination with the electric drives and the advanced refuelling system developed by Messer Griesheim GmbH in conjunction with the cryogenic onboard fuel storage.

[Partners: Ansaldo Ricerche s.r.l. (ARI); Azienda Servizi Municipalizzati (ASM); Messer Griesheim GmbH (MG); De Nora Permelec (DNP)].

2.6.1.4. *City Bus with a Stirling Engine; Gaseous Hydrogen Storage (Ireland).* A mid-size urban bus will be powered by a hybrid system, consisting of an external combustion engine operating on the Stirling cycle (AB 4-95 Stirling ECE, Kockums Marine) and fuelled with gaseous hydrogen, together with an electrical drive and energy storage.

The fuel supply system will comprise a storage system for compressed gaseous hydrogen operating at 13,7 MPa and accommodated in a manifold cylinder pallet assembly as well as necessary supply infrastructure (e.g. piping system) and relevant control and safety components (e.g. valves, sensors, actuators and protective devices).

The electric drive train, including storage batteries, control, drive motors and the structural components of the bus vehicle body will be adapted to the propulsion system which is an innovative 2-phase DC motor. This motor provides higher efficiencies, smaller sizes and higher starting torques than traditional electric motors.

The concept under development involves a series of accumulators charged on the one hand by the Stirling-alternator engine and on the other by energy recovered during breaking. It will power the main motor mounted over the rear axle. The project includes the design of the overall electrical systems (control, accumulation and propulsion) and the coupling of the two propulsion systems within the same vehicle.

Two engines units will be tested: one of 30 kW and the other of 80 kW. They will enter service in late 1994 and will be operated in stationary mode at first, fuelled by hydrogen from a methanol reformer. Present plans call for only one engine to be mounted on a mid-size bus for local transportation in 1995 in a demonstration that will take place in Ireland, in the city of Dublin.

[Partners: ESAT Motor Co. Ltd. (ESAMCO); Irish Industrial Gases; Kockums Marine AB].

2.6.1.5. *City Bus with Internal Combustion Engine Operating on Hythane (hydrogen-natural gas mixture) (Montreal).* This project aims at developing the particular advantages of natural gas and hydrogen mixed in a fuel called hythane (typically 15-20% hydrogen and 85-80% natural gas), a fuel mixture capable of meeting the strictest environmental standards, namely the California ultra low emission vehicle requirements, at an acceptable price level.

It is estimated that for public transit vehicles hythane provides better potential for reducing pollution emissions than does natural gas alone or the combination of catalytic converters and particle traps recommended for buses using diesel fuels.

Adapting bus motors for hythane involves designing of a complete technology for the efficient use of hythane as a fuel for urban buses. In this demonstration project, the tasks are oriented towards the development of a system of fuel storage and fuel supply, the choice of appropriate engines and their modification and the integration of these various techniques within a bus to be used for public transit. Personnel training and maintenance procedures as well as safety measures and standards applicable to this technology will be defined.

In this project, which has started in October 1993, three buses to be operated with hythane will be engineered, fabricated, tested and demonstrated in Montreal (Québec), by the end of 1995.

[Partners: Andiax Conseil inc.; Ecole Polytechnique; Industries Electrolyseurs du Québec (IEQ); FRE Composites inc.; Gaz Métropolitain inc.; Groupe informatique en recherches opérationnelles (GIRO); Hydrogen Consultants inc.; Hydro-Québec; NOVABUS; SNC-Lavalin inc.; Société de transport de la Communauté urbaine de Montreal (STCUM); Université Concordia].

2.6.2. *Aviation Propulsion.* Aircraft emit pollutants at altitudes where the atmosphere is extremely sensitive. Thus, these emissions may have a high specific effectiveness.

The elimination of CO, CO_2, SO_2 and VOC by using hydrogen not only is beneficial for the atmosphere but reduces considerably airport ground pollution. For example, the NO_x emissions resulting from the daily average 1259 landing/take-off cycles at Los Angeles International Airport are equivalent to the operation of about 1 Mio passenger cars.

2.6.2.1. *Adaptation of a Jet Engine Combustion Chamber for Hydrogen Burning with Low NO_x Emissions.* This experimental project will establish the NO_x reduction potential in an aircraft combustion operated with hydrogen chamber and provide design guidelines for later development of a real flight engine combustor.

NO_x result from the combination of atmospheric oxygen with atmospheric nitrogen in the combustion chamber. The amount produced is largely a function of the combustion temperature and pressure, as well as the "dwell time" i.e. the time the air remains in the zone of high temperature and pressure. A combustion chamber designed for kerosene, if operated with hydrogen, could produce increased NO_x emissions. Therefore, an optimized combustion chamber must be designed to take advantage of hydrogen's favourable properties: (1) as a gas it mixes well with air during combustion (no "hot spots" and corresponding high NO_x production rates), (2) the combustion is possible with a large air or fuel surplus (low temperatures) and (3) the rate of combustion is extremely high (short dwell times). The overall engine configuration for combustion with kerosene can remain unchanged when operated with hydrogen.

While the reaction zone of a conventional kerosene combustor is operated virtually stoichiometrically at full load - this is deliberate to provide a wide range of engine control - the wide flame stability limits of hydrogen allow the shifting of the operating point into the lean or rich region. Considering the limitations of current technology the preferred approach for attaining an extremely low emission target is to concentrate on a fuel injection scheme intended to produce lean combustion. In this case, it is very important that the combustion mixture be homogeneous, otherwise local hot spots occur, which result in the formation of NO_x.

Based upon available literature five candidate fuel injection concepts for the lean combustion of hydrogen have been defined. The challenges are to provide for rapid homogeneous fuel/air mixing, burning and quenching. This can be achieved with the use of swillers or by many strongly interacting jets. A different approach is related to a lean premixed combustion. The fundamental question here is whether adequate premixing prior to the introduction of the mixture into the combustor can be achieved without the danger of pre-ignition and flash back.

Different fuel injectors have been investigated in a comprehensive generic test campaign until the end of 1993. Two of these concepts identified as the most promising will be further pursued in a research type combustor section under all critical operating conditions of a modern turbofan engine. An appropriate test program has been defined and will be executed in 1994 and 1995.

After steady state testing tests are planned to study the transient behaviour of the combustion systems.

[Partners: Allied Signal Aerospace GmbH; Deutsche Aerospace AG (DASA); Deutsche Aerospace Airbus GmbH (DAA); Deutsche Forschungsanstalt für Luft und Raumfahrt (DLR); Deutsche Aerospace Motoren- und Turbinen-Union (MTU); Pratt & Whitney Canada Inc. (PWC)].

2.6.3. *Public Transport on Domestic Waters.* Transportation on domestic waters (lakes, rivers, canals, lagoons) has a negative impact on the environment due to the pollution caused by exhaust gases and fuel leaks; in the case of canals (in Venice, for example), noise pollution and vibrations are added sources of environmental stress. Another associated environmental hazard is the danger of collision or the possibility of fuel spillage into the water, particularly in areas of heavy traffic. For some small European lakes there are already cases in which diesel-powered boats are prohibited from navigating.

Through the use of hydrogen-powered boats, particularly with noise-free systems powered by fuel cells, these sources of pollution and hazard could be eliminated.

2.6.3.1. *Passenger Boat Powered by a Fuel Cell.* A passenger boat, originally powered by a diesel engine, is converted having the same propulsion system i.e. a hybrid fuel-cell with traditional batteries as the Italian bus project (2.5.1.3).

The boat has an electric power output of about 100 kW, of which approximately 30-50 kW will be covered by a PEM fuel cell to be supplied by De Nora Permelec, whereas the peak demand will be provided from a battery pack. In operating modes with low power demand the fuel cell will charge the batteries. The exact definition of the nominal power output is being determined in a cooperation between Ansaldo and Gestione Governativa Laghi Maggiore-Garda-Como which is the operator of the boat. The electric conditioning unit includes a converter which adjusts the fuel cell voltage to the battery voltage and controls the fuel cell power output. The fuel cell system consists of the fuel cell stacks, the heat rejection system, the gas humidification system and the necessary equipment. The hydrogen fuel supply system developed by Messer-Griesheim GmbH, includes onboard LH_2 storage vessels and vaporizers in order to provide gaseous hydrogen at a pressure suitable for the operation of the fuel cell system. The same mobile fuel filling station designed for the fuel cell bus demonstration project will supply hydrogen to the boat, on a weekly basis.

A half year demonstration period of the fuel cell boat will start in the fall of 1995 and will be carried out in Italy on the Lago Maggiore, by Gestione Governativa Navigazione Laghi Maggiore-Garda-Como.

[Partners: Ansaldo Ricerche S.r.l.; Ansaldo Industrie S.p.a.; Messer-Griesheim GmbH (MG); Registro Italiano Navale (RINA); Gestione Governativa Navigazione Laghi Maggiore-Garda-Como (GNL)].

2.6.4. *Hydrogen in Steelmaking.* Traditional iron ore processing and refining techniques release some two tonnes of CO_2 into the atmosphere for every tonne of steel produced. On the planetary level, the steelmaking industry is responsible for more then 10% of the CO_2 emissions associated with human activities.

The introduction of hydrogen in the steelmaking process with no CO_2 emissions can contribute to an important reduction of the pollution caused by traditional iron ore processing and refining techniques.

2.6.4.1. *Demonstration of the Utilization of Hydrogen to reduce CO_2 Emission In Steelmaking.* This project aims at demonstating a technology that is destined to replace the conventional methods of iron ore blast furnace reduction by an advanced method based on hydrogen and plasma arc technology.

In a melter-reducer, processing a mixture of scrap iron ore, fusion-reduction is accomplished by means of plasma generators fuelled by hydrogen which convert the electrical energy into a plasma at very high temperatures with optimum yield. The overall power consuption for crude steel production is minimised, the energy requirements of the endothermic reactions involved in hydrogen reduction of iron ore being supplied by the plasma-arc device which is a convenient, compact and highly efficient way of transferring electrical into thermal energy. A test furnace was used in 1992 in Sweden to confirm the approach. The output of liquid metal exceeded the estimates because of the higher-than - expected energy level of the process and the absence of oxidizing agents. The life of the modified plasma arc electrodes fell well short of that expected and a modified approach will be necessary to overcome the arc behaviour when hydrogen only is the plasma gas. Tests of the revised units will start in 1994 at a new test centre at Arigna in Ireland.

An iron ore and scrap melter-reducer component with two plasma arc injectors of 0,8 MW at an output of 0,5. To steel/h are being built and tested, demonstrating the direct reduction of iron ore fines with hydrogen as reductant yielding crude steel (Kent Steel). The goal is to commence demonstration in a steel plant (preferably at a larger output capacity) as soon as possible and to build a commercially operating plant rapidly. An optimized plant of larger scale shall include additional CO_2 neutral reductants such as forest wastes.

Provision is made for an alternate supply of synthesis gas derived from biofuel, thereby preserving the CO_2 neutrality from an environmental viewpoint. The CO and H_2 derived from gasification of the biofuel is achieved by plasma-arc devices, with steam injection and its water splitting in the plasma arc providing for additional hydrogen.

[Partners: Arigna Catchment Area Ltd.; Arigna Enterprise Management Ltd.; EMEC Arigna Ltd.; Industrial Consultants International Ltd. (ICIL); Kent Steel Ltd.; Westgate Engineering Ltd.].

2.6.5. *Storage and Transportation Technology.* The best way of transporting large quantities of hydrogen over long distances and also the use of hydrogen in mobile applications is in its liquid form given the high gravimetric energy content of liquid hydrogen. Two projects will be carried out in countries of the European Union and in Québec, concerning the storage of large quantities of hydrogen and also the storage of hydrogen inboard of vehicles.

2.6.5.1. *Development and Testing of Large Storage Tanks.* This project aims at optimizing the transport of large quantities of liquid hydrogen by sea and its safe storage. A reduced scale model of the large tank (see 2.3.4) will be built and subjected to laboratory testing.

In addition to the usual phenomena related to heat transfer, to the effects of pressure variations and to mechanical limitations, tasks will focus on the configuration of the structures supporting the inner shells of the tanks and connecting them to their outer shells. Specific attention will be paid to optimizing these structures so that they can withstand the accelerations and jolting associated with transporting the large scale tanks of 3000 m^3 by sea. Finally, an evaluation will be made of how such a transport system could comply with the current shipping regulations and the related safety requirements.

A test tank of 180 m^3 was designed by Air Liquide SA for demonstration purposes. The tank represents a 1:20 scale model of the barge tank foreseen in the previous phase of the EQHHPP. It incorporates novel design features and satisfies both German and Canadian rules and regulations for storage and transportation tanks (Germanischer Lloyd, ASME, AD Merkblatt).

Between 1992 and 1995, such vessels are engineered, built and tested in Germany. An experimental program will simulate different operating conditions. It will yield the temperature development at selected points in the inner tank, at tank walls and in the vacuum space. The experimental data will be used to validate a numerical model of the tank.

[Partners: Air Liquide SA; Bundesanstalt für Materialforschung und- prüfung (BAM); Germanischer Lloyd AG (GL); Holinger; Staatliche Materialprüfunganastalt der Universität Stuttgart (MPA); Thyssen Nordseewerke GmbH (TNSW); Uhde GmbH].

2.6.5.2. *Development and Testing of Hydrogen Tanks for Busses and Automobiles.* This project consists of two tasks concerning metallic and composite materials LH_2 tanks for vehicle application.

The first task, as a part of research activities at BMW, is to investigate the safety of LH_2 in a program carried out in conjunction with Messer-Grieshiem GmbH and Linde AG aimed at estimating the hazard potential of LH_2 metallic tanks when used in vehicles. Consequently, the design criteria for vehicle tanks will be established. For this purpose, cylindrical LH_2 tanks with a capacity of 125 l have been engineered, built and tested. Burst tests with 10 LH_2 tanks under various levels of pressure and filled to varying degrees have been carried out. A subsequent series of test tanks will be exposed to extreme conditions (fire), to accidents caused by the malfunctioning of safety systems (valves, overflow pipes) and to damage resulting from violent impact.

[Partners: Bayerische Motoren Werke AG (BMW); Fraunhofer Institut für Chemische Technologie (FHG-ICT); Linde AG; Messer-Griesheim GmbH (MG); Technische Universität München (TUM); Technische Überwachungsverein Bayern-Sachsen (TÜV)].

The second task within this project is to demonstrate the suitability of composite materials tanks for vehicle applications. The first step is to demonstrate that this material, which can already be used to contain mixtures of hydrogen and natural gas, can also be used to contain liquid hydrogen.

A number of tanks with capacities of 100 to 350 litres will be designed, manufactured and equipped with instruments for measuring pressure and temperature at various critical points, both inside the tank and on its surface.

The tanks will be manufactured in Quebec, by FRE Composites inc. , and then transported to Europe where destruction tests will be carried out, simulating the same type of foreseeable accidents.

[Partners: Bayerische Motoren Werke AG (BMW); Bundesanstalt für Materialforschung und- prüfung (BAM); Forschung für Luft- und Raumfahrt (DLR); Fraunhofer Gesellschaft, Institut für Chemische Technologie (FHG-ICT); FRE Composites Inc.; Hydro Québec; Linde AG; Messer-Griesheim GmbH (MG); Staatliche Materialprüfungsanstalt der Universität München (TUM); Technische Überwachungsverein Bayer-Sachsen (TÜV)].

2.6.6. *Cogeneration*. The cogeneration of electricity and heat is by now a well known industrial process finding increasing favour because of its energy saving advantages.

When operated with hydrogen, cogeneration exhibits further advantages: the low or zero emissions allow combined electricity/district heating plants to be moved closer to urban centres; the cryogenic properties of liquid hydrogen providing a heat sink has the potential of high conversion efficiencies.

2.6.6.1. *Fuel Cell Cogeneration Plant*. A 200 kW_{el}/220kWth phosphoric acid fuel cell (PAFC) will be operated with both natural gas and hydrogen in parallel and hydrogen alone providing data for later full-sized cogeneration plants in the 10 mW range.

The cell operates at ~ 200°C and has an efficiency of ~40%.

An urban location is chosen in order to clarify the licensing and liquid/gaseous hydrogen transport issue and to provide connections to existing networks of electricity and district heating.

The fuel cell activity will last until 1998.

[Partners: CONSULECTRA GmbH; Hamburger Electricitätswerke AG (HEW); Hamburger Gaswerke AG (HGW)].

2.6.6.2. *Piston Engine Cogeneration Plant*. The aim of the project is to demonstrate that hydrogen, produced from wind energy offers the potential to substitute the imported energy requirements of remote communities.

The project consists of the design, construction and testing of a cogenerator unit based on a water cooled gas engine powered by hydrogen gas and coupled to an asynchronous generator. The unit will be coupled to a stand-alone wind energy converter.

The engine is a modified 4 cylinder diesel engine, type Valmet-420 D. It has a displacement of 4.4 liters and a maximum power of 64 kW at 2400 rpm.

The plant will be operated at 1500 rpm. It then has an efficiency of about 90% having an output of 30 $kWk_{el}/70kW_{th}$ removing the heat from the cooling water of the engine and from the exhaust gas.

The unit will be coupled to an electrolyser manufactured by Hydrogen Systems N.V. (registered trademark IMET), having a ~90% thermal efficiency. The required electricity is produced by a windturbine (type Windmaster-200), manufactured and commercialised through a sister company of Hydrogen Systems N.V., i.e. H.M.Z- Belgium.

Endurance tests are envisaged at the Belgian coast near Zeebrugge.

[Partner: Hydrogen Systems NV (HS)].

2.6.7. *Safety and Standards*. Broader use of hydrogen implies new safety measures and regulations that will have to be harmonized at the international level. The objective of this study is to determine the measures to be taken in this area.

Even though the issues of safety measures and procedures for using hydrogen are covered in the framework of the various demonstration projects it will be considered in a more global manner.

The properties of hydrogen, hythane, propane, gasoline and gas-oil will be analyzed and compared in relation to safety aspects. This will be followed by a more specific assessment of the potential risks associated with the handling of hydrogen and hythane. Attention will also be paid to compiling and comparing the existing codes and regulations at the local, national and international level, particularly in North America and Europe. Recommendations will be made with a view to harmonizing the existing codes and regulations and to fostering the use of hydrogen for energy purposes.

The study will also address the question of social acceptability of using hydrogen and the perceptions that the public at large or certain groups might have about the dangers related to hydrogen. Widespread dissemination of the findings shall be made to curb the circulation of erroneous and unfounded information on hydrogen safety aspects.

It is intended to push world standards through the Technical Committee on Hydrogen Energy Technologies ISO TC 197 of which the Bureau de normalisation du Québec is assuming the secretariat.

[Partners: Bureau de normalisation du Québec (BNQ); Continental Contractors GmbH (CONOC); Conseil de l'industrie de l'hydrogène (CIH); Germanischer Lloyd AG (GL); Holinger GmbH; Linde AG; SNC-LGL; Société Européenne de Propulsion S.A. (SEP); Staatliche Materialprüfunganastalt der Universität Stuttgart (MPA); Tektrend inc.].

2.6.8. *Operational Management System for Integration of Results of Urban Bus Demonstrations*. Analysis will make it possible to integrate the results of the work carried out in Europe and in Québec to develop urban buses adapted for hydrogen. On the basis of the results of these complementary projects, it will be possible to develop common standards with the objective of optimising the use of this kind of technology.

Analysis activities will identify the requirements that operators of such systems will have to meet with regard to infrastructures and operational procedures for using gaseous or liquid hydrogen. Then modes of operation and characteristics will be defined for fleets of public transit vehicles consisting of hydrogen-fuelled buses. Optimisation of the systems will guarantee the highest possible energy efficiency and maximum environmental gain, and also to foster strong public acceptance of this means of transportation.

Analysis will be carried out aiming at the adaptation of the HASTUS (transit scheduling system) and GEOROUTE (street network database system) software packages to schedule the use of hydrogen and hythane fuelled city buses based on their specific operating characteristics and their estimated emissions.

In Belgium, the site selected for the implementation of the adapted software packages is the Brussels Transit Authority, that already uses the HASTUS system for its conventionally fuelled transit systems. It will be the demonstration site for the EQHHPP Belgian hydrogen bus project. Similar adaption of the HASTUS and GEOROUTE systems will be performed in Québec at the STCUM in Montréal where the hythane project will be demonstrated and where the HASTUS system is also used.

[Partners: Groupe Informatique en Recherches Opérationnelles (GIRO); Hydrogen Systems NV (HS); Société de transport de la Communauté urbaine de Montréal (STCUM); Tele Atlas].

2.6.9. *Economic Analysis*. As it stands today, external costs of environmental damages resulting from the use of energy are not established and not attributed to the users. A socio-economic study is to assess the relative advantages of hydrogen in the field of public transport by evaluating the pollution emissions and evaluating the reductions in social costs that could be attained by replacing conventional transport systems by hydrogen-based technologies.

The study involves the investigation of the following emissions: NO_x, HC (with separate consideration of CH_4,), SO_2, CO, CO_2 and particulates, as well as secondary ozone formation from NO_x, and HC. Investigation of direct burdens will be given highest priority, but also indirect burdens derived from the fuel supply chain (extraction, processing, transportation, distribution) will be covered. The various impacts on humans and on the environment will be described, quantified, and as far as possible monetized as social costs. A number of scenarios will be analyzed combining various technologies based on conventional hydrocarbons, on electric streetcars, on natural gas, on hythane and on pure hydrogen. The various effects brought to light will then be described, quantified and, as far as possible, compared to each other, taking into consideration the overall social costs.

The tasks are to evaluate the emissions, burdens, and attributable social costs in the urban areas of London, Los Angeles and Munich, as well as to assess the possible social cost reductions derived from replacing parts or all of the conventionally fuelled transportation systems (namely actual buses and aircrafts) by advanced conventional or advanced hydrogen fuelled systems.

[Partners: Ecole des Hautes Etudes Commerciales (GERAD-HEC); Fraunhofer Institut für Systemtechnik und Innovationsforschung FHG-ISI); Ludwig-Bölkow-Systemtechnik GmbH (LBST); University of Northumbria of Newcastle (UNN)].

2.6.10. *Study on Distribution Systems*. A study will be made to define the storage and distribution systems best adapted to the transportation and delivery of hydrogen to specific markets, and to evaluate the costs of such systems.

Based on the various modes of transportation available, such as ship, rail and road the characteristics of the best hydrogen delivery and distribution systems will be examined with a view to limiting the costs, while guaranteeing the reliability of the equipment. The types of tanks as well as the systems of transportation by ship, by train or by truck will be reviewed and analyzed. Cost estimates will be made, taking into account the distances, volumes and frequencies of delivery defined in the various scenarios corresponding to the target market.

This study should lead to the design and manufacture of specialized containers in Québec and their testing in Europe. Variants of the containers are to be adapted to different conditions prevailing in the target markets.

[Partners: Econoden; FEDNAV ltée; Hydrogenal; Hydro-Québec; Navtech].

2.7. FUNDING

With the financial participation of the Commission of the European Communities (19.7 MECU), the Government of Quebec (6.1 MECU), European industry (16 MECU) and Quebec industry (3 MECU), the EQHHPP's total funding up to now is 45 MECU.

3. Conclusion

With the EQHHPP, the European Commission and the Government of Québec are carrying out an unique experiment in international partnership with committments from political entities and universities, as well as from public corporations and private industries. This collaboration promises to generate technological, environmental and economic benefits stemming from the promising hydrogen technology.

The demonstration projects undertaken in the field of hydrogen within the EQHHPP are well in line with other technological efforts undertaken worldwide in the field of hydrogen such as: ultra low/zero emission vehicles; advanced storage concepts for LH_2, for gaseous hydrogen and hythane; fuel cell development and application in vehicles; the use of hydrogen in steelmaking; the identification and harmonization of safety relevant aspects and of rules and regulations; considerations of socio-economic aspects of hydrogen use and the fostering of the acceptability of hydrogen. It contributes in a broad scope to the advancement of hydrogen as a future fuel option with a favorable overall cost-benefit for society.

Whilst the EQHHPP for the moment, due to the lack of financing for the realisation of the 100 MW project, focuses on the pilot demonstration of the entire hydrogen supply chain, hydrogen supply on a much smaller scale of approx. 1 MW, is under investigation. It uses advanced containerized transport concepts for container ship and rail transport of liquid hydrogen. The supplied LH_2, will mainly be used to run the EQHHPP demonstration activities. The EQHHPP thus assists on a small scale but already involving hardware components to provide a possible option for an energy usage and trade based on renewable energy sources via a new clean energy vector in the 21st century. Given the long lead times and the dynamics of the introduction of energy technologies, time is likely to become a precious raw material in view of the menacing deterioration of the environment.

In the past it took about 100 years for a market penetration from 10 to 50%, which means that solar hydrogen should cover 10% of the energy market by the year 2000 if by the year 2100 it is assumed to cover 50% and if things would go at the same pace as in the past. Even if things move faster now, solar hydrogen will not cover 10% of the market by the year 2000. Work on solar hydrogen should - by judiciously analyzing the penetration curve - have started before the beginning of the century in order to have 50% of mankind's energy consuption covered by solar hydrogen by the year 2100. The least one can say is that it is not too early to start working on it now.

4. References

[1] Dechema, A. study for the Generation, Inter-Continental Transport, and Use of Hydrogen as a Source of Clean Energy, on the Basis of Large-scale and Cheap Hydro-Electricity (Hydrogen Pilot Project - Canada), Dechema, Frankfurt 1987.

[2] The 100 MW Euro-Quebec Hydro-Hydrogen Pilot Project, J. Gretz, J. Baselt, O. Ullmann, H. Wendt - International Journal of Hydrogen Energy, Vol. 15, No.6, pp. 419-424, 1990.

[3] The Euro-Quebec Hydro-Hydrogen Pilot Project Phase II, Executive Summary, Joint Management Group Hydro-Quebec and Ludwig-Bölkow-Stiftung, D. Kluyskens, O. Ullmann, R. Wurster, Montréal and Ottobrunn, March 31, 1991.

[4] Phase II and Phase III.O of the 100 MW Euro-Quebec Hydro-Hydrogen Pilot Project, J. Gretz, B. Drolet, D. Kluyens, F. Sandmann, O. Ullmann - proceedings of the 9th World Hydrogen Energy Conference, Paris, 1992.

[5] The Euro-Quebec Hydro-Hydrogen Pilot Project: Demonstration Phase, B. Drolet, J. Gretz, D. Kluyskens, F. Sandmann, R. Wurster - 10th World Hydrogen Energy Conference, Cocoa Beach, Florida, 1994.

HYDROGEN ENERGY SYSTEM IN CALIFORNIA

R. M. ZWEIG
2936 Mc Allister Street
Riverside, CA 92503
U.S.A.

ABSTRACT. Hydrogen use as a clean burning fuel in California and the South Coast Air Quality Management district tests using hydrogen as a clean burning environmentally safe fuel is given. Solar Hydrogen Projects in California and recent medical data document human lung damage of patients living in air polluted urban areas are summarized.

1. Hydrogen Integration into Community Energy Systems

California has been a pioneer in the use of hydrogen as a clean burning fuel. The South Coast Air Quality Management district (SCAQMD) is particularly suited as a living laboratory to test the benefits of using hydrogen as a clean burning environmentally safe fuel. It is interesting to recollect the history of hydrogen in California with several historic high lights:

-1970, Development of a hydrogen vehicle at UCLA under the direction of Dr. Van Vorst in which a Gremlin car was converted to utilize cryogenic hydrogen in 1972.
-1973, Dr. James Pitts who director of the California State Air Pollution Research Center at UCR mentioned that the use of hydrogen as clean burning fuel would eliminate the problems found in California resulting from fossil fuel combustion air pollution.
-1973, A clean fuel rally for all of the engineering schools in California demonstrated the various alternate fuel vehicles at the Riverside International raceway. Hydrogen was found to be the cleanest and therefore, a new organization known as Clean Air Now was established to promote hydrogen as the perfect fuel.
-1975, A Helioscience conference was held in Palm Springs to investigate the feasibility of utilizing solar energy to produce various energy forms including hydrogen.
-1976, The City of Riverside operated a 19-passenger hydrogen bus, which was evaluated by the California Transportation Authority (CAL TRANS). Their conclusion:
1. The hydrogen bus utilized its fuel very efficiently.
2. Hydrogen fuel could reduce air pollution in the South Coast Air Basin.
3. Hydrogen vehicle technology should be pursued further.

Mechanical problems to be corrected:
1. Back firing.
2. Increase power for climbing hills.
3. Speed of refuelling procedure. Increase the efficiency for refuelling hydride tanks.

Y. Yürüm (ed.), Hydrogen Energy System, 317–327.

-1980, A solar hydrogen conference was held at University of California (UCR) where a new vehicle, a Dodge D50 truck was presented at the conference. The Dodge D50 was converted by Clean Air Now, University of Denver and Ergenics. It was driven from Denver to Riverside utilizing hydrogen from a tube trailer supplied by Air Products Corporation.

-1990, A solar hydrogen program was established at Riverside Community College, which eventually was transferred to the University of California at Riverside. The project will engineer and convert up to there vehicles to operate on hydrogen fuel. UCR's prototype vehicles include two pickup trucks and a passenger car. The staff is implementing testing modifications and performing conversions on remaining vehicles. They will then conduct engine analysis and perform vehicle demonstrations for the SCAQMD, American Lung Association, and other organizations.

This program is also developing a curriculum to train key personnel and students, who will apply mechanical and technological expertise for the production, distribution, and conversion of these and future hydrogen vehicles. This project creates one of the first hydrogen fuels student training centers in the United States. Currently, UCR faculty are preparing operating manuals for classroom use in technical retrofitting. An additional handbook is being developed for fleet conversion.

-1991, Dr. Alan Lloyd, the chief scientist for the SCQAMD, presented the report for "California Plan," which outlined the potential role of hydrogen in California as follows:
"In order for the District to achieve its air quality goals, all available clean technologies need to be implemented. The future potential for hydrogen is significant. The lack of sufficient technological research and development efforts, coupled with limited funding resources, and an absence of a determined, focused national energy agenda has provided a limited exploration of hydrogen potential to date. For hydrogen to become a viable energy option in the United States, progress in the production, handling, and storage of the fuel are essential."

South Coast Air Quality Management District (SCAQMD) has been financially supporting hydrogen programs for the past five years.

A basic conceptual hydrogen infrastructure is in process of development, demonstrating solar electrolysis production, transmission, and end use in internal combustion engine (I.C.E.) vehicles and eventually hydrogen fuel cell electric vehicles, including buses, trucks and locomotives.

Other programs supported by the SQAMD demonstrate that hydrogen can be produced using solar electrolysis, be transmitted and stored at various locations in California, i.e. Humboldt, UC Davis, SMUD, UCR, Xerox - El Segundo and at Kaiser Hospital. These projects will eventually be integrated into a master plan for hydrogen utilization as a common fuel for I.C.E. engines, stationary fuel cells, or fuel cell electric vehicles.

National and international scientist and consultants will cooperate in integration of these various systems. At the present time, the South Coast has been working with the Electrolysis Incorporated from Toronto, Canada, the Ballard Fuel Cell Corporation from Toronto, the Ontario energy Department from Ontario, Canada, and Ontario Quebec for integrating hydrogen into the fuel system.

One recent hydrogen program breakthrough will integrate programs utilizing the "hythane" systems as promoted by Hydrogen Consultants Incorporated (H.C.I.). Other programs promoting hythane have been utilized in Canada, Pennsylvania and in the Florida areas. The South Coast would utilize this process by these pioneer programs and develop any future refinements required to make hythane an early interim step in the hydrogen economy. Hythane offers a great benefit for reducing air pollution and provides a valuable interim toward using hydrogen as a pure fuel. The scientists at the South Coast have offered to use the basin as a test bed to demonstrate the various technologies and demonstration, programs, which will be promoted within the next five years. NATO should offer the date derived from these programs to other international cooperatives including those of the third world countries.

1.1. THE MASTER PLAN FOR HYDROGEN IN CALIFORNIA

California offers an excellent test bed for a complete infrastructure for the hydrogen economy. All aspects for the future infrastructure are in place. The following is a list of the potential participants. In the overall, this "blueprint" study will list all the institutions now in existence, which could be integrated for the master plan. We will discuss production, distribution, and end use for the hydrogen.

1.1.1. *Production*

1.1.1.1. Fossil Fuels

1. Coal -There is a coal gasification plant in Barstow, California, which could be upgraded to produce hydrogen without environmental degradation at a cost competitive with the present prices of gasoline.
2. Oil-Oil refineries, i.e. Arco, Texaco and Unocal in the Long Beach area, which produce hydrogen in excess of the refineries' present needs.
3. Natural Gas-Praxair Corporation is now producing hydrogen at its plant in Fontana, California. The capacity presently can produce several tons a day more than the demand requirements using a steam reforming technology.

1.1.1.2. Renewables

1. Wind-There are three primary locations in California where wind generation would make hydrogen production economically feasible, namely the Altamont Pass in Northern California, Tehachapi Pass in Central California and the Banning pass in Southern California. These three areas could be integrated into one massive hydrogen pipeline.
2. Solar-There are several locations where solar electrolysis would be economically feasible. The most extensive solar collector is in Kramer's Corner area where LUZ Corporation has many square miles of parabolic collectors producing electricity at the present time. There are several solar electrolysis plants now in operation, i.e. Humboldt State College and UCR.

1.1.1.3. Kaiser Plasma Pyrolysis Plant is an excellent example of reduction of medical and municipal waste produce a high percentage hydrogen fuel gas.

1.1.2. *Distribution*

1.1.2.1. Transmission-present modes of transmission could be upgraded utilizing the system Praxair from Fontana plant whereby they distribute the hydrogen by tube trailer trucks, cryogenic trucks and rail lines. Upgrading natural gas pipelines could also be employed for long distance transmission.

1.1.2.2. Storage-All present forms of hydrogen storage have been utilized in California and there is a wealth of experience utilizing the compressed gas cryogenic and improvements for materials and concentrated supplies of hydrogen could easily be implemented in the very near future. Of special importance is modification of high-pressure tanks utilizing new materials, i.e. kevlar and carbon fibers.

1.1.3. *Utilization*

1.1.3.1. Transportation

1. For internal combustion engines (I.C.E.)-several programs are utilizing hydrogen and I.C.E.s, namely UCR and Xerox. Future use of hydrogen in the area would consist of representative vehicles from Mazda, BMW, Mercedes and Canadian Nova buses.

2. Use in locomotives-There is an ongoing study for the use of hydrogen in locomotives to replace diesel either as diesel engines or fuel cells. This study is by JPL will make the necessary recommendations for long-term use of hydrogen in locomotives.

3. Transportation use of hydrogen in fuel cell electric vehicle. Several programs are underway, i.e. Georgetown Bus, Ballard bus, and the green car all have been demonstrated in ,California and future vehicles demonstrating the fuel cell electric vehicle concept is ongoing.

1.1.3.2. Stationary Fuel Cells

Presently, the Southern California Gas Company has installed several fuel cell demonstration plant throughout the basin to establish electrical generation using their natural gas reformation to hydrogen on site for on peak needs for electricity. This program will be magnified in the near future and will benefit greatly from improvement of the infrastructure for hydrogen pipeline direct supply.

California, especially the South Cost Air Basin with their many universities and engineering schools, as well as private industry, including Electrolysis Incorporated, Ballard Fuel Cell Copporation and Xerox, could develop these programs to develop a package available for other needy countries. Moving directly into a solar hydrogen energy system would eliminate many health problems found at the South Coast Basin and would be a benefit to all NATO nations and mankind in general.

2. Solar Hydrogen Projects in California

2.1. PHOTOVOLTAIC ELECTROLYSIS PRODUCTION OF HYDROGEN
DR. PETER LEHMAN
HUMBOLDT STATE UNIVERSITY, ARCADIA, CALIFORNIA

This project has been ongoing for approximately five years and Dr. Lehman reports excellent results of photovoltaic electrolysis production of hydrogen. The use of hydrogen eventually will provide off peak fuel cell electricity for heating the fish hatchery. On campus, the results for the first part of the program have been satisfactory.

Dr. Lehman is now in the process of developing a hydrogen fuel cell electric program manufacturing his own membrane fuel cell funded by the Schatz Foundation.

2.2. HYDROGEN FUEL CELL VEHICLES
DR. DAVID SWAN, MARK DE LUCCI AND MARSHALL MILLER
UNIVERSITY OF CALIFORNIA, DAVIS, CALIFORNIA

This program is a continuation project first developed at Texas A&M by Dr. Swan. Dr. Swan has had the experience of racing electric vehicles at the Pheonix, Arizona raceway. He has also driven the electric vehicle in Mexico City with very great success.

At the present time, the program at UC Davis is to test and install the hydrogen fuel cell electric technology into the ongoing electric vehicles already produced from Texas A&M. The static state test will prepare the program to study the load on a dynamometer. The future of this program will be to install the fuel cell on electric buses to be run in the vicinity of the University of California at Davis. This program will encompass five or six years of testing and demonstration.

In addition to this research and development, the university has been charged with writing a complete manual for the California Energy Commission having to do with any future hydrogen programs in the state of California. This manual will consider all the safety factors involved with production, transmission, storage and utilization in transportation.

2.3. HYDROGEN AS A FUEL FOR THE METROPOLITAN FLEETS
ROBERT WICHERT
SACRAMENTO METROPOLITAN UTILITIES DEPARTMENT

This program is developing hydrogen as a fuel for the Metropolitan Fleets in the Sacramento Utilities area.

2.4. SOLAR HYDROGEN RESEARCH FACILITY
JOE NORBECK AND JIM HEFFEL
UNIVERSITY OF CALIFORNIA, RIVERSIDE

The SCAQMD supported program is located at UCR where some of the highest smog levels in the United states have been recorded.

The facility is a project of the College of Engineering center for Environmental Research and Technology (CE-CERT), who will install, operate and evaluate a solar-hydrogen production facility; review and evaluate hydrogen as a fuel for transportation; and develop and evaluate hydrogen vehicles. The program is utilizing the latest in solar powered electrolysis and hydrogen technologies to demonstrate a virtually zero emission, renewable energy source. In so doing, UCR will also train graduate and undergraduate students (engineers of tomorrow) in the production and utilization of hydrogen.

The facility uses 368 square feet of photovoltaic modules (solar panels) to produce up to 150 amps of electricity. Electric current travels through a control room to 12 electrolysis cells where water is split into hydrogen and oxygen.

The photovoltaic array at UCR has a maximum power output of 3.45 KW, 150 amperes and 24 volts. In one hour of production, this would consume about .1 gallon of fresh water to produce 28 SCF of hydrogen. The rate limiting factor of the current system is the number of solar cells. The system can also be run at night using a rectifier that uses commercially produced electricity to split the water.

Electrolyser Corp. Ltd. has installed an integrated, automatic hydrogen generation and vehicle fuel supply system which is designed to deliver electrolytic hydrogen at a rate of 9.5 cubic meters /hr. at a purity of 99.9%, and 4.75 m^3/hr. electrolytic oxygen at a purity of 99.7%

The oxygen is released and hydrogen is initially stored in a low pressure gas holder. When the gas holder reaches its maximum storage capacity, a high pressure compressor is automatically activated. The hydrogen is compressed to 5,000 psi and transferred to a high pressure storage station. The gas can then be transferred to a filling station for hydrogen fuelled vehicles.

A computerized system monitors temperature, voltage, current, solar irradiance wind speed and direction, allowing for quantitative analysis of the system. The electronic controller starts and stops the electrolyzer equipment and notifies operators of system status.

The system at UCR includes the entire process of hydrogen utilization from production to transportation. Staff, faculty and students will evaluate the efficacy of each phase of the process, including the photovoltaic electrolysis production, and the performance and emission characteristics of hydrogen powered vehicles.

Additionally, a comprehensive literature survey of hydrogen fuelled transportation has been conducted. Students have compiled the information and evaluated hydrogen-fuelled transportation power plants and requirements for a hydrogen-fuel economy. The results will be published.

CE-CERT researchers are working on two hydrogen fuel vehicles. The first vehicle is a 1992 Ford Ranger truck with a modified 2.3 liter four-cylinder engine. Modifications include reworking the cylinder heads and camshaft, addition of a turbocharger and aftercooler and the installation of a electronically controlled hydrogen fuel injection system supplied by the Hydrogen Consultants Inc. This vehicle has been designed to take advantage of the unique flammability characteristics of hydrogen to utilize a "lean burn" (excess air) combustion and therefore minimize the NOx level to near zero.

The second vehicle is a hybrid electric with an on-board hydrogen fuel cell. This vehicle will utilize electricity from the hydrogen fuel cell to keep the batteries charged, offering a much greater range than a pure electric vehicle. Fuel cells, which in the past have been mainly used in space applications, promise to be about 2.5 times as efficient as the internal combustion engine. Because the fuel cell converts the hydrogen to water electrochemically, no NOx is formed. The exhaust from the fuel cell is pure water.

2.5. HYDROGEN PRODUCTION UTILIZING HYDROPOWER
RAMBACH AND SMITH
LAWRENCE LIVERMORE LABORATORY , CALIFORNIA

These projects are looking into various forms of hydrogen production utilizing hydropower. They will also be demonstrating a hybrid/hydrogen electric vehicle utilizing a steady state internal combustion engine on board an electric vehicle.

2.6. XEROX-CLEAN AIR NOW (CAN) SOLAR HYDROGEN FLEET VEHICLES
ROBERT M. ZWEIG
RIVERSIDE, CALIFORNIA

This project is similar to 2.4. above and will demonstrate a practical application of solar generated hydrogen as fuel for utility vehicles at the Xerox facility in El Segundo, California. Clean Air Now (CAN), will oversee and manage all aspects of this program through agreements and contracts with the participating entities. The project will utilize state-of-the-art technology including photovoltaics, hydrogen production through water electrolysis, and hydrogen internal combustion engine retrofits. The project also includes comprehensive educational and promotional activities. The results of this project will be expandable on the Xerox site, replicable at other sites, and will lead to jobs and increased clean fuel technology transfer to other private, commercial and public entities.

CAN has organized a partnership consisting of Xerox, Proxair/Linde and Solar Engineering Applications, Inc., three major corporations with vision to the future. In addition, other companies, top educators, scientists and engineers are cooperating with CAN to produce a first of its kind-private sector application of solar hydrogen.

There are three key technical elements to the project:

1. SEA Corp. an innovative photovoltaic manufacturer located in Sunnyvale, California. will install a Linear Focus Fresnel Lens solar tracking photovoltaic array.
2. Praxair/Linde Industrial Gases will provide a slow-fill fuelling system which will later be upgraded to a "fastfill" system that system that will have the capability of fuelling multiple vehicles in succession in less than seven (7) minutes.
3. Advanced Machining Dynamics (AMD), of Highland, California will convert three vehicles (1993 Ford Rangers) to operate on hydrogen.

The project will consist of a photovoltaic array producing 50 KWH of power, which will drive the electrolytic generation of approximately 370 scf of hydrogen gas. This generating system will also supply an on-site fuelling station to serve ULEV/ZEVs (Ultra Low Emission and Zero Emission Vehicles).

In conclusion, missions and goals of these programs are to :

1. Produce hydrogen from solar energy.
2. Prove the efficiency of hydrogen as a clean transportation fuel.
3. Demonstrate fleet vehicles conversion technology.
4. Demonstrate the viability of hydrogen fuel cell electric vehicles.

Upon completion of these goals, the first of the converted vehicles will be delivered to the Xerox facility for full systems testing. Energy Technology Engineering Center (ETEC) and UCR/CECERT will conduct system testing and analysis of all four completed, SEA Corp., Electrolyser and Praxair/Linde will train Xerox and can personnel for systems operation and maintenance.

This project will demonstrate a low environmental impact, "closed-loop" system that will mitigate stationary and mobile source air emissions. The twenty-four (24) months program will include ongoing and continuous pubic and private outreach and educational activities. The first year will consist of installation and initial testing phase; the second year, start -up, operational and further educational phases.

2.7. PLASMA GENERATION OF HYDROGEN
KAISER HOSPITAL
SAN DIEGO, CALIFORNIA

This program demonstrates the use of waste products from hospital utilization including bacteria, hospital waste etc. using a high plasma pyrolysis system to produce a 40% hydrogen fuel which can be utilized in fleet vehicles or fuel cells.

The product gas may be used in may various energy intensive applications. A cost benefit analysis would require a torch large enough to make it cost effective. Municipal landfills would be prime candidates.

The hydrogen-rich fuel gas produced by the plasma torch in this San Diego project will be used as feedstock for fuel cell demonstration projects. Kaiser Permanente and San Diego Gas & Electric entered into an agreement to demonstrate a 250 KW molten carbonate fuel cell (MCFC)

The fuel cell will use all fuel gases produced by the plasma torch. The fuel cell shares many characteristics of a battery-no mowing parts, silent operation and electrochemical reaction to generate energy. Chemical reaction in the cell is the combination of hydrogen ions and oxygen ions interacting to from water, producing electrical and thermal energy. The hydrogen-rich fuel produced in the plasma torch is an ideal match for the fuel cell.

The combination of two emerging technologies used to recycle hospital waste is recycling in its highest form. The medical center's converted waste stream will produce two forms of energy from the fuel cell. This electrical and thermal energy will be used at the medical center. The MCFC is scheduled to be installed in 1994.

Plasma pyrolysis, an electrotechnology at work, is an environmentally benign answer to the waste disposal dilemma. A new application of a proven technology, the plasma torch reaches instantaneous arc temperatures of 21 000°F. It converts all organics into clean fuel gases and vitrifies inorganics into an inert glassy slag. All by-products are recyclable. There are a variety of uses for the vitreous slag. It can be used as concrete filler in roadbed construction, in composition roofing and even for insulation; The product gas can be used as a boiler gas or for vehicular fuel.

Environmental pollution is eliminated as well as the long term liability associated with disposal of medical waste at municipal solid waste sites. Landfills will no longer be considered "dumps" but sources of transportation fuel. Plasma pyrolysis of both medical and municipal solid waste is an innovative, environmentally sound solution to a complex and growing problem.

2.8. CONCLUSION

Mounting evidence that FCC causes inanimate material loses, damage to flora, fauna and human health. Long term exposure to present FFC and byproducts results in environmental health problems. Expenditures for treating these pollution related diseases could be better spent developing the HE infrastructure where solar electrolysis hydrogen could replace fossil fuel and offer economic, environmental and energy benefits to global populations. The South Coast Air Quality management District has been very generous with their support of these programs in order to demonstrate the availability of hydrogen to reduce air pollution in the South Coast Air Basin. As a result of this reduction of air pollution related diseases and disease aggravation.

3. Public Health Needs for Hydrogen

Recent medical data document human lung damage of patients living in air polluted urban areas. Nowhere in the United States are pollution control policies and strategies needed more than in the South Coast Air Basin (SoCAB). Health costs from fossil fuel combustion (FFC) effects mount steadily in spite of billions of dollars invested in presently outmoded pollution control measures. Therefore, SoCAB has been strategically targeted by medical researchers, who found increasing evidence of:
-permanent lung scarring in patients residing in the SoCAB for fifteen years.
-chronic lung disease due to long–term ambient concentrations of total suspended particulates (TSP) and oxidants (Ox) in California Seventh Day Adventists.
-childhood asthma incidence, severity, and deaths.

Without a Hydrogen Economy (HE), air pollution health problems for improving air quality, such as "add–on smog devices" and scrubbers are inadequate. All obsolete vehicles and plants should be replaced with new vehicles and plants utilizing hydrogen fuel; thus entirely eliminating toxic pollutants from FFC.

3.1. COST

If hydrogen were utilized as an alternative to fossil fuels, non–energy costs could be reduced by 50%. Logically, to begin paying for pollution, 30 cents a gallon should be added to the price of each of the roughly 100 billion gallons of gasoline sold in the U.S.A. each year. Other nations pay for clean air. Cities such as Zurich, Switzerland, change $20 an hour to park cars, effectively forcing the use of public transit. We in America prefer the flexibility of car to public transit, which costs billions of dollars to build. We want both mobility and clean air– and we must think of how to pay for it. Health costs related to disease management caused by, or aggravated by FFC continue to increase. These health care dollars spent in impacted areas could better serve society as the immediate investment for HE infrastructure.

Air quality conditions in SoCAB are among the worst in the nation. Health–based standards for ozone, (Particulate Matter (PM–10) (less than 10 microns in diameter) and carbon monoxide (CO) are exceeded frequently. Elimination of health problems associated with these elevated levels would result in a 14.3 billion U. S. dollars annual benefit in the SoCAB. We can have a new worldwide industrial fuel base and clean air too! In addition, to direct health costs from FFC, many indirect costs are identified (i. e. oil spills, military defence of oil production facilities, and international warfare to protect these sources). Congressman Bill Alexander (Democrat–Arizona) estimates that when the Desert Storm military costs are included in the cost of fuel, gasoline's true cost is $5.50 per gallon. Energy economists estimate hydrogen production would cost far less per equivalent BTU.

3.2. FOSSIL FUEL COMBUSTION SEQUELAE

3.2.1. *Carbon dioxide/greenhouse phenomenon.* A U. S. National Academy of Science report describes decreased food production and other adverse effects of increased CO_2 production. A "worst–case" greenhouse scenario justifies immediate action. In order to stabilize atmospheric concentrations of long–lived greenhouse gases, emissions must be cut by 60 percent or more. Suggested strategies for reducing the greenhouse effect would:

-switch coal– and oil–fired power plants to natural gas
-promote energy efficiency
tax energy to reflect its environmental damage
-phase out chlorofluorocarbons (CFCs)
-plant trees
-support commercialization of high–efficiency fuel cells
-Hydrogen would satisfy 5 of recommendations.

Adopting these suggestions, the United States could reduce the 1990 levels of greenhouse gases by 10–40%.

3.2.2. *Acid precipitation.* The United States Environmental Protection Agency (EPA) reports high acidity levels are found in many national lakes and streams. The study analyzed pH 1,180 lakes and 4,670 streams in the Adirondack Mountain area. 75% of these lakes and rivers are acidic as a direct result of sulphuric acid produced from non–hydrogen FFC.

3.3. EPIDEMIOLOGICAL EFFECTS FROM NON–HYDROGEN FOSSIL FUEL COMBUSTION BY PRODUCTS

3.3.1. *Recent health studies*

3.3.1.1. Lung biopsies from healthy volunteers before and after exposure to 0.2 parts of ozone per million parts of air (0.2 ppm) for four hours of moderate exercise showed a six–fold increase in inflammatory cells in the airways of ozone–exposed volunteers. Damage from ozone is similar to a sunburn except the burn occurs in the airways of the lung and it may linger for several days.
3.3.1.2. American Lung Association (ALA), researchers warned that there is a poor correlation between ozone–induced lung injury and outward signs of damage such as cough, chest pains, or reduced lung function. Therefore, individuals may suffer from ozone damage without being aware of the harm.
3.3.1.3. Symptoms and decreased lung function are two parameters normally used to assess pollution impacts on the human body and establish public health standards for safe levels of exposure. "This study suggests that these methods used to assess the risk from ozone and other pollutants may be too crude to protect public health, "says Dr. Alfred Munzar, president of the ALA.
3.3.1.4. Harvard Six Cities Study–Statistically significant and robust associations between air pollution and mortality were observed. Air pollution was positively associated with death from lung cancer and cardiopulmonary disease but not with death from other causes. Mortality was most strongly associated with fine particulate (PM–10) air pollution.

In this study, a well–characterized cohort of adults participating in the Harvard Six Cities Study of Health effects of air pollution was followed prospectively, beginning in 1974.

Elevated levels of particulate air pollution have also been associated with declines in lung function or with increases in respiratory symptoms such as cough, shortness of breath, wheezing, and asthma attacks. Other studies found associations between particulate air pollution and rates of hospitalization, chronic obstructive pulmonary disease and restricted activity due to illness.

The compatibility of the effects of air pollution on mortality in this study with those observed in population–based cross–sectional studies and daily time–series studies provides further evidence for concluding that exposure to air pollution provides additional impetus for development of strategies to reduce urban air pollution.

3.3.2. *Previous Health studies*

3.3.2.1. University of Southern California (USC)- researchers found that children living in Houston, Texas (The smoggiest city in Texas). California children contract more lower respiratory tract infections (colds and influenza) than children who breath clean air.
3.3.2.2. A newborn infant breathes there times as much air per pound of body weight as an adult; therefore, increased pollution will be there times as toxic for a newborn who traps particulate pollution ten times greater than adults.
3.3.2.3. Lungs of teenagers living in SoCAB had 80% "notable" lung abnormalities versus 27% similar respiratory lesions in teenagers residing in cleaner areas.

3.3.3. *Asthma*

3.3.3.1. A Natural Resources Defence Council (NRDC) report endorsed by the ALA and the American Academy of Pediatrics, concludes that federal air–pollution laws and regulations do not take into account "compelling medical evidence" of increased vulnerability of children to $O3$, PM-10 and NOx.
Since children spend more time outdoors than adults, absorb proportionately more pollution (based on their body weight) and have bronchial systems that trap more particles than fully developed lungs and airways, they tend to be more susceptible to airborne pollutants than any other segment of the population.
The EPA must draft a California clean-air implementation plan which could include control strategies advocated by the NRDC and other groups.
3.3.3.2. Peter J. Barnes, Chairman of the Department of Thoracic Medicine at the National Heart and Lung Institute, Bronpton Hospital London, states that asthma affects over 5% of the total population in industrialized countries.
3.3.3.3. Prevalence, severity, and deaths from asthma are rising. It has been postulated that the increased severity of asthma is aggravated by increased inflammation and bronchospasm and limits protective functions of pulmonary macrophages.
3.3.3.4. Both asthma and chronic bronchitis result in airway inflammation, and the additional burden posed by 03 may provoke or exacerbate symptoms. We do not believe that 03 causes asthma or bronchitis, but we do believe it makes these afflictions worse,. There are so many people with chronic lung disease that even if 03 were hurting only these groups, it may have widespread impact.
3.3.3.5. Asthma causes a general thickening of airway walls. Airway inflammation, rather than "twitchy airways" is now seen as the root cause of asthma.

3.3.4. *Effects of particulate matter (PM-10)*

3.3.4.1. EPA concludes that PM-10 accounts for higher mortality associated with heavy episodes of smoky pollution. As many as 60,000 Americans will die annually from inhaling particles of soot at levels that do <u>not</u> exceed the maximum federal standards.
3.3.4.2. Dr George Thurston, researcher, New York University School of Medicine mentions that acidic particles are especially toxic for all humans and cause increased mortality rate of "wheezers" (i.e. asthmatics or emphysema patients). Particulates also contribute to other illnesses, (i.e. lung cancer) in previously healthy people. It is suggested that ultimately we will be mandated to revise federal standards for "safe" levels of particulate matter.
3.3.4.3. PM-10 reduction of U.S. federal standards would result in 1600 fewer premature deaths SoCAB.

3.4. SOLUTION

Hydrogen is the ultimate healthy fuel. It is the cleanest and most efficient way to use and store solar energy. It is abundant and renewable. Technology now exists for practical applications of hydrogen as fuel. Replacement of FFC to H_2 will solve health problems. Hydrogen as a fuel produces no carbon monoxide, carbon dioxide, reactive organic gases or oxides of nitrogen or sulphur. H_2 projects in California will demonstrate safe, clean the use of this technology and will advance the use of this truly clean fuel.

ROLE OF RENEWABLE HYDROGEN IN GLOBAL ENERGY-ENVIRONMENT-ECONOMIC SYSTEM

F. BARBIR*, H.J. PLASS, Jr., and T.N. VEZIROGLU
Clean Energy Research Institute, University of Miami
P.O. Box 248294, Coral Gables, FL 33124, U.S.A.

Abstract. The energy language method has been used to develop a dynamic model which generated intersubstitution of energy sources and energy currencies in the energy market, and provided information on complex interactions between energy, environment and economics. Substitution of the fossil fuel system by the solar hydrogen energy system has been analyzed, with particular attention given to effects on economic growth and on the environment. The results have shown that an early transition to the solar hydrogen energy system would provide long term benefits to economy and to the environment.

1. Introduction

Because the energy system is only a part of the economic system, which in turn is only a part of the environment, establishment of a new energy system, such as the solar hydrogen energy system, will inevitably have impacts on both the socio-economic system and the environment. These impacts can be analyzed only by taking into account the complex interactions between energy, environment and economy. A dynamic model has been developed, which included these complex interactions, although in a simplified way, and which enabled simulation and analysis of the long term global trends resulting from the replacement of the global energy system.

2. Modeling Technique

Odum (1983) has developed the energy circuit language for analysis and modeling of various systems. Although initially developed for ecological systems, this method has been successfully applied for modeling and simulation of human economic systems on regional, national and global levels (Odum, 1987; 1988; 1989). Society can be modeled as an energy system because the major socio-economic patterns are the ultimate result of energy usage. All primary energy forms have their origins in the environment, i.e., outside the socio-economic system. There is always an exchange of energy and materials between the socio-economic system and the environment. The dynamics of the economic process is therefore ultimately constrained by the availability of energy in the environment and its accessibility for economic consumption.

* *Current address: Energy Partners, 1501 Northpoint Pkwy, #102, West Palm Beach, FL 33407, U.S.A.*

Y. Yürüm (ed.), Hydrogen Energy System, 329–338.

Energy circuit language is a simple but flexible way to visualize the structure and dynamics of the systems being analyzed. A special set of symbols has been developed in order to describe various components of the system and interactions between them (Odum, 1983). By combining the appropriate symbols, and following some specific but simple rules, it is possible to construct the energy diagram of the system. The symbols have rigorous kinetic and energetic definitions so that they not only provide graphic visualization of the model, but they also provide the equations for simulation.

This modeling technique provides a tool for the analysis of the behavior of the economic systems. Such models provide important physical information on the structure from which the behavior evolves. The models can be set to replicate patterns of real society, while at the same time they give an insight into the system structure which produces these patterns. In constructing a model of an energy-economic system, only long term changes in process dynamics are considered, such as consumption of resources from the environment and their use in building and maintaining a socio-economic infrastructure. It is not possible to account for the short term fluctuations.

Bodger and Baines (1988) have applied this technique to develop a model which simulates the penetrations of the new forms of primary energy sources, each having successively greater accessibility and availability. They managed to replicate some historical trends in energy consumption and performance of the socio-economic sector.

3. Global Energy-Economy-Environment Model

The world energy-economy-environment system has been modeled with energy circuit language method. The graphic diagram of the model is given in Fig. 1. The model relates total world assets and the global economy to the availability of renewable and nonrenewable resources in a very aggregate way. Boundaries of the system were selected at the global level. In that case the only energy flows crossing the boundary are global external energy sources coming into the system and degraded energy leaving the system.

The system draws energy from two kind of sources:

- Renewable or flow limited sources. Incoming solar energy is a flow limited source because the system can extract only that much energy that is flowing in. The unused (untapped) portion leaves the system.
- Non-renewable, i.e., fossil fuels. Fossil fuel reserves are drawn at the boundary because they could be considered either as a part of the system or as an outside energy source. They are represented as a tank, from which the system draws as much energy as needed, but only until there is enough reserves in the tank.

The second law of thermodynamics requires that any process disperses some of its available energy in degraded form, which leaves the system unable to do more work. This law also requires that any stored material, energy, goods, or information, depreciate, losing its concentrations, because its energy is dispersing. The flows of used energy leaving the system are shown in the diagram by pathways that go downward into the heat sink. These pathways represent only degraded heat energy. Any dispersing materials, commodities, services, or energy still concentrated enough to do work, form the network of pathways amongst the sources and storages.

The world energy-economy-environment system has been reduced to only three stored values:

- Fossil fuel reserves are aggregated into a single tank, "F", i.e., coal, oil and natural gas are represented as a single source.
- World economic assets, which in the model are represented as a physical storage "A", include all man-made physical structures, e.g., buildings, transport, communication networks, technological hardware, as well as the accumulation of skills, experience, knowledge and culture. It is the accumulation of hardware and software which is involved in the economic process and continuously being maintained, replaced and added to. In the dynamics of the process the assets have a twofold role: (i) they generates a demand for consumption of environmental resources, and (ii) provide the economic feedback necessary for generating goods and services.
- Pollution tank, "P," stores pollutants being released by combustion of fossil fuels. Carbon dioxide has been selected as a measure of pollution, since the quantity of pollutants is proportional to the fossil fuels consumed, and CO_2 generation is also proportional to the fossil fuels consumed. Since carbon dioxide is a natural constituent of the atmosphere, only excess of CO_2 (above pre industrial level) is assumed to have negative effects on economy.

Interactions aggregate the entire economic process of creating goods and services. The flows coming into an interaction are environmental flows (in this case simply aggregated to only energy flows) and feedback of already stored goods and services. The result of this interaction is the economic product (goods and services).
The lines between the sources, storage tanks and interactions represent the flows of energy, materials, goods, services and information. The flow always has the same unit as the tank from which it has originated or to which it is headed.

In the model, there are three pathways of resource use that generate economic assets, each slightly different. Each pathway actually represents a different stage of economic development:

(1) The lower pathway represents utilization of renewable resources in building and maintaining the world assets (e.g., agriculture, fishery, forestry, etc.). It actually represents undeveloped, non-industrialized economies.

(2) The next pathway represents developed economies that have tapped vast storages of mineral and fuel reserves to power their economies. Utilization of fossil fuel reserves has provided the growth of the economic activities and increased the world assets dramatically, particularly in the second half of this century. However, as already mentioned, there are at least two kinds of problems related to the utilization of the fossil fuels, i.e., (1) their reserves are finite and therefore sooner or later will be depleted, and (2) they create pollution on a local, regional and global scale.

(3) The solar hydrogen economy is considered as a new stage in the evolution and development of human civilization. A new source is actually added to the model, which represents solar energy available over vast desert areas of the world. This solar energy has not been used by human economic systems either directly or indirectly. Through hydrogen and electricity, it can be used in the economic process to provide essentially the same services as fossil fuels are providing today.

Once the graphic model has been completed, its relationships have been translated into mathematical equations by following the rules of the energy circuit language. Each flow, is represented by an equation, which are combined in differential equations, one for each storage tank. Table 1 provides the list of all the flows, sources and storages, as well as associated equations. In order to solve the resulting set of differential equations, the initial values and coefficients are required for each flow and storage. Year 1950 has been selected as the initial year. Most of the initial values (such as energy reserves, energy consumption, CO_2 content in the atmosphere, CO_2 emissions, etc.) were taken from the available statistical data (United Nations, 1991; Smil, 1991; White, 1987; German Bundestag, 1989). Selected initial values are shown in Table 1.

Because of their highly aggregated form, assets were taken as a normalized value (1 in 1990). Initial values for the resulting flows of assets (i.e., economic production, natural depreciation, and depreciation due to use of assets in production) were set by trial and error, so that model simulation replicates historical trends in energy consumption and economic production for the period 1950-1990 for which statistical data were available. World Gross Product, measured in constant 1990 U.S. dollars, is actually proportional to the economic production. Using statistical data for the World Product (World Bank, 1990), a conversion factor of $268 trillion per normalized unit of assets is found to be applicable.

Utilization of solar energy through hydrogen as an energy carrier has been modeled to start in the year t_i ($t_i \geq 2000$). Penetration into the energy market has been assumed in logistic manner, according to Marchetti's (1973) observations. The time period required for the solar hydrogen to take 50 percent of the energy market (t_h) is selected to be equal to the time required to go from 10% to 90% of the market, so called "takeover time" (t_o). This time has been arbitrary selected to be 40 years, which is shorter than "takeover times" for most of the fuels or energy technologies in the past. However, it is longer than the lifetime of energy equipment and infrastructure (which is usually considered as a minimum time required for painless transition). Penetration in a shorter time period is not likely, because of the economic stress which would occur since the new energy infrastructure would have to be built even before than the old one has been depreciated.

Initial values of storages and flows were used to determine the coefficients in the differential equations. The resulting set of the first order nonlinear differential equations has been solved by simple Euler's numerical integration, using one year as the time step.

4. Results of Simulation

Computer simulation of the described model of energy-environment-economy interactions offers opportunity to examine a variety of different scenarios. The time period which has been selected for simulation is from 1950 to 2100. The period between 1950 and 1990 was used to verify the model and to determine missing coefficients. The real simulation starts in 1990. The time span of about one century is enough to find out long term trends. Short-term variations would be impossible to detect with such a simplified model.

Model simulation for the period between 1950 and 1990 has shown good agreement with the historical records for energy consumption, world gross product (as a measure of economic activity) and carbon dioxide in the atmosphere, as shown in Figs. 2, 3 and 4 respectively. If the world economy is based solely on utilization of fossil fuels (which means that the solar hydrogen sector is not activated at all) the consumption of fossil fuels will start to decline some time in the middle of the next century, after reaching a peak at about three times today's level. Decline in energy consumption will be accompanied by similar decline in economic activity. Carbon dioxide content in the atmosphere would continue to grow until the end of the century reaching 700 ppm - almost double what it is today.

In the case where utilization of solar energy and hydrogen is introduced in the energy market, energy consumption continues to increase, although at a lower rate than that between 1950 and 1990 (Fig. 2). Consequently, economic activity (Fig. 3) also continues to increase and would eventually reach a steady state some time in the future. Hydrogen introduction dramatically affects carbon dioxide in the atmosphere, which reaches the maximum before 2050 at 520 ppm (Fig. 4). From 2050, the carbon dioxide level in the atmosphere continues to decrease and eventually it would reach the pre-industrial revolution level.

Figures 5 and 6 show what would happen if transition to the solar hydrogen energy system is delayed 25 years. Energy consumption and economic activity would be higher than in "no hydrogen" scenario, but much lower than the case when hydrogen is introduced in the year 2000. Carbon dioxide would continue to increase until approximately 2070 reaching 620 ppm. If transition starts at 2050 there would be almost no positive effects. This suggests that an early transition to the solar hydrogen energy system would benefit the economy and the environment in the long run.

Historically, in the context of the longer time span, the fossil fuels era could be considered as a short interlude between the solar past and the solar future. In that short period (about three hundred years) fossil fuels made possible a tremendous development of the human civilization. However, if they are used to support the establishment of a permanent energy system such as the solar hydrogen energy system, they could be considered as a spark which provided a transition from the low level solar energy past to the higher level solar energy future. Solar energy is steadily available in a quantity which exceeds human needs by several orders of magnitude.

Any delay in action would make a transition more difficult, and the point of transition to hydrogen and solar energy could be another "lost moment in history" as feared by Winter (1990). If transition starts when economy begins to decline, it may be too late, because economy would not be able to afford investing in such a long term project as establishing of the new energy system.

As any system based on finite sources cannot grow forever, a system based on utilization of constant flow of incoming energy should reach a steady state. A steady state economy is defined as an economy with constant stocks of people and artifacts, maintained at some sufficient levels, by the lowest feasible flows of matter and energy. If economic development means getting more services per unit of throughput, and economic growth is defined as the increase of services by increasing the size of stocks, then it is clear that the steady state economy would force an end to pure growth but would not curtail, and in fact would stimulate, development (Daly, 1977). The steady-state level, which the human civilization can obtain, depends on the rate of utilization of available solar energy and the effort required to convert solar energy into more useful forms of energy.

Similar results were obtained by other researchers and modelers, using different mental and mathematical models. Hubbert (1973) concluded that exponential growth in fossil fuel consumption is not possible because fossil fuel resources are finite. After cessation of exponential growth, there is only one possible future – steady state, but different levels are possible: (1) stabilization at maximum point, (2) overshoot and stabilization at some intermediate level capable of being sustained, (3) decline to zero or to some very low value. However, he has not discussed in detail the instruments for obtaining these steady states, although he did mention solar energy.

Odum (1983, 1987, 1988, 1989) argues that utilization of dispersed renewable energy sources, such as solar energy, yield a little if any net energy, and therefore could not provide a steady state at a level higher than it used to be before the industrial revolution started.

Similar prediction of world collapse, has been provided by Meadows, et al. (1972). Their results (using the model World3) have shown that population and industry output grow until combination of environmental and natural resource constraints eliminate the growth. Industrial capital begins to depreciate faster than the new investment can rebuild it, and the result is the collapse of the world economic system. They investigated other possible scenarios, and found that it would be possible to reach the steady state if sustainable policies (such as population control, technologies to reduce emissions, erosion and resource use) are implemented in the year 1995. If the same policies are implemented with a 20 years delay, there will be an overshoot and the steady state will be reached at some lower level. These results are qualitatively similar to those obtained in the present study.

5. Conclusions

Energy language method, originally developed for analysis and simulation of the ecological systems, is applicable for large-scale, long-term modeling of the socio-economic system. It is easy to understand and convenient to use. A model of global energy-environment-economy interactions has been developed and used for simulation of future trends, including the transition to the solar hydrogen energy system.

The results of the model simulation have shown that the present energy system will not be able to support economic growth in the next century. However, if the fossil fuel system is gradually replaced by the solar hydrogen energy system, the growth could continue (although at a slower rate) until a steady state is ultimately established.

The results has also indicated that the transition to the solar hydrogen energy system should start as soon as possible. Any delay could have a negative effect, i.e., increase in energy consumption, accelerated depletion of natural resources and a bigger burden on the already damaged environment.

References

Bockris, J. O'M., and T.N. Veziroglu. 1985. A solar-hydrogen energy system for environmental compatibility, *Environmental Conservation*, Vol. 12, No. 2, pp. 105-118.

Bodger, S., and J.T. Baines. 1988. Dynamics of an energy-economic system subject to an energy substitution sequence, *Energy Systems and Policy*, Vol. 12, pp. 167-178.

Daly, H.E. 1977. *Steady State Economics*, W.H. Freeman and Co., San Francisco.

German Bundestag, 1989, *Protecting the Earth's Atmosphere: An International Challenge*, German Bundestag, Bonn.

Hubbert, M.K. 1973. Survey of world energy resources. *Can. Mining and Metallurgical Bull.* Vol. 66, pp. 37-53,

Marchetti, C. 1977. Primary energy substitution models: On the interaction between energy and society, *Technol. Forecast. Soc. Change*, Vol. 10, pp. 345-356.

Meadows, D.H., D.L. Meadows, J. Randers and W.W. Behrens, III. 1972. *The Limits to Growth*, Universe Books, A Potomac Associates Book, New York.

Odum, H.T. 1983. *Systems Ecology*. John Wiley & Sons, New York.

Odum, H.T. 1987. Models for national, international, and global systems policy, in L.C. Braat and W.F.J. van Lietrop, eds., *Economic-Ecological Modeling*, Elsevier Science Publishers, Amsterdam, Holland.

Odum, H.T. 1988. *Energy, Environment and Public Policy: A Guide to the Analysis of Systems*, UNEP Regional Seas Reports and Studies No. 95, United Nations Environment Programme, Nairobi, Kenya.

Odum, H.T. 1989. Simulation Models of Ecological Economics Developed with Energy Language Methods. *Simulation*. August 1989: 69-75

United Nations Organization. 1991. *Energy Statistic Yearbook 1989*, United Nations, New York.

Smil, V. 1991. *General Energetics: Energy in the Biosphere and Civilization*. John Wiley & Sons, New York.

White, D.A. 1987. Conventional oil and gas resources, in D.J. McLaren and B.J. Skinner, eds., *Resources and World Development*. Wiley, Chichester, pp. 113-128

Winter, C.-J. 1990. Hydrogen and Solar Energy - Ultima Ratio Avoiding a "Lost Moment in the Hystory of Energy," in T.N. Veziroglu and P.K. Takahashi, (eds.), Hydrogen Energy Progress VIII, Vol. 1, pp. 3-47, Pergamon Press, New York.

World Bank, 1990, *World Development Report 1990*, Published for the World Bank by Oxford University Press, New York.

Table 1. Storages, Sources and Flows, and Corresponding Equations for the Model

	Description	Equation ([1]	Initial Value ([2]
F	Fossil fuels reserves	$dF/dt = -f_4$	170×10^{21} J ([3]
A	Economic assets	$dA/dt = f_2+f_5+f_9-f_3-f_6- -f_{10}-f_{11}-f_{12}$	0.236 (normalized) ([4]
P	Carbon dioxide in the atmosphere	$dP/dt = f_7-f_{13}$	0.310×10^{21} kg ([5]
I	Environmental resources	I = const.	0 ([6]
N	Unused (untapped) environmental resources	$N = I/(1+k_1A)$	
J	Solar energy available for hydrogen production	J = const.	25.1×10^{21} J ([7]
R	Untapped solar energy	$R = J/(1+k_8A)$	
f_1	Utilization of environmental resources	$f_1 = k_1AN$	0 ([6]
f_2	Production of assets resulting from f_1	$f_2 = k_2AN$	0 ([6]
f_3	Assets depreciation through use in production f_2	$f_3 = k_3AN$	0 ([6]
f_4	Fossil fuels consumption	$f_4 = k_4AF$ or $D(1-\xi)$ ([8]	0.082×10^{21} J/yr ([9]
f_5	Production of assets resulting from f_1	$f_5 = (k_5/k_4)f_4$	0.0149 nau/yr ([10]
m_5	Gross World Product	$m_5 = p_5(f_5+f_9)$	4.0×10^{12} \$/yr ([11]
f_6	Assets depreciation through use in production f_5	$f_6 = (k_6/k_4)f_4$	0.0028 nau/yr ([10]
f_7	Carbon dioxide emissions resulting from f_4	$f_7 = k_7f_4$	0.00615×10^{15} kg/yr ([12]
D	Energy demand	$D = f_4(1-\xi) + k_8AR(\eta_{S-h}/\alpha_u)\xi$ ([13]	0.082×10^{21} J/yr
ξ	Hydrogen's share in energy market	$\xi = \{1+\exp[-\beta(t-t_i-t_h)]\}^{-1}$ ([14]	0
h	Hydrogen consumption rate	$h = \xi D$	0
f_8	Utilization of solar energy	$f_8 = (\alpha_u/\eta_{S-h})h$ ([13]	0
f_9	Production of assets resulting from h	$f_9 = (k_5/k_4)h$	0
f_{10}	Assets depreciation through use in production f_9	$f_{10} = (k_{10}/k_4)h$	0
f_{11}	Assets depreciation	$f_{11} = k_{11}A$	0.00236 nau/yr ([10]
f_{12}	Assets degradation due to pollution	$f_{12} = k_{12}AP$	0.00059 nau/yr ([10]
f_{13}	Carbon dioxide absorption from the atmosphere	$f_{13} = k_{13}P$	0.00372×10^{15} kg/yr ([15]

Footnotes for Table 1:

1. Coefficients, $k_{i=1 \text{ to } 13}$ and p_5, are determined from the initial values.
2. All initial values are for the year 1950.
3. Fossil fuel resources, [United Nations, 1991; Smil, 1991, White, 1987):
   coal 150,000 EJ, oil and gas 20,000 EJ, total 170,000 EJ = 170×10^{21} J.
4. Assets were normalized so that A=1 normalized asset units (nau) in 1990.
5. CO_2 content in the atmosphere before industrial revolution (German Bundestag, 1989):
   $CO_2^{1850} = 275$ ppm $= 2.17 \times 10^{15}$ kg
   CO_2 content in the atmosphere in 1950 (ibid): $CO_2^{1950} = 315$ ppm $= 2.48 \times 10^{15}$ kg
   Excess of CO_2 in the atmosphere: $P = (2.48 - 2.17) \times 10^{15} = 0.31 \times 10^{15}$ kg
6. Environmental resources, other than those used to generate hydrogen and electricity, have been neglected, because their contribution in building and maintaining the assets is much smaller than the contribution from fossil fuel powered economic process.
7. Solar energy available for conversion into hydrogen is assumed to be 10% of solar radiation over the world's desert areas:
   World deserts: 3,140 million hectares = 31.4×10^{12} m^2; solar radiation in deserts: 8×10^9 J/m^2/yr (Smil, 1991)
   Available radiation: $0.1 \times (31.4 \times 10^{12}$ m$^2) \times (8 \times 10^9$ J/m^2/yr) $= 25.1 \times 10^{21}$ J/yr
8. Share of fossil fuels in energy consumption starts to diminish after solar hydrogen has been introduced.
9. Fossil fuel consumption in 1950 (United Nations, 1991): 82 EJ/yr = 0.082×10^{21} J/yr
10. nau = normalized asset units; initial values are determined by model simulation between years 1950 and 1990.
11. Gross world product in 1950 (World Bank, 1990): $4,000 billion/yr
12. CO_2 emissions in 1950 = (fossil fuel combustion)x(fossil fuel carbon content) = $(0.082 \times 10^{21}$ J$) \times (75$ kg/10^9 J$) = 0.00615 \times 10^{15}$ kg/yr
13. α_u = utilization efficiency advantage of hydrogen defined as $\alpha_u = \eta_f/\eta_h$, where η_f is the overall end-use conversion efficiency of fossil fuels, and η_h is the overall end-use conversion efficiency of hydrogen; $\alpha_u = 0.73$ (Bockris and Veziroglu, 1985).
    $\eta_{s\text{-}h}$ = overall conversion efficiency of solar energy to hydrogen (assumed to be 0.05).
14. β= penetration rate = $4.4/t_O$, where t_O = 40 yr = takeover time defined as the time required to go from ξ=0.1 to ξ=0.9).
    t = time; t_i = initiation time; t_h = half time (time required to reach half of the market) = $t_O/2$ = 40 years;
15. It takes about 100 years until a quantity of CO_2 released into the atmosphere is reduced to about 30% of its original volume (German Bundestag, 1989). Therefore, the initial value for the flow f_{13} is:
    $$f_{13} = [1 - 0.3^{(1/100)}]P = 0.00372 \times 10^{15} \text{ kg/yr}.$$

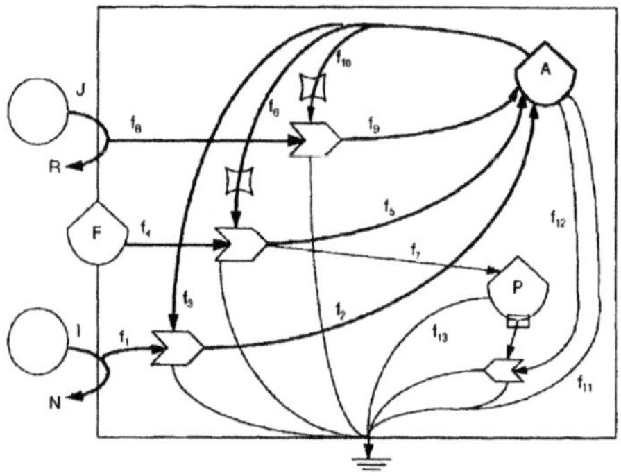

Figure 1 Global energy-economy-environment model diagram

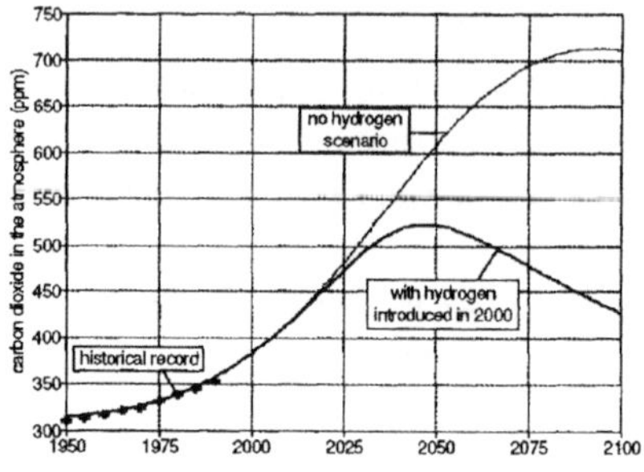

Figure 4. Carbon dioxide in the atmosphere (base case scenario)

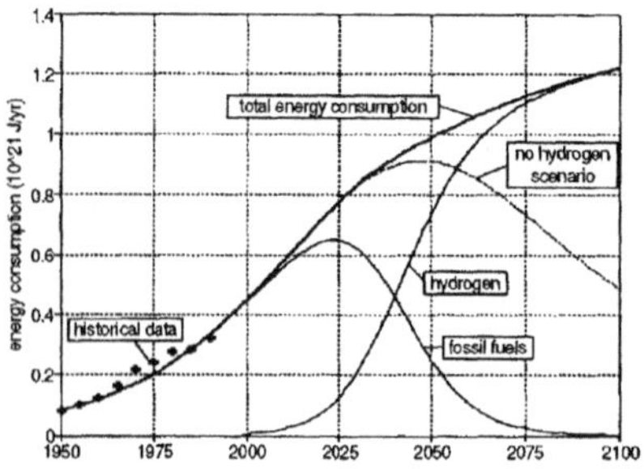

Figure 2. Energy consumption (base case scenario)

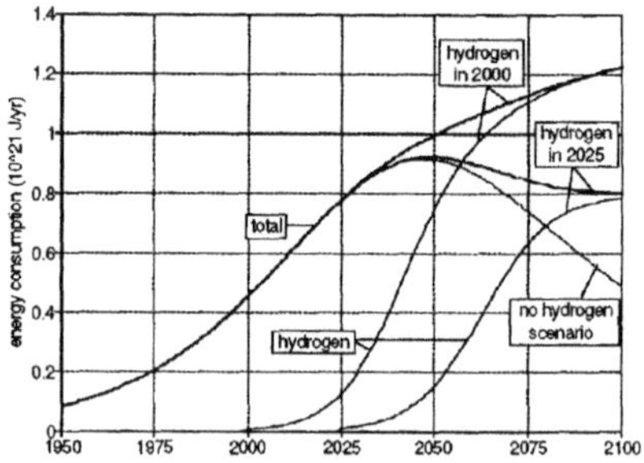

Figure 5. Energy consumption (initiation of hydrogen energy system is delayed 25 years)

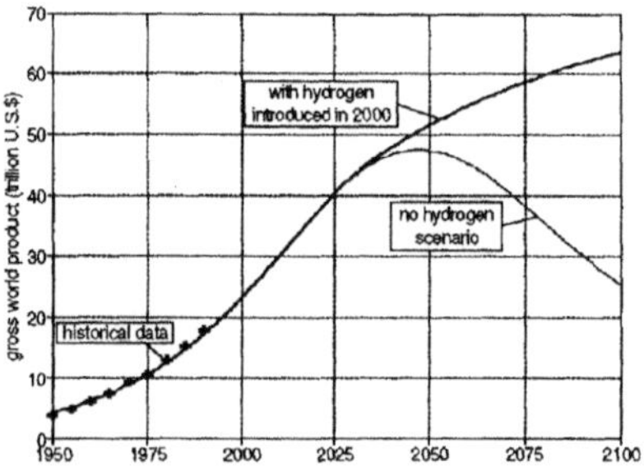

Figure 3. Gross world product (base case scenario)

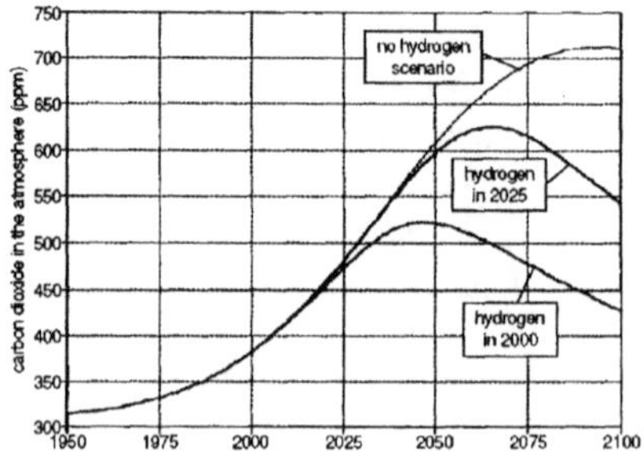

Figure 6. Atmospheric carbon dioxide (initiation of the solar hydrogen energy system is delayed 25 years)

INDEX

H

I

J

L

M

N

O

P

R

S

T

U

V

W

Z

GPSR Compliance

The European Union's (EU) General Product Safety Regulation (GPSR) is a set of rules that requires consumer products to be safe and our obligations to ensure this.

If you have any concerns about our products, you can contact us on ProductSafety@springernature.com

In case Publisher is established outside the EU, the EU authorized representative is:

Springer Nature Customer Service Center GmbH
Europaplatz 3
69115 Heidelberg, Germany

Batch number: 10816188

Printed by Printforce, the Netherlands